# Scientific Elite

*Foundations of Higher Education*
David S. Webster, Series Editor

*Academic Freedom in the Age of the College,* Richard Hofstadter,
with a new introduction by Roger L. Geiger

*The Academic Man,* Logan Wilson,
with a new introduction by Philip G. Altbach

*Centers of Learning,* Joseph Ben-David,
with a new introduction by Philip G. Altbach

*The Distinctive College,* Burton R. Clark,
with a new introduction by the author

*The Future of Humanities,* Walter Kaufmann,
with a new introduction by Saul Goldwasser

*The Higher Learning in America,* Robert Maynard Hutchins,
with a new introduction by Harry S. Ashmore

*The Ideal of the University,* Robert Paul Wolff,
with a new introduction by the author

*The Impact of College on Students,*
Kenneth A. Feldman and Theodore M. Newcomb,
with a new introduction by Kenneth A. Feldman

*Making the Grade,* Howard S. Becker, Blanche Geer, and
Everett C. Hughes, with a new introduction by Howard S. Becker

*Mission of the University,* Jose Ortega y Gasset
with a new introduction by Clark Kerr

*The Organization of Academic Work,* Peter M. Blau
with a new introduction by the author

*Rebellion in the University,* Seymour Martin Lipset,
with a new introduction by the author

*Reforming of General Education,* Daniel Bell,
with a new introduction by the author

*The Rise of the Meritocracy,* Michael Young,
with a new introduction by the author

*Scientific Elite,* Harriet Zuckerman,
with a new introduction by the author

*Universities,* Abraham Flexner,
with a new introduction by Clark Kerr

Harriet Zuckerman

**WITH A NEW INTRODUCTION BY THE AUTHOR**

# Scientific Elite

**NOBEL LAUREATES IN THE UNITED STATES**

**TRANSACTION PUBLISHERS**
New Brunswick (U.S.A.) and London (U.K.)

New material this edition copyright © 1996 by Transaction Publishers, New Brunswick, New Jersey 08903. Originally published in 1977 by The Free Press.

This book is printed on acid-free paper that meets the American National Standard for Permanence of Paper for Printed Library Materials.

Library of Congress Catalog Number: 95-45046
ISBN: 1-56000-855-5
Printed in the United States of America

Library of Congress Cataloging-in-Publication Data

Zuckerman, Harriet.
    Scientific elite : Nobel laureates in the United States / Harriet Zuckerman ; with a new introduction by the author.
        p.  cm. — (Foundations of higher education)
    Originally published : New York : Free Press, c1977.
    Includes bibliographical references and indexes.
    ISBN 1-56000-855-5 (pbk. : alk. paper)
    1. Scientists—United States. 2. Nobel prizes. 3. Science—Social aspects—United States. 4. Scientists—United States—Biography. I. Title. II. Series.
Q149.U5Z8   1996
509.2'273—dc20                                                    95-45046
                                                                      CIP

*For my parents*
*HZ and AWZ*

Science today must search for a source of inspiration above itself or it will perish. There are just three reasons for doing science: 1° technical applications; 2° chess game; 3° the way toward God. (The chess game is embellished with competitions, prizes and medals.)

Simone Weil
*Le pesanteur et la grâce*

[I am grateful to Erwin Chargaff for having led me to this quotation.]

# CONTENTS

Introduction to the Transaction Edition    xiii

Preface and Acknowledgments    xliii

Chapter 1    Nobel Laureates and Scientific Elites    1

Chapter 2    The Sociology of the Nobel Prize    16

Chapter 3    The Social Origins of Laureates    59

Chapter 4    Masters and Apprentices in Science    96

Chapter 5    Moving into the Scientific Elite    144

Chapter 6    The Prize-Winning Research    163

Chapter 7    After the Prize    208

Chapter 8    The Nobel Prize and the Accumulation of Advantage in Science    243

Appendix A    Interviewing an Ultra-elite    256

Appendix B    Nobel Laureates in Science, 1901-76    282

Appendix C    Prize-Winning Research: Specialty and Year of Award    292

Appendix D    Official Occupants of the Forty-first Chair: "Honorable Mentions" for Nobel Prizes    296

Appendix E    Age-Specific Annual Rates of Productivity of Laureates and a Matched Sample of Scientists Who Survived to Each Age    302

Bibliography    304

Index of Names    327

Index of Subjects    332

# LIST OF ILLUSTRATIONS

**Figure 2–1**  Purchase Value and Cash Value of Nobel Prizes (1901–75)  21

**Figure 2–2**  Total Number and Last-Round Candidates for Nobel Prize in Physiology/Medicine (1901–71) Compared with Number of Publications in *Biological Abstracts* (1926–71)  45

**Figure 4–1**  Masters and Apprentices Among Laureates in Physics and Physical Chemistry (1901–72), According to Decade of Apprenticeship  101

**Figure 4–2**  Masters and Apprentices Among Laureates in Biological Chemistry and in Biological Science (1901–72), According to Decade of Apprenticeship  102

**Figure 4–3**  Laureate Masters and Apprentices Associated with J. J. Thomson and E. Rutherford (1901–72)  103

**Figure 4–4**  Age Differences between Masters and Laureate Apprentices (1901–72)  119

**Figure 6–1**  Percent of Collaborative Papers on Which Laureates Are Highly Visible Authors  181

# LIST OF TABLES

Table 3–1    Elite Origins of American-Reared Laureates (1901–72) and Other Elites    64

Table 3–2    Socioeconomic Origins of American-Reared Laureates (1901–72), Scientists Receiving Doctorates (1935–40), and Employed Males    66

Table 3–3    Religious Origins of American Professoriate and Laureates (1901–72) in Fields of Science Eligible for Nobel Prizes    75

Table 3–4    Percent of Jews Among American-Reared Laureates (1901–72) and American Professoriate    79

Table 3–5    Socioeconomic Origins of American Professoriate and of American-Reared Laureates (1901–72), According to Religious Origins    80

Table 3–6    Baccalaureate Origins of American-Reared Laureates (1901–72) and Male Graduates (1924–34)    84

Table 3–7    Baccalaureate Origins of American-Reared Laureates (1901–72), According to Socioeconomic and Religious Origins    87

Table 3–8    Doctoral Origins of Laureates (1901–72), Members of National Academy of Sciences, and Other Scientists (1920–49)    90

Table 3–9    Socioeconomic and Baccalaureate Origins of American-Reared Laureates (1901–72) Holding Elite Doctorates    94

Table 4–1    Laureate Master-Apprentice Pairs (1901–72): Status of Masters at Time of Apprenticeship    109

Table 4–2    Mean Age at Prize of Laureates (1901–72), According to Doctoral Origins and Master's Status    114

Table 4–3    Mean Age at Prize of Laureates (1901–72), According to Master's Status at Time of Apprenticeship    114

Table 4–4    Mean Age of Laureates (1901–72) at Prize-Winning Work and at Prize, According to Master's Status    117

Table 4–5    Laureate Apprentices Whose Awards Preceded Their Master's    121

Table 5–1  Doctoral Origins and First Jobs of Laureates (1901–72) and Matched Sample of Scientists  150

Table 5–2  First Jobs of Foreign-Trained Laureates (1901–72)  155

Table 5–3  Age at Full Professorship: Laureates (1901–72), Members of National Academy of Sciences, and Sample from *American Men of Science,* According to Prestige of Institution Making Appointment  159

Table 6–1  Mean Age of Laureates (1901–72) at Time of Prize-Winning Research, According to Year and Field of Prize  166

Table 6–2  Age Distribution of Laureates (1951–72) at Time of Prize-Winning Research and of American Scientists (1970)  169

Table 6–3  Sites of Prize-Winning Research: Laureates (1901–72)  171

Table 6–4  Percent of Laureates (1901–72) Cited for Collaborative Research and Multi-Authored Papers Published in Comparable Years  177

Table 6–5  Percent of Collaborative Papers on Which Laureates Are Highly Visible Authors  182

Table 6–6  Organizational Mobility and Promotion of Laureates (1901–72) Five Years After Prize-Winning Research  194

Table 6–7  Mean Age of Laureates (1901–72) at Time of Prize-Winning Research, at Time of Election to National Academy of Sciences, and at Prize  197

Table 7–1  Mean Age of Laureates (1901–72) at Time of Prize, According to Field and Year of Prize  217

Table 7–2  Mean Interval between Prize-Winning Research and Prize, According to Field and Year of Prize (1901–72)  218

Table 7–3  Percent Change in Publications Five Years Before and Five Years After Prize, According to Age and Eminence at Prize  225

Table 7–4  Affiliations of Laureates (1901–72) Five Years After Prize  241

Table C–1  Laureates in Physics, According to Specialty and Year of Award (1901–72)  292

**Table C–2** Laureates in Chemistry, According to Specialty and Year of Award (1901–72)    292

**Table C–3** Laureates in Medicine or Physiology, According to Specialty and Year of Award (1901–72)    293

**Table E–1** Age-Specific Annual Rates of Productivity of Laureates and Matched Sample of Scientists Who Survived to Each Age    302

# INTRODUCTION TO THE TRANSACTION EDITION

Now almost a century old, the Nobel prize remains the supreme symbol of accomplishment in science. Greatly esteemed by the general public, the prize evokes far more complicated responses from scientists; disdain, veneration, desire, and ambivalence are standard reactions, sometimes separately and sometimes in tandem. Indeed, few scientists are indifferent about the prize. This was so when *Scientific Elite* was published and it continues to be so now. The aura of the prize endures as does the impulse of free-spirited scientists to poke fun at it. No other scientific prize, as far as I know, has spawned its own anti-prizes. Twenty years ago, a band of oceanographers, statutorily ineligible for the prize, formed the "Laureates of the Albatross" and since 1991, an assortment of M. I. T. and Harvard Nobelists have been ceremoniously awarding Ig-Nobel Prizes to recognize "individuals whose achievements cannot or should not be reproduced."[1]

And, as if reading from a common script, the newest Nobel laureates react to the call from Stockholm much as they have for decades. They manifest surprise that the great event has occurred at all while confessing that they thought it might. Phillip Sharp, the laureate in bio-medicine for 1993, "could hardly believe the news. 'I was shaking.'"[2] Yet his secretary had kept a bottle of champagne on hand for fifteen years. E. Donnall Thomas, the laureate in medicine for 1990, was incredulous, "I really never felt the prize would go for patient-oriented research"[3] Clifford Shull, the laureate in physics for 1994, thought he might win the prize "throughout the 1950s, 60s, 70s, and 80s.... 'Well yes,...I guess that's inbred in any scientist.'"[4]

*Scientific Elite* is about three related subjects: Nobel laureates in the United States, the Nobel Prize as an institution, and the stratification system of science. It draws on biographical studies of all American laureates chosen between 1901 and 1972 and on interviews with some four-fifths of those living in the United States at the latter time, while investigating their careers and showing how their research accomplishments are connected with opportunities and obstacles shaped by the stratification system of science. It also investigates the extent to which merit and privilege are connected throughout the laureates' lives. Along the way, it details social processes of accumulation of advantage in science, whereby resources required for further accomplishment accrue to those who are already accomplished, and it earmarks a spate of unintended consequences of the Nobel prizes, particularly those the prizes have had for their winners. Finally, the book explores "the prize" as an evolving social institution: how it has become the "gold standard" for measuring all other awards in science; how it has been appropriated as a symbol of prestige by individuals, organizations, and institutions associated in any way with the prize-winners; and finally, how it has became an ever more blurred social metaphor for supreme accomplishment of any kind.

The book was first published twenty years ago. Yet, with predictable regularity, its contents are reactivated by the press and by scientists when the prizes are announced each autumn, with a focus above all else on the effects of winning the Nobel prize. Alfred Nobel had hoped that his prizes would honor great scientific accomplishment, and provide enlarged incentive and wherewithal for new research. However, it has become increasingly clear that the prize has been more effective in achieving Nobel's first objective than his second. For the laureates, the great honor has proved a decidedly mixed blessing. Every laureate I interviewed derived satisfaction from both the honor and the not inconsiderable cash. (No scientist has ever turned down a Nobel prize.) But, as chapter 7 shows, the socially induced demands that come with the prize divert scientists from their research (and greatly reduce their rate of publication), not only in the short run but in the long run as well.[5] Moreover, these unintended consequences are most severe for young laureates and for others who, not having achieved eminence before the prize, have not learned to cope with fame. All this has become so much a part of scientists' thinking about the prize that it has been elevated into "the not-so-amusing metalaw... [which] states that the receipt of the Nobel Prize marks the end of one's productive career."[6] Sociologically evocative and theoretically interesting as this part of the study may be—reaffirming as it does—Émile Durkheim's observation that abrupt upward mobility can produce virtually as unsettling effects as abrupt downward mobility—it is not a subject I want to develop here. Instead, I touch on other themes and questions in the book that have become subjects of continuing discussion, focussed scholarship and occasional controversy.

First, what are the limitations of the prize and how have these affected its legitimacy over the decades?

Second, and closely related, what effects has the prize had on the behavior of scientists and on the development of scientific knowledge?

Third, what consequences derive from the prize having become the gold standard for awards in science and how is this linked to the proliferation of prizes in science?

Fourth, do the laureates actually constitute an ultra-elite in science? Is the sharply graded stratification system described in *Scientific Elite* inherent in modern science or is it only distinctive of American science in the late twentieth century?

And last, do the marked changes occurring in American science over the last two decades call for changes in the principal conclusions of the study?

This inventory of renewed questions omits much else that is treated in *Scientific Elite* and most of the research the book has evoked. Thus, I do not treat ongoing studies of accumulation of advantage and disadvantage in science and outside, in domains as varied as gender and ethnic stratification and the stratification of organizations generally.[7] Nor do I take up studies of the still poorly understood nature of scientific creativity and ability,[8] including those centered on the ages at which scientists make contributions of enduring significance;[9] how it happens that Jews are greatly "overrepresented" among winners of the science prizes and women are greatly "underrepresented";[10] whether, and if so, in what ways, the Nobel prizes reflect the health and illnesses of national scientific establishments[11] and the effectiveness of peer review systems;[12] how the reward system of science compares with reward systems in the arts, politics, corporations, and the professions;[13] and the extent to which the discipline of economics and its relatively new Nobel award resemble the original disciplines capable of being awarded the prize—physics, chemistry, and the Nobel composite of physiology or medicine.[14] Each of these subjects has its own substantial literature which plainly cannot be considered here.

## LIMITATIONS OF THE PRIZE:
## RULE-GOVERNED CONSTRAINTS AND SELECTION ERRORS

As the significance of the prize has grown and become almost universally recognized, so has recognition of its limitations, especially those deriving from the distinctive rules and practices that govern the selection of laureates. Though treated in *Scientific Elite,* the limitations of the prize did not get the systematic attention they deserve. These have since been examined more fully in a paper published in *The American Scientist,* a paper reprinted often enough both here and abroad to indicate at least a degree of interest in the subject.[15]

A primary question is whether the prize recognizes surpassing scientific excellence.[16] This deceptively simple question elicits two seemingly contradictory answers. The first is yes, of course it does. In the aggregate, the laureates have made contributions to science of the highest order. To be sure, stronger cases can be made for some awards than for others but, on balance, few prizes have been awarded for less than excellent work and few outright errors have been made. As I noted in *Scientific Elite,* the prize committees have been decidedly risk-averse and future-oriented. Rolf Luft, one-time chairman of the Committee on Medicine put it this way, "The Nobel committees prefer deliberate conservatism to displays of daring."[17] In statistical terms, this means that the committees have decided that if errors are to be made, they had better be "errors of the first kind," passing over worthy candidates, than "errors of the second kind," awarding prizes for work that would later on prove worthless. This decision rule has little to do with whether the prize selections are just or fair, but it has much to do with the fact that practically all prizewinners have done truly excellent if not always the most original work. It is a rule that has minimized embarrassment and enshrined safe choices.

## Types of Limitations

The second answer to the question whether the prize recognizes scientific excellence is no. Much scientific excellence of the highest order has not been and cannot be recognized by the awards; the rules governing the eligibility of contributions greatly limit the range of research that can be considered for the prize and this is probably further circumscribed by the limits placed on eligible nominators. Nor has the prize unfailingly recognized all those who contributed in a major way to the cited research. The requirement that only a limited number of prime contributors to a given research be identified makes for selection errors as, of course, do lapses in committee judgment about the identity of major contributors.

## Rule-Governed Constraints

From the beginning, Nobel prizes have been awarded in accord with rules spelled out in the Foundation's Statutes, rules laid down in 1900 and little changed since. To be sure, the rules limiting the fields in which prizes are given, the number of laureates chosen each year, and the groups eligible to nominate have, in important respects, contributed to the luster of the prizes by producing a scarcity of laureates. But they have also limited the capacity of the prize to identify contributions to science of the highest order.

*Limits on Fields.* The primary and enduring constraint has stemmed from the fact that prizes can be given in just three fields—physics, chemistry, and physiology-or-medicine (the composite field which has come to cover parts of the biological and biomedical sciences). Only one more discipline has been tacked on since. The Alfred Nobel Memorial Prize in Economic Science was added in 1969, thus extending the disciplinary bounds somewhat, but not much. The limitation of disciplines plainly excludes a significant fraction of the current scientific enterprise—the earth and marine sciences (the revolution of plate tectonic theory has won "no Nobel notoriety"[18]), mathematics (neither Gödel nor von Neumann could have received Nobels), much biological research, and all of the behavioral and social sciences, except economics.[19]

Tradition has further circumscribed the specialties the committees consider eligible for awards even within the four specified fields. Astrophysics, for example, was ruled out of bounds for the physics prize for decades,[20] and evolutionary biology continues to be out of bounds for the biomedical prize. The Nobel committees have occasionally relaxed the customary disciplinary boundaries but they have not done so often and show no signs of doing so in the future. The committees have also evolved preferences for certain kinds of research within the disciplines in which prizes are awarded. Early on, fundamental research came to be favored over applied research and, in spite of Nobel's will explicitly including inventions and "improvements" in the purview of the prizes, the committees seemed disinclined to honor this kind of contribution.[21] Further, in accord with their positivist perspectives, the committees more often opted for experiment than theory (though theory, especially physical theory, has come to be more acceptable);[22] and when theory was under consideration, they opted only for theory that had been thoroughly supported by experiment.[23] These preferences have ruled out awards for grand theoretical formulations. Most important of all is the Committees' preference, indeed insistence, that work be seasoned enough to be practically incontrovertible.[24] In Stockholm, safety clearly outweighs timeliness in deciding on prize-winning research. It seems evident then that restrictions on fields, specialties, and the kinds of research considered eligible for awards "reinforc[e] a narrow and conventional stereotype"[25] of the broad spectrum of science and, in some quarters, make the prize appear anachronistic and unjust.

*Limits on Numbers.* Disciplinary limits aside, scarcity also derives from the rule of three; according to the statutes, no more than three scientists may share an award. This would-be anti-inflationary rule assumes that it is possible to identify no more than three prime contributors to most research (especially important research) and in effect celebrates individual scientists in an era of increasingly collaborative research. This, too, causes the awards to sometimes be regarded as anachronistic and unjust. These rule-governed constraints ensure that many more scientists have done research of Nobel quality than can become Nobel laureates. They produce an accumulation of "uncrowned laureates," who are peers of prize-winners in every relevant respect except that of having won the prize. Chapter 2 likens these uncrowned laureates to the "immortals" who were not elected to one of the forty seats in the French Academy; those the French like to call "occupants of the forty-first chair." With regard to the Nobel prize, occupants of the forty-first chair include all past scientists of genius or great talent who never gained the prize; scientists who, despite their great accomplishments, will never do so; and of course, those eventual Nobelists whose work has not yet been honored.

I made much in *Scientific Elite* of that concept of the occupants of the forty-first chair since it serves as an important corrective to the notion that Nobel laureates constitute "the world's best scientists." Knowledgeable scientists are abundantly aware that occupants of the forty-first chair exist not only in the disciplines in which the prizes are awarded but in all the sciences.[26] The concept of the occupants of the forty-first chair also holds considerable sociological interest. It not only calls attention to the limitations of the prizes as mechanisms for exhaustively identifying the scientific elite but also to the widely experienced sense of injustice created by reward systems marked by great scarcity.[27] I shall return to the occupants of the forty-first chair when I take up questions about the ultra-elite in science.

*Limits on Nominators.* The rules governing eligible nominators produce their own variety of constraints. To become candidates for the prize, scientists must be formally nominated and nominators can come only from two cadres: those with permanent nominating rights and those invited to nominate in a given year. The cadre of permanent nominators consists of members of the Royal Swedish Academy of Sciences, members of the Nobel committees, professors at selected Swedish and Danish universities and the Caroline Institute (which oversees the prize in biomedicine) and all prior Nobel laureates. Temporary nominators are described as "other scientists" from whom the Academy and the Caroline Institute see fit to invite proposals.[28] In principle and perhaps in practice, these arrangements could variously skew nominations in a number of ways:

• By specialty—since permanent nominators represent only a fraction of specialties in fields in which prizes awarded, this increases the prospects of scientists who work in these selected specialties and reduces the prospects of others even though their work fits the general guidelines.

• By nationality—since Northern Europeans and now Americans predominate among prior laureates, this, combined with a tendency toward chauvinism in nominations, may have contributed to the concentration of Northern European laureates for a time and of Americans later on.

• By institutional affiliation—since laureates tend to cluster at particular institutions, this may reinforce the tendency for new laureates to come from these same institutions,[29] notably the great research universities and institutes; chapter 7 details such marked clustering of American laureates.

• By association with Nobel prize-winners—to my mind, this is especially consequential in light of the pattern of "laureates breeding laureates," described in chapter 4; associates of the laureates may have a better chance of being nominated than other equally qualified scientists.

In practice, such skewing of nominations is important only if the committees fail to invite a sufficient number of yearly nominations from a broad and knowledgeable group and if committee decisions are influenced by the sheer number of nominations candidates receive. The existence of effects of this kind have often been the subject of speculation but, owing to the much vaunted secrecy attending Nobel deliberations, it has not been possible to examine the matter systematically throughout the near-century of the prize. However, in 1974, the statutes were amended to allow scholarly examination of the archives on prize selections made fifty or more years before and thus has opened up new archival research on the early years of the prize.[30]

Archival studies to date have been of three sorts: historical inquiries into particular awards, chiefly, those to Einstein, Arrhenius, and Nernst; studies of the role played by the scientific "interests" of committee members in their prize decisions and quantitative analyses of nominators, candidates, and prize-winners. The archival studies, which happen to have focused on the physics and chemistry prizes more than on biomedicine, find a marked consensus in nominations early on.[31] As I assumed in *Scientific Elite*, there was a substantial queue of nineteenth-century giants to be honored. Nominators soon proved more enthusiastic about candidates from their own countries than elsewhere (despite Nobel's fiat that nationality not be considered in selections for the prize). Four-fifths of the British nominators proposed British candidates between 1901 and 1929; three-fourths of the French nominators, French candidates; three-fifths of the Germans, German candidates; and half of the Americans, American candidates.[32] Furthermore, national loyalties were soon compounded by institutional loyalties. Crawford's detailed analysis of German nominees and nominators affiliated with the elite Kaiser Wilhelm Institutes shows a clear pattern of such loyalties, a pattern that in the course of time became reinforced by the concentration of laureates associated with these same institutes.[33]

Not surprisingly, the Nobel committees continue to be alert to such nationalist tendencies in nominations and express concern when they occur. Indeed, Bo G. Malmstrom, until recently, head of the chemistry committee, has publicly complained that "members of the committee have been unhappy for some years about the fact that Americans always seem to nominate Americans even though they are supposed to be nominating 'the most important chemists in the world.'" Malmstrom went on to observe "that in the great majority of cases, Americans nominate chemists from their own departments.... Americans are not deficient in their knowledge of work by foreigners but are just 'more chauvinistic in this regard.'...the problem is particularly evident at big universities like Harvard and the University of California at Berkeley." Further, he observed that "it is obvious from the pattern of nominations that Americans discuss them among [themselves]. 'We take confidentiality very seriously...apparently this is not in the American tradition.'"[34]

Since knowledge about colleagues' work and nationalist bias press in the same direction, skewed nominations seem all but inevitable. Yet the central issue is whether there is evidence showing that such skewing did in fact affect the selection of laureates. Marc Friedman, a historian, and Elisabeth Crawford, a historical sociologist, say that it did not, while a group from the Science Studies Unit at the University of Bielefeld, Gunther

Küppers, a physicist, and the sociologists Norbert Ulitzka and Peter Weingart, disagree. Friedman and Crawford base their conclusions mainly on archival data (on the physics and chemistry prizes 1901-31) and on their studies of the Swedish scientific community, particularly of the committee members.[35] The Bielefeld group, in contrast, draw principally but not exclusively on quantitative data they assembled from the archives on nominees and nominators.[36] Friedman and Crawford conclude that choices made in the early years (particularly the delay in the award to Einstein) depended much less on numbers and sources of nominations than on the scientific and personal predilections of the committee members, particularly their inclinations to use the prizes to advance the directions they wanted Swedish science to take. The awards, they say, cannot be understood without placing the committees' actions within the context of the Swedish scientific community at the time.[37] On this hypothesis, neither the numbers of nominations candidates received nor the sources of their nominations were decisive. Neither academy members nor Scandinavian professors nor laureates prevailed as Friedman observes in a separate paper on research in the Nobel archives.[38]

However, the quantitative data on the nationality of nominators and candidates reported by Küppers, Ulitzka and Weingart apparently indicate otherwise.[39] To their mind, nationalist tendencies in nomination did matter. "The Nobel committees and the Academy in their prize decisions...did not level the national differences that exist[ed] in the nominating procedure." Only the British received substantially more prizes in both physics and chemistry than predicted from the sheer number of British nominators or nominees.[40] Drawing on slightly different but nonetheless consistent data, Crawford agrees that "the nominating system...became infused with chauvinism" in this period but concludes that the committees' choices were unaffected by the nominating system.[41] The committees' deliberations may not, in fact, have been affected by the number of nominations candidates received, but the quantitative evidence is not consistent with this conclusion. In the early period, at least, the number of laureates chosen from particular countries closely parallels the national distribution of nominees.

Furthermore, Küppers, Ulitzka, and Weingart contend that laureate *nominators* in chemistry and physics had "extraordinary influence" and that their nominations held sway "twice as often as those of other nominators" between 1901-29. A striking conclusion but one that must remain provisional absent information about how the calculation was made.[42]

Even so, these data do not indicate whether laureates typically pressed the causes of their own students and whether this accounts for the, to me, still arresting finding reported in chapter 4 that slightly more than half the American laureates were trained by prior laureates. Of the 92 whose careers I studied, 48 had worked as a graduate student, post-doctoral fellow, or a junior collaborator with older laureates. Much the same pattern holds for the foreign-born and foreign-trained laureates who came to the United States as for the home-grown products. As the laureate in economics, Paul Samuelson, noted in his own acceptance speech, "I can tell you how to get a Nobel prize. One condition is to have great teachers."[43]

Having had great teachers helps explain the observed pattern of "laureates having trained laureates." Another, complementary, explanation is self-selection. The ablest young scientists are tuned into important developments in their fields and know with whom to study. (There is support for this interpretation inasmuch as some two-thirds—69 percent—of the laureates' apprentices studied with the senior scientists before those seniors got their prizes.)

A third proposed explanation, and the one at issue here, is whether the laureates are quite literally a "self-perpetuating elite," as Crawford puts it.[44] If so, the data should show that the laureates nominate their own students more often than they do others, that their students get little or no support from other nominators, and that the laureates' nominations hold sway, thereby accounting for the filiation of laureates reported in *Scientific Elite*. At the time the monograph was published, it was not possible to say; relevant archival data were not available for any period, nor are they now available for the last half-century.

However, data on nominators and candidates in the years 1901–37 have now been published for the physics and chemistry prizes.[45] Only eight Americans were selected for prizes in that early period,[46] and seven of these were apprenticed early in their careers to Nobel laureates. Of the seven, just two were nominated by their laureate "masters." Carl Anderson, laureate in physics for 1936, was nominated by R. A. Millikan, who supported Anderson's candidacy for three years running until he got his prize. And Clinton H. Davisson, laureate in physics for 1937, was nominated by O. W. Richardson but not until the second year of his candidacy, having first been proposed by the laureate, A. H. Compton, and the laureate-to-be, James Franck, among others. As it turns out, between 1934 and 1936, Anderson received 20 nominations, most coming from outside the United States. In addition to Millikan's nomination, these came from the laureates A. H. Compton, Perrin, von Laue, Planck, and de Broglie. Between 1929 and 1937, Davisson was proposed by 44 nominators, most of them also not American. At various times, in addition to Richardson, Davisson was nominated by the laureates A. H. Compton, Franck, Millikan, de Broglie, Schrödinger, and Appleton, this a laureate-to-be who would not get his own prize until 1947. Presumably each of these laureates could have touted his own apprentices rather than Anderson or Davisson but did not. As evidence becomes available for the more recent period, we will be able to determine whether the laureates really are a self-perpetuating elite in the narrow sense.[47]

So much for the effects of rule-governed constraints on the number of recipients and nominators limiting the capacity of the Nobel prizes to identify the most significant contributions to world science.

## Errors in Selection

As I noted at the outset, the Nobel establishment has long preferred orthodox to venturesome decisions. Contributions that seem at all controversial—either scientifically or statutorily—have generally been passed over in favor of those that are rock solid. It has been more important to the committees to make incontrovertible choices than to make inspired ones. Since there have always been far more qualified candidates for the prize than can possibly receive it, this preference has allowed the committees to compile an extraordinary record and has resulted in more questions about errors of exclusion than of inclusion.

This does not mean, of course, that all research cited for the prize has stood the test of time. It has not. There was, of course, the award in biomedicine to Johannes Fibiger in 1926 which cited his discovery that a cancer could be induced experimentally, in this instance by the parasitic worm *Spiroptera*. If valid, this obviously would be a contribution of great consequence. But that work proved to be mistaken. Perhaps as a result, the Nobel committees have avoided giving any prizes for research on cancer for 40 years until they finally got round to recognizing Peyton Rous for his discovery of viral-induced sarcoma.

In any event, the Nobel committees cannot be faulted for what they could not know when their prize decisions were made. Errors established after the prize testify to the continuing revision of the scientific corpus. That so few *ex-post* errors have been identified in the history of the prize indicates that the committees' preference for safe decisions has had its intended effect; the Nobel awards prove sound over the long term.

A quite different problem is posed by Svedberg's award in 1926 for his work on colloids and the nature of Brownian movement.[48] At the time, Einstein was only primus among several who were saying that the Svedberg model was simply wrong. Yet Svedberg's work on the ultracentrifuge, before and after the prize, is widely recognized as consequential not just in chemistry but in medicine and physics as well. When judgments of prize worthiness are made, the particular contribution cited for the prize evidently carries less weight than the corpus of the recipients' work.[49]

In short, the committees have made few errors of commission. It is the errors of omission that are troublesome, those in which they passed over extraordinary contributions and those in which they failed to identify all the principal contributors to work they elected to honor. In the latter instance, the rule limiting each award to no more than three recipients surely creates difficulties when reality exceeds that rule-governed maximum. But especially when all the major contributors could have been honored without breaking the rule of three, the committees have been faulted both for excluding scientists who knowledgeable experts judge were critical to the research[50] and those who made the same discoveries independently.

As *Scientific Elite* notes in describing the controversy surrounding Selman Waksman's award in 1952 for the discovery of streptomycin, errors of exclusion involving collaborators almost but not quite always involve those junior to the prize-winner.[51] In addition, such historians of science as Margaret Rossiter have begun to document instances of exclusion of women collaborators, junior and senior. However, in the absence of evidence on the gender distribution of "excluded" contributors, it is not possible to say whether women are disproportionately represented among them.[52] In any case, authentic exclusions both of junior collaborators and of women would be prime examples of the Matthew Effect, which, according to Robert Merton, "consists of the accruing of greater increments of recognition of particular scientific contributions to scientists of considerable repute and the withholding of such recognition from scientists who have not yet made their mark."[53]

Rossiter, only partly in jest, suggests that credit is so often denied women collaborators that the Matthew Effect should, in fact, be named the "Matilda Effect," after Matilda Joslyn Gage, the American "feminist, suffragist... and early sociologist of knowledge, who glimpsed what was happening, perceived the pattern, deplored it, but herself experienced some of the very phenomena described here."[54]

The most persuasive evidence of unwarranted exclusion comes from laureates themselves when they declare that others should share their credit. Rossiter notes, for example, that G. H. Whipple, laureate in medicine in 1934, recognized his debt to his collaborator Freida Robscheit-Robbins, lavishly praised her and divided his prize money with her and two women assistants.

Lise Meitner did not fare so well. Widely recognized by the scientific community for her significant role in the work, her exclusion from Otto Hahn's prize in chemistry in 1944 had quite a different outcome. Hahn mentions in correspondence having had an "unhappy" conversation with Meitner after his award and, goes on to say that the Nobel was "given to me for work I had done alone or with my colleague Fritz Strassmann" and adds that he "really had no responsibility for the course events had taken."[55] Quite so with regard to responsibility. But, according to the census of Nobel nominations, Meitner was nominated for the prize thirteen times in nine different years, between 1924 and 1937, and as far as one can tell from the published sources, almost always jointly with Hahn.[56]

The pattern recurs (in part) some fifty years later, in 1985, when the Nobel prize in chemistry was awarded to Jerome Karle and Herbert Hauptman.

Isabella Karle, her husband's longtime collaborator, did not share in it. Her contribution to the research "is said to have been considerable... for fifteen years, scientists did not appreciate [the Karle-Hauptman work] until [she] immersed herself in the mathematics of the research and pointed out to others its potential applications in the understanding of crystals and the development of drugs." One of the Karle daughters, a chemist, put it this way, "My mother worked very closely with my father all these years. . . if it was not for her, it would have taken much longer for the methods that he and Herb Hauptman developed to be accepted."[57]

The central issue, of course, is what constitutes an unambiguous and critical contribution either to a particular research or to a larger research program. The exclusion of Jocelyn Bell from the 1974 physics prize puts this issue sharply. That prize cites Anthony Hewish for research in radio astrophysics and "particularly for his role in the discovery of pulsars." Bell, a graduate student at the time, was indeed the first to observe pulsars, but scientists continue to disagree about whether that contribution was sufficient to merit the prize.[58] Not surprisingly, senior scientists are more apt to confine responsibility and credit to the originators and planners of research and to exclude "those who merely carry out orders."[59] Still, the question remains whether the research would have gone as it actually did if others had carried out the orders. Would someone in Hewish's laboratory other than Bell have observed pulsars and might Hewish, in fact, have gotten the award for work other than the pulsar discovery? One cannot say. Such hypotheticals are no substitute for the implied experiment. What one can say is that neither Meitner nor Karle were mere recipients of "orders" and their being excluded from Nobel awards inevitably raises the question of the part played by gender in committee decisions.

Nobel awards for contributions involving multiple independent discoveries raise different problems. Here the issues are whether discoveries made independently are really "the same" and, if so, whether the committees have included all the major contributors or only some of them. The issue of "sameness" is at the center of the exclusion of William H. Oldendorf from the 1979 prize in medicine for the development of the CAT scan which went to Godfrey Hounsfield and Allan Cormack. Oldendorf, a neurologist at U.C.L.A., published the first paper on radiographic tomography, received the earliest patent for a scanner, and shared the Lasker award with Hounsfield for the "conception of a scanning system." As Oldendorf put it when the prize was announced: "Anybody who goes into science expecting to win the Nobel Prize is about as realistic as a person going to Las Vegas to get rich. But my feeling is that I should be preparing to go to Stockholm."[60] Why Oldendorf was passed over may be revealed when the archives for 1979 are finally open. Speculation at the time had it that Oldendorf's work did not provide the mathematical foundation for scanning and, as a consequence, did not satisfy the fundamental science "wing" of the Nobel As-

sembly, the body charged with accepting the Medicine Committee's recommendations.[61] This accords with Bettyann Kevles' assessment after careful study of the history of the CAT scan, as part of her history of medical
imaging.[62]

In review, though it seems clear that the prizes honor great scientific
accomplishment, formal constraints imposed by rules governing the awards
(primarily the limits on eligible fields and the scarcity of prizes) and informal boundaries imposed by committee members on contributions considered capable of receiving prizes have had various unintended consequences.
They have created the occupants of the forty-first chair, increased skepticism about the award, occasionally disputed its legitimacy and raised questions about whether the image of science fostered by the prizes has enough
to do with the reality of contemporary scientific practice.

## EFFECTS OF THE PRIZE

Some observers worry that the very existence of the prize has put such
a premium on "winning" in science that they doubt it should be continued.
Long after his own award, Salvador Luria, laureate in medicine for 1969,
gave voice to this concern and provided a textbook example of normative
rejection of "goal displacement," the socio-psychological process wherein
secondary or illegitimate goals are substituted for primary institutionalized
ones: "The goal in science should be to find out things—not to win a
prize.... Yet in conversations with some of my younger colleagues, I get
a sense that it has become a goal, and that is not good. I think it would be
better if there were no prizes."[63]

But few would adopt the solution proposed by the physicist, Tullio
Regge: "abolishing the [Nobel Prizes] would only make them more valuable. Better to inflate them—give out five, ten, a hundred a year until no
one wanted them anymore."[64] Yet since the prize is not about to disappear
or to become irretrievably inflated, what can be said about its effects on
scientists and on scientific practice?

To no one's surprise, receiving the prize exhilarates and gives laureates
a sense that their work has value. As laureate in medicine Peter Medawar
put it: "scientists are most eager for the high opinion of their peers. The
effect of gaining an award [and Medawar should know] is a great moral
boost—this expression of the confidence and esteem of others will...
perhaps help them do better than before. Very likely, too, the prizewinner
will want to show everyone that it wasn't all a fluke."[65]

And yet, as *Scientific Elite* noted, the laureates' enthusiasm is not unre-
served. Most of them have rather more complicated reactions to the prize.
They recognize in cost-free fashion that the work of peers could just as
well have been chosen, acknowledge that others could reasonably have

shared in their prizes, and, after the first flush of public admiration, often retreat from the limelight. In the plainspoken language of the irreverent physicist Richard Feynman: "the Nobel Prize has been something of a pain in the neck."[66]

In combination with those mandatory reiterated assertions of humility, such responses reflect layers of ambivalence about the prize as well as a paradox.[67] Michael Mulkay notes in his astute analysis of the Nobel lecture as a rhetorical form that central features of the ceremonially required lecture are "scrupulous avoidance of self-praise, the downgrading of their own achievements,... [and] the reassignment of praise by the laureates to their colleagues and families."[68] But, of course, if their achievements are as insignificant as they claim, they surely should not have been enNobeled and this, few are ready to admit. This "structural joke," Mulkay observes, tells us much about the inauthenticity of ceremonial claims to humility.

The prize, as I have noted, has a good many disagreeable effects on its winners. It disrupts research, unsettles collaborative relationships[69] and, as Medawar implies, reinforces the sense that one must continuously demonstrate through further accomplishment of a high order that the award was no mistake. It also has the effect on some winners of inflating already substantial egos beyond what is good for their work or for those in their vicinity.

Unwanted effects of the prize are not limited to the laureates. "Contenders" also feel the Nobel's weight. Since research capable of being awarded a prize is not hard to spot, contenders generally know who they are. Their prospects are often reviewed and ranked, fueling Nobel ambitions in all but the most practical or remote of scientists. Such ambitions can spur both superb science and dubious behavior, as J. D. Watson's now classic and still engrossing scientific autobiography, *The Double Helix,* makes plain.[70]

The same conclusion about effects of questing for the prize can be drawn from the decades-long competition between Roger Guillemin and Andrew Schally to isolate peptide hormones produced in the brain. Their research which would lay the groundwork for the field of neuroendocrinology led ultimately to the award of the prize in biomedicine in 1977. Over the years, their work was marked by extraordinary persistence in the face of failure and a striking array of competition-induced behaviors—insistence on priority, secrecy, and premature publication.[71]

Did Nobel ambitions play a role? Guillemin says no. After the fact, he maintained that he never expected the prize. "It has been a long road, an arduous road...but there was nothing conceptually revolutionary in this field which made me think a Nobel prize had to be awarded for it."[72] However, Schally confessed, "It was a dream, I can't deny it." An anonymous member of one research team puts it this way: "Despite what either may say about not caring about the prize, they've been after it for years...."

Both of them would have been happier people if the prize didn't exist. It has its negative features and the world might be better off without it."[73]

Receiving the prize is one thing, losing out is quite clearly another. Some obvious prospects for the prize left unlaurelled become envious and embittered, forever forced to explain to themselves, and others, why they "failed" to get the prize. Apropos Oldendorf's loss, Rosalyn Yalow, 1977 laureate in medicine, observed: "'The only thing that will make it up is if he gets another money award,'...[she] then pause[d] for a second. 'But you don't go down in history with money. You go down with a Nobel.'"[74]

Mixed as the effects of the prize often are on laureates and laureates manqués, the broader sociological question concerns its diverse effects on the community of science and scientific practice, involving such matters as recruitment to the scientific career, the perpetuation of elites, sex stratification in science, civility in science, and, hardly least, the advancement of scientific knowledge. This question has its complexities too.

Joshua Lederberg, laureate in medicine in 1958, has expressed a firm sense of the shortcomings of the prize. Yet he continues to think that the prize has a beneficial effect by "conferring an aura on science," making it more important to young people considering careers and even to those in fields for which prizes are not available, those who cannot and do not aspire to the prize. Evidence for this being the case turns up in autobiographies of scientists and in writings about the prize.[75]

Correspondence with George Stigler, who would be awarded the Nobel prize in economics in 1981, reveals a distinctly economic take on these matters:

> [I]f Adam Smith is right, that men overestimate their chances of winning great prizes in lotteries of all sorts, the Nobel Prize may even have served to increase slightly the number of able young scholars entering the covered fields...and to lower slightly the average earnings in these fields. It would be difficult to argue that this redistribution of talent is socially useful, however: there is no reason to believe that the covered fields were drawing less talent, and uncovered fields more talent, than served to equalize prospective marginal products.[76]

It is more difficult to assess the effects the prize has on perpetuating elites and on gender stratification in science. To be sure, the Nobel prize reinforces and legitimates elite status, both of individual scientists and scientific organizations. To the extent that it also reinforces informal networks that exclude women scientists, it reinforces sex stratification. However, the dominance of elites and of elite institutions in science are endemic outcomes of the process of accumulation of advantage, through which resources for future work and rewards for past work are allocated on the basis of merit assessed in peer evaluations of prior contributions to science. Such a process, as *Scientific Elite* spells out, is bound to enlarge differences in achievements and ultimately to create elites. The perpetuation of elites based on such unequal access to resources seems to me to be affected far more by long-standing processes of assessment and allocation in science than by the existence of the Nobel prize. The prize may reinforce these large-scale existing patterns but not by much.

Its impact on gender stratification in science is also marginal and for much the same reason. Stratification in science based on the functionally irrelevant attribute of gender rather than on the functionally relevant criterion of scientific merit begins early in careers and, of course, such discrimination systematically limits access of women scientists to resources and rewards. The prize has little independent impact on this continuing pattern in science; women do not do better in the sciences sans Nobel prizes than in those with them. As for the effects of the prize on the advancement of scientific knowledge, what was noted in *Scientific Elite* is still the case: little is actually known, as distinct from being suspected, about the effects of the Nobel or any other prize on problem choice and foci of attention in science. Some observers think that the prize, with its emphasis on rewards, distorts problem choice, that it leads scientists to "swarm" like bees into new areas where prizes may be awarded, speeds their development and, in the process, diverts attention from other equally significant problems that are less likely to attract Stockholm's attention.[77] Qualitative research on problem choice among men and women scientists by Jonathan Cole and myself shows that first-class investigators are indeed motivated to take up important problems but, contrary to this imagery, many are competition-averse and loathe the idea that others might be "breathing down their necks."[78]

Prizes aside, we do know that the reward system accords more peer recognition to scientists who open new fields than to those who elaborate old ones, to those whose work illuminates fundamental and long-standing problems than more peripheral ones, to those whose contributions have ramifying implications for many fields, and, most of all, to successful outcomes rather than important failures. These broader features of the reward system affect problem choice by attracting scientists to some problems and diverting them from others. The choices may be somewhat reinforced by the Nobel prize in the fields in which prizes are available but it is these broad patterns of incentive in the reward system far more than the rare Nobel prize itself that significantly affect patterns of problem choice and foci of attention. However, systematic evidence on the determinants of problem choice and, by extension, the effects of the prize on the pace and directions of science, is in self-exemplifying style still absent.

I am inclined to think that the principal effect of the prize on science in the large is indirect; its influence on the public's image of science probably counts for more than its function as incentive for scientific accomplishment. Decades-long reiterated attention to the prizes and the laureates in the public press, to their great achievements and to the ceremony honoring them, announces to the public that great things are stirring in science, things worthy of public admiration and public support. The laureate in economics, George Stigler, goes farther. Even though the average educated citizen has "no possibility of understanding the work that won the prize, or [even] of tracing any connection between that work and contemporary well-being,...even the uneducated citizen knows that the laureate is the Life Baron of science." The public, Stigler holds,

> has good reasons for what it does.... [It] wishes to admire superior performance in every legitimate calling—athletic and military...as well as scientific. If there were a good objective measure of scientific performance...they would use this basis for selecting champions, rather than the more fallible choices of the Swedish Academy.... But presently it is the best ranking they have, and so they heap their kudos on the laureates.[79]

The public esteem accorded the laureates indicates that great scientific work is also esteemed, even as public worries endure about the diverse social impacts of science on everyday life.

## THE PRIZE AS THE GOLD STANDARD AND THE PROLIFERATION OF PRIZES

Defined as the gold standard by which all other scientific awards are judged, the Nobel prize has become a universal and instantly understood metaphor of supreme achievement, often in odd applications. The review of a novel whose central figure is "despicable...and consumed by self hatred" is headlined as "The Nobel Prize in Misanthropy,"[80] while an article in the sports pages about the cantankerous owner of the New York Yankees, George Steinbrenner, begins "There goes the Boss's Nobel Prize."[81]

The symbolic uses of the prize hardly stop with such oddities. Enlisting the prize and recruiting its laureates has become standard practice among universities and colleges, journals, publishers, industrial firms, laboratories, and, for that matter, other awards for scientific achievement. All are bent on increasing their derivative prestige. Universities and colleges publicize the counts of laureates associated with them in one capacity or another.[82] Since there is no statute of limitations on claiming laureates and since, as chapter 7 shows, Nobel prizes are typically given for research far from new, these claims usually say more about the past than the present. "Counting the number of Nobel Prizes at an institution," Jacques Barzun observed, "may be only recording extinct volcanoes."[83] Publishers, journals, and business firms also announce—usually in full-page advertisements—the laureates associated with them, while the guardians of other awards declare their prescience by enumerating the number of Nobelists they identified before the committees in Stockholm got round to doing so. All this results in emphasizing the acquisition of laureates for the prestige they confer rather than improving cognitive environments for new potentially prize-winning contributions, another classic instance of goal displacement.

Yet "having" a laureate is not always what it is cracked up to be. In one of the more ironic episodes of laureate recruitment, the University of Houston announced that it would not renew the appointment of A. J. P. Martin, the British chemist laureate who had been given its Welch chair, "because he had not published enough scientific papers in the five years that he has been a member of its faculty." According to Houston's dean, Martin had not "done all that much for the University." For his part, Martin countered that, "I certainly got the impression that I had this chair for as long as I was able to stagger into the laboratory."[84] Evidently, not having had a laureate before (or, one gathers, since), the university must have had grand expectations when it acquired the 64-year-old prize-winner. Other prize-winning acquisitions might take note.

The prize as universal metaphor for supreme achievement has also encouraged the more outré elements to seek a Nobel association. Think only of the astrological charts of 572 prize-winners in *The Astrology of Genius: A Study of Nobel Prize Winners*[85] or, better still, take note of the Nobel Sperm Bank (formally titled The Repository for Germinal Choice). Recent reports claim that the bank now contains deposits from Nobelists' sons as well as from prize-winners themselves.

The Nobel, having become the gold standard for awards, is no mere metaphor. Along with the great increase in the number of awards in science, the honoraria attached to the more munificent of them have been growing rapidly. Successive editions of the Gale directory, *Awards, Honors and Prizes*, lists some 3,000 prizes now available in the sciences in North America alone, five times as many as two decades ago.[86] Many are evidently designed to honor the donors quite as much as the recipients. After all, the greatest honoree of the Nobel prizes has surely been Alfred Nobel himself.

The most conspicuous two dozen or so of the new awards are rich by any measure. Among the richest of these are the Japan Prize (50 million Yen or about $500,000), the Kyoto Prize for basic science (45 million Yen or $460,000), the Fiuggi International Prize for medicine and biology, established at 500 million lire (about $300,000), the Bowen Awards of $300,000 each, and two large awards in technology, the Lemuelson Prize for Invention at $500,000 and the Draper Award for engineering at $375,000. Then there are the new awards that provide support for research such as the Donald Bren Fellowships at the University of California Irvine with grants of $1 million and the Prix Louis Jeantet, with grants of $1.36 million plus an award of $60,000. These awards fit comfortably with the MacArthur Fellows awards which vary with the recipients' ages and run between $150,000 and $375,000. These may not seem large when compared to the scale of some government research grants but coming as they do relatively free of strings attached, they provide a greater measure of freedom to the recipients. They also set new standards of opulence for science prizes, exceeding all but the Nobel Prizes, each of which came to $1 million in 1995, down from the peak of $1.2 million in 1992. The plethora of new rich awards inevitably raises the question: what, if anything, have they to do with the Nobel prize?

Often, a great deal. The most opulent new awards are self-defined offspring of the Nobel prizes that divide into two classes: Nobel complements and Nobel surrogates. Nobel complements are explicitly described as correlatives in such diverse spheres of accomplishment as architecture, religion, and education. Nobel surrogates have a narrower intent, being aimed specifically at filling gaps left in the reward system of science, gaps left, as noted earlier, by rule-governed constraints. In both types, the rich prizes (especially the eponymous ones) merge altruism and self-interest. Most of the time, in providing the wherewithal (of which they usually have an abundance), donors ask only to bask in the prestige of the recipients.

The Pritzker Architecture Prize of $100,000, financed by the Hyatt Foundation and the Pritzker family, neatly exemplifies a Nobel complement. As it happens, this award is directly descended from the Nobel prize, having been established at the suggestion of Sweden's King Gustaf VI Adolf who observed that Alfred Nobel's prizes were confined to too few fields of accomplishment. Architects have come to take the new prize seriously. As Philip Johnson put it in so many words, "It's to us, the Nobel."[87] Considerably more lavish but just as much a Nobel complement, the Templeton Prize for Progress in Religion "grew out of Sir John's [that is, the founder, John Marks Templeton's] dissatisfaction with the Nobel Prizes. He believed they failed to recognize spiritual achievements." Determined to pay tribute to the spiritual, Templeton has done so in substance. Since founding the awards in 1972, he has raised the honorarium seven times—in 1992, to a round million—expressly to keep pace with the Nobels.[88] And to go no further, there are the Right Livelihood Awards, the self-defined "alternative Nobel prizes," founded by Jacob von Uexhull, a Swedish-German writer and member of the European Parliament. When the Nobel Foundation proved unwilling to create an award to "honor work which is ecologically responsible and does not ignore the traditional wisdom of mankind, especially the knowledge of the Third World," Uexhull established his Right Livelihood Foundation which literally anticipates the Nobels by making its awards of $50,000 to $100,000 in Stockholm a day before the Nobel ceremonies.

The severe limits on fields in which prizes are awarded in science plainly leaves ample room for Nobel surrogates, prizes expressly intended to fill

Nobel gaps. The Crafoord Prizes, founded in 1980 by Anna-Greta and Holgar Crafoord, were explicitly designed to honor great achievement in sciences ignored by Nobels—the geosciences, biosciences emphasizing ecology, mathematics, astronomy, and the investigation of arthritis, the last a disease which afflicts Holger Crafoord.[89] The honoraria are ample (running upwards of $250,000) but plainly do not match the Nobels. But their kinship with the Nobels is direct; the Crafoords persuaded the Royal Swedish Academy of Sciences, the body charged with selecting Nobel laureates in physics, chemistry, and economics to award the Crafoord prizes as well.

There are also surrogate offspring in engineering. The Nobel neglect of this field led directly to the establishment in 1985 of the Draper Prize awarded by the U.S. National Academy of Engineering. According to the Academy's president, "We hope that in years to come the...Draper Prize will be just as well known and respected an award in engineering as the Nobel Prizes are today."[90] Its not inconsiderable honorarium of $350,000 and its sponsorship by the National Academy give it a certain authority. But it remains to be seen whether it will accumulate a roster of recipients sufficient to provide Nobel-like prestige.

Like the Nobels, the new rich awards evoke powerful criticism from those who believe that they, too, reinforce the wrong motives and celebrate the wrong values. This was the leitmotif of the decision taken by the French mathematician, Alexandre Grothendieck to decline the 1988 Crafoord Prize. In a letter to the Academy, the one-time Fields Medalist wrote,

> The work which...[has been honored] goes back 25 years...in the past two decades, the ethics of the scientific profession (at least among mathematicians) have become so degraded that wholesale plundering of ideas (and particularly at the expense of those in no position to defend themselves) has become almost the general rule.... It is, at any rate, tolerated by all.... Under the circumstances, agreeing to play along with the practice of granting prizes and rewards would also be endorsing a spirit and a development in the scientific world that I see as deeply unhealthy...it is...suicidal spiritually as well as intellectually and materially.[91]

The proliferation of prizes has enlarged and diversified the reward system of science but has not fundamentally changed its character.

## THE LAUREATES, THE ULTRA-ELITE, AND THE STRATIFICATION SYSTEM IN SCIENCE

*Scientific Elite* is centrally concerned with the stratification system in science, its shape, maintenance and consequences. Describing the laureates' place in that system, chapter 1 notes:

[T]he Nobel prize elevates its recipients not merely to the scientific elite
but to the uppermost rank of the scientific ultra-elite, the thin layer of
those at the top of the stratification hierarchy of elites who exhibit especially great influence, authority or power and who generally have the highest
prestige with in what is a prestigious collectivity to begin with. (p. 11)

This characterization has proved more controversial than I expected.
One of the foremost analysts of the Nobel archives, Elisabeth Crawford,
has written at length about what she takes to be the shortcomings of both
the analysis of the ultra elite and of the stratification system of science. I
want to address both issues here to set the scholarly record straight. Crawford
holds that the laureates do not constitute the ultra-elite and, in any event,
that the stratification system described in the monograph is not characteristic of science as a social institution but only represents "the high point of a
particular historical development."[92]

First in a paper[93] and then in her monograph, *Nationalism and Internationalism in Science, 1880–1939,*[94] Crawford contends that the laureates
cannot be considered an ultra-elite in science because the career histories
of American laureates selected between 1901 and 1939 are much the same
as those of unsuccessful candidates for the prize. Drawing on data about
the careers of the 37 American candidates nominated at least twice for the
physics and chemistry prizes (these include three physicists who did not
meet this criterion and excludes the candidates Thomas A. Edison and the
Wright brothers), Crawford reports that the undergraduate origins and graduate education of laureates and candidates in the period under consideration
were much the same, that laureates were only slightly more often trained
by prior laureates, and that both laureates and candidates were equally apt
to have been elected to the National Academy of Sciences. However,
Crawford does report that the laureates as a group were more often academics than were candidates, more apt to have become full professors at
elite institutions, had acquired awards earlier and had been accorded more
international awards than candidates not selected for the prize.

But, of course, Crawford's data which demonstrate that the laureates
named in the first third of the century were much like other highly esteemed
scientists do not counter the analysis in *Scientific Elite;* they simply confirm
it. That as careful an analyst as Crawford can so egregiously misread *Scientific Elite* makes it clear that I must review and clarify the attributes both of
the ultra-elite and the occupants of the forty-first chair, the "'uncrowned'
laureates who are peers of the prize-winners in every sense except that of
having the award" (p. 42). For one thing, *Scientific Elite* seems to make it
plain that the laureates are not the only members of the ultra-elite:

There are some [laureates] whose contributions in the judgment of many
qualified scientists have not advanced the frontiers of their fields as much
as others who do not wear the Nobel laurels. These uncrowned laureates

either have not yet or will not win Nobel prizes. Nevertheless, in the
aggregate, the Nobelists, taken together with the publicly less visible
scientists of comparable stature, take their place in the highest stratum of
the loosely formalized hierarchy of science. (p. 11)

So, too, it is made plain that the early laureates probably are likely not
to differ greatly from other candidates for the prize, since both are members of the ultra-elite. Laureates are not "a breed apart"[95] nor does the composition of the ultra-elite depend "on successive prize decisions by scientific
corporations in Sweden,"[96] as Crawford mistakenly claims I say.

Crawford's data on early candidates for the prize are especially useful
for identifying distinctive attributes of the occupants of the forty-first chair,
those peers of the laureates in every respect but that of having received the
prize. When *Scientific Elite* was written, the archives on the early prizes
were not yet open but it was evident that major but unsuccessful candidates
for the prize would provide a "strategic research site."

> In the strict sense, occupants of the forty-first chair at a given time are
> scientists who are known to have been serious contenders for Nobel
> prizes—that is, those especially worthy candidates selected by the com-
> mittees for painstaking "special investigation" that precedes the final
> decision...but who have not won Nobel prizes—and their peers in fields
> not included within the purview of the prizes. (p. 42)

Had the archival data been available then (and had I command of Swedish), I would have been tempted to undertake the kind of analysis Crawford
actually did. That the findings turned out as they have, that the laureates'
careers and those of scientists nominated more than once for the prize are
largely similar is precisely what was, in effect, predicted about the occupants of the forty-first chair.[97]

Further inquiry into the stratification system in science would do well
to examine the functions served by elite, in certifying scientific contributions, in representing science to the public and in organizing scientific establishments; on this last, Crawford and I agree. But we do not agree that
research on elite formation "is not very interesting."[98] The long tradition of
research in the social sciences on the origins of elites suggests otherwise.

Turning to the stratification system of science, Crawford insists that
the sharply graded pyramid described in *Scientific Elite* and the processes
of accumulation of advantage which reinforce it are not "immanent
features...of science—that have determined the organizational structure
of American science in the twentieth century"[99] but simply the outcome of
my having treated the Nobel population as "an ahistoric entity, view[ed] in
a socioinstitutional context restricted to the time [the] study [was] carried
out."[100] She argues that had the data on the laureates been disaggregated
into finer-grained historical periods, differences would have become evident between the organization of science in the earlier and the later de-

cades and I would then not have concluded that the patterns I observed are endemic to American science.

Crawford's insistence that *Scientific Elite* applauds the stratification system as both "immanent" and as "functional to progress in science" is, shall we say, perplexing since it does neither.[101] That a critical and nuanced review[102] of *Scientific Elite* made much of its analysis of dysfunctions flowing from the Nobel prize and the stratification system in which it is embedded suggests that this is covered in the text had Crawford wished to find it.

## SCIENTIFIC ELITE: TWENTY YEARS LATER

Two decades have passed since the research for *Scientific Elite* was completed. In the interim, new fields and institutions of science have emerged and old ones have receded. The scientific enterprise continues to grow, though more slowly. The Cold War over, national security provides a far less popular and often extraneous rationale for science. Scientists can no longer count on backing for research which aims to guarantee American dominance in science. Indeed, the necessity for such dominance is now put in question, as the collapse in 1993 of the hugely expensive Superconductor Super Collider project clearly testifies. Nevertheless, American scientists continue to gain more Nobel awards than ever. The proportion of prizes going to American citizens has risen steadily from 42 percent in the first post-World War II decade (1945–1954)[103] to 65 percent in the decade 1985–1994, and that proportion would be greater had the attributions centered on the location of the research rather than the citizenship of the laureates.

There is also reason to think that the stratification system described in *Scientific Elite* remains largely unchanged even though the great research universities have become somewhat less dominant in garnering research support than they were two decades ago.[104] There are, for example, few signs that rewards in science are being allocated differently. Indeed, women scientists, particularly, hold that the system has not changed enough as they tellingly justify their claims to resources and rewards on grounds of merit, not entitlement. There are also few signs that disparities between the most advantaged and the least advantaged scientists are narrowing significantly. If some details of the stratification system described in the monograph need revision, this is not true of its main contours.

So, also, with the Nobel prize and its distinctive role in science. The gold standard may have become somewhat tarnished and its laureates' charisma faded, but its symbolic significance has not diminished and, despite the great proliferation of prizes, its preeminence remains unchallenged.

So much for empirical description. The principal ideas developed in *Scientific Elite* continue to be confirmed by research. The process of accumulation of advantage enters into stratification—in science as in other departments of social life. All reward systems, including the reward system of science, however apt when first instituted, acquire dysfunctional features when historical contexts change and they do not. The phenomenon of occupants of the forty-first chair continues to remind us that the prize is not a unique or exhaustive indicator of supreme accomplishment in science. The last chapter of *Scientific Elite* earmarked these themes as central. They remain so now.

H.Z.

New York
November 1995

## NOTES

1. The "Igs" aim particularly to identify "those who persistently work at convincing the world and themselves of something that apparently just isn't so." One paradigmatic Ig-Nobel prize-winner is Jacques Benveniste whose 1988 paper in *Nature* "claimed that water had a memory for materials dissolved in it" (*Scientific American December* [1991]: 26). See also Craig Lambert, "The Ig-Nobel Prizes," *Harvard Magazine* (January-February 1994): 12–13, especially p. 12. The Ig-Nobel Prizes, once sponsored by the M.I.T. Museum and the *Journal of Irreproducible Results*, are now overseen by the *Annals of Improbable Results*. On "Laureates of the Albatross," see *Science* 181 (7 September 1973): 926.
2. Lawrence K. Altman, "Surprise Discovery about 'Split Genes' Wins Nobel Prize," *New York Times*, 12 October 1993, p. C3.
3. Joseph Palca, "Overcoming Rejection to Win a Nobel Prize," *Science* 250 (19 October 1990): 378.
4. Elizabeth Karagianis, "Profile," *SPECTRUM: MIT Newsletter* (Winter 1995): 16.
5. Reduced publication is not entirely unintentional, as Jerome Karle (laureate in chemistry in 1985) observed in a talk he gave to the American Association for the Advancement of Science. My notes on that session paraphrase Karle's own words: "After the prize, paper writing will go down. But that's a decision. One thinks, are there really interesting problems that haven't been broken? I'd rather fail at solving very hard problems and if so, I'll not write many papers. But if I do solve them, there will be many papers" (14 February 1987).
6. James Gleick, *Genius: The Life and Science of Richard Feynman* (New York: Pantheon, 1992), 382.
7. Harriet Zuckerman, "Accumulation of Advantage and Disadvantage: The Theory and Its Intellectual Biography," in Carlo Mongardini and Simonetta Tabboni, eds., *L'Opera di Robert K. Merton e la sociologia contemporanea* (Genova: Edizioni Culturali Internazionali, 1989), 153–76. See also Jonathan R. Cole and Burton Singer, "A Theory of Limited Differences: Explaining the Productivity Puzzle in Science," in H. Zuckerman, J. R. Cole, and J. T. Bruer, eds., *The Outer Circle: Women in the Scientific Community* (New York: W.W. Norton, 1991), 277–310.
8. R. S. Albert, ed., *Genius and Eminence* (New York: Pergamon, 1983); D. K. Simonton, "Developmental Antecedents of Achieved Eminence," *Annals of Child Development* 4 (1987): 131–69; R. D. Clark and G. A. Rice, "Family Constellations and Eminence: The Birth Orders of Nobel Prize Winners," *Journal of Psychology* 110 (1982): 281–87; Albert Rothenberg, "Psychopathology and Creative Cognition: A Comparison of Hospitalized Patients, Nobel Laureates, and Controls," *Archives of General Psychiatry* 40 (September 1983): 937–42.
9. Paula E. Stephan and Sharon G. Levin, "Age and the Nobel Prize Revisited," *Scientometrics* 28, no. 3 (1993): 387–99.
10. See Margaret Rossiter, *Women Scientists in America: Struggles and Strategies to 1940* (Baltimore: Johns Hopkins Press, 1982); Sharon B. McGrayne, *Nobel Prize Women in Science: Their Lives, Struggles and Momentous Discoveries* (New York: Birch Lane Press, 1994); Evelyn Fox Keller, *A Feeling for the Organism: The Life and Work of Barbara McClintock* (New York: Freeman, 1983); Helena M. Pycior, "Reaping the Benefits of Collaboration While Avoiding Its Pitfalls: Marie Curie's Rise to Scientific Prominence," *Social Studies of Science* 23 (1993): 301–23; and Harriet Zuckerman, "The Careers of Men and Women Scientists: A Review of Current Research," in H. Zuckerman, J. R. Cole, and J. T. Bruer, eds., *The Outer Circle: Women in the Scientific Community* (New York: W. W. Norton, 1992), 27–56.

11. Congressional Research Service, Library of Congress, "The Nobel-Prize Awards in Science as a Measure of National Strength in Science." Report Prepared for the Task Force on Science Policy, Committee on Science and Technology, U. S. House of Representatives, Washington: GPO, September 1986; Daniel S. Greenberg, "Nobels: The Embarrassment of Repeated Wins," *Washington Report* 301, no. 21 (22 November 1979): 1191-92; B. R. Martin, "British Science in the 1980s—Has the Relative Decline Continued?" *Scientometrics* 29, no. 1 (1994): 27-56.
12. National Science Foundation, Sources of Financial Support for Research Prize Winners, Evaluation Staff Report 2-87, Washington: National Science Foundation NSF87-87, 1987.
13. Walter Broughton and Edgar W. Mills, "Resource Inequality and Accumulative Advantage: Stratification in the Ministry," *Social Forces* 58 (1980): 1289-1301; Kathryn M. Moore, "Women's Access and Opportunity in Higher Education: Toward the Twenty-first Century," *Comparative Education* 23, no. 1 (1987): 23-34; Patricia A. Taylor, "The Celebration of Heroes Under Communism: On the Reproduction of Inequality," *American Sociological Review* 52 (April 1987): 143-54; Emanuel Levy, *And the Winner Is—The History and the Politics of the Oscar Awards* (New York: Ungar, 1987); Michael Useem and Jerome Karabel, "Pathways to Top Corporate Management," *American Sociological Review* 51 (April 1986): 184-200.
14. Assar Lindbeck, "The Prize in Economic Science in Memory of Alfred Nobel," *Journal of Economic Literature* 23 (March 1985): 37-56, and Arthur Diamond, "Citation Counts for Nobel Prize Winners in Economics," *H. E. S. Bulletin* 10 (Spring 1988): 67-70.
15. Harriet Zuckerman, "The Sociology of the Nobel Prize: Further Notes and Queries," *American Scientist* 66 (July-August 1978): 420-25.
16. One might also ask whether, as Alfred Nobel stipulated, the prize also recognizes contributions which "confer the greatest benefit on mankind." This aspect of the awards fell by the wayside quite early as prize-committees elected to equate contributions to fundamental knowledge with benefiting mankind—one way out of the difficulty of making Nobel's contradictory wishes accord with one another.
17. Quoted in *Science* 211 (27 March 1981): 1404.
18. S. J. Gould, "Balzan Prize to Ernst Mayr," *Science* 223 (20 January 1984): 255-57, especially p. 255.
19. As Irving Louis Horowitz has observed, "The Nobel Prize for economics in one fell swoop disenfranchises all other social sciences, rewards a special variety of business analysis as it uniquely pertains to market systems, and ultimately negates the scientific aspect of social research apart from such heuristic values," *PS* 16 (Winter 1983): 57-58, especially p. 57. Horowitz's diagnosis is the same as Gould's, differing only in the detail of the fields excluded but agreeing that the prize provides a narrow notion of science.
20. The archives, open since 1974 for materials fifty years old or more, suggest that this was the outcome of efforts by members of the Physics committee to restrict competition for funds then available for specialties in which prizes may be awarded. Robert Marc Friedman, "Nobel Physics Prize in Perspective," *Nature* 292 (27 August 1981): 793-98; Elisabeth Crawford and Robert Marc Friedman, "The Prizes in Physics and Chemistry in the Context of Swedish Science," in C. G. Bernhard, Elisabeth Crawford and Per Sörbom, *Science, Technology and Society in the Time of Alfred Nobel* (Oxford: Pergamon, 1982), 311-31.
21. Elisabeth Crawford observes that after 1972, the cut-off date for the analysis reported in *Scientific Elite,* "a string" of awards in physics have gone to scientists working in industrial laboratories and asks if other honors were extended to scientists outside academe. Elisabeth Crawford, "Scientific Elite Revisited: American Candidates for the Nobel Prizes in Physics and Chemistry 1901-1938," in Stanley Goldberg and Roger Stuewer, eds., *The Age of Michelson in American Science* (New York: American Institute of Physics, 1988), 258-71, especially p. 267. A number of recent prizes in physics have gone to "industrial scientists" (Arno Penzias, Robert W. Wilson, and Philip Anderson of Bell Laboratories, Ivar Giaever of General Electric, and Leo Esaki of IBM). However, these awards were for fundamental research and neither signal a new interest in honoring inventions nor a basic change in the Nobel reward system. There are no data on whether the proportion of prize-winning scientists affiliated with non-academic institutions is increasing or decreasing overall.
22. James MacLachlan, "Defining Physics: The Nobel Prize Selection Process, 1901-1937," *American Journal of Physics* 59 (February 1991): 166-74, especially p. 170, notes the marked increase in the number of prizes in physics being awarded to theorists and goes on to say that physics has simply become "more theoretical."
23. Certain theoretical positions have found favor at particular times. Claire Salomon-Bayet concludes that microbial theory dominated the prizes in medicine between 1901-20, sanctioning the "already accomplished revolution" in the field of bacteriology. "Bacteriology and Nobel Prize Selections, 1901-1920," in C. G. Bernhard, Elisabeth Crawford, and Per Sörbom, eds., *Science, Technology and Society in the Time of Alfred Nobel* (Oxford: Pergamon, 1982), 377-400.
24. In combination with the scarcity of prizes, this preference leads the committees to put off making decisions for some time.
25. S. J. Gould, "Balzan Prize to Ernst Mayr," *Science* 223 (20 January 1984): 255-57, 255.

26. Eugene Garfield has been using citation analysis for some time to identify scientists he calls "of Nobel class," scientists including laureates-to-be and other highly cited authors. Early studies at the Institute for Scientific Information by Sher and Garfield began by analyzing laureates' citations before the prize, then turned to "forecasting" Nobel selections and then to examining the work of highly cited scientists more generally. For a review, see E. Garfield and A. Welljams-Dorof, "Of Nobel Class: A Citation Perspective on High Impact Research Authors," *Theoretical Medicine* 13 (1992): 117-35.

27. As noted in *Scientific Elite,* the selection of prize-winners is neither wholly universalistic (meritocratic) nor wholly particularistic. Both criteria operate. Decision making for the prize involves successive phases of selection from pools of decreasing size. Early on, universalistic criteria are applied to reduce the number of candidates but there comes a point when those remaining are judged to be equally meritorious. At this point, additional bases of selection are required if a choice is to be made. Secondary criteria are then called into play, some functionally relevant for the advancement of science and some particularistic and functionally irrelevant. Particularistic criteria include nationality, candidates' ages or other personal characteristics and committee members' preferences. Jonathan Cole has gone on to develop the implications of "individual particularism and institutional universalism" in science and the law. "The Paradox of Individual Particularism and Institutional Universalism," *Social Science Information* 28, no. 1 (1989): 51-76.

28. Statutes of the Nobel Foundation and the Prize-Awarding Institutions, in Nobelstiftelsen, ed., *Alfred Nobel: The Man and His Prizes* (Amsterdam, London, and New York: Elsevier, 1962), 653 and 659.

29. Rosalind Yalow, a laureate in medicine and never a faculty member of a major university, is especially sensitive to the effects of the institutional skewing of nominations. She remarks that "it's much easier to be nominated if you are from an institution that already has a laureate." Gordon L. Weil, "Nobel Prize Politics," *Politics Today* 5 (1978): 50-53, especially pp. 52-53.

30. See Elisabeth Crawford, "The Secrecy of Nobel Prize Selections in the Sciences and Its Effect on Documentation and Research," *Proceedings of the American Philosophical Society* 134, no. 4 (1990): 408-19. The archives are less instructive than one might hope since the committees did not keep full records of their deliberations.

31. G. Küppers, N. Olitzka, and P. Weingart, "The Awarding of the Nobel Prize: Decisions About Significance in Science," in C. G. Bernhard et al., eds., *Science, Technology and Society,* pp. 332-51, p. 342. The consensus was more pronounced in chemistry than in physics, and it declined in both fields after the first fifteen years.

32. Küppers et al., in Bernhard et al., *Science, Technology and Society,* 337 and 339.

33. Tendencies toward chauvinism in nominations are evident in the early years in physics and chemistry. The extent of chauvinism varied by country of nominators and as a result of complicated international politics in the post-World War I period. See E. Crawford, *Nationalism and Internationalism in Science, 1880–1939: Four Studies of the Nobel Population* (Cambridge: Cambridge University Press, 1992), chapters 3 and 5.

34. The paraphrases are Constance Holden's, "Chauvinism in Nobel Nominations: American Chemists Nominate Other Americans," *Science* 243 (27 January 1989): 471.

35. Elisabeth Crawford and Robert Marc Friedman, "The Prizes in Physics and Chemistry in the Context of Swedish Science: A Working Paper," in Bernhard et al., eds., *Science, Technology and Society,* 311-31.

36. Küppers et al., "The Awarding of the Nobel Prize," see fn 32.

37. Diana K. Barkan's recent study of Walther Nernst's award in chemistry in 1920 leads to the same conclusions; she recounts the bitter rivalry between Nernst and Svante Arrhenius, a laureate and committee member for physics. Arrhenius stood in the way of Nernst getting a prize for more than a decade. "Simply a Matter of Chemistry: The Nobel Prize for 1920," *Perspectives on Science* 2, no. 4 (Winter 1994): 357-95.

38. "Statistics provide few clues about what persuaded committee members in any given year. Rarely did any candidate receive a clear mandate from the nominators and rarely did the candidate who received the most nominations get the prize." R. M. Friedman, "Text, Context, and Quicksand: Method and Understanding in Studying the Nobel Science Prizes," in *Historical Studies in the Physical and Biological Sciences* 20 (1989): 63-77, p. 75. As it happens, the physics committee managed to turn back Poincaré in 1910 even though he received half of all the nominations made that year. See Elisabeth Crawford, J. L. Heilbron, and Rebecca Ullrich, "The Nobel Population: 1901-1937." Berkeley Papers in History of Science, 11, and Uppsala Studies in History of Science, 4; Office for History of Science and Technology, University of California at Berkeley and Office for History of Science, Uppsala, 1987, pp. 46-50.

39. Friedman is highly critical not only of the interpretations the Bielefeld group makes from the archival reports but also of their data. "Unfortunately, errors abound in the lists, revealing fundamental problems in the tabulation and subsequent analyses of the data." Friedman in *Historical Studies,* p. 75. Although Friedman identifies the character of errors in the data, he does not indicate the extent to which they bias the conclusions.

40. Küppers et al., "The Awarding of the Nobel Prize," 338-40, especially p. 340. The French also received slightly more awards relative to nominees in chemistry: 17.4 percent of the French nominees received awards as against their having 12.4 percent of the nominations while the Germans received fewer awards in chemistry (19 percent) as against having 36.2 percent of the nominators and 34.1 percent of the nominees.

41. It may be that the committee members disregarded the nominations system as Crawford asserts (p. 44). But the chances of their having done so is only about 2 percent, according to a chi-square test of the independence of the distributions of candidates and winners she provides. Furthermore, the committees' tendencies to favor candidates from Britain and Sweden and to disfavor those from Italy are not large. Stephen Stigler of the University of Chicago was kind enough to provide to the chi-squares and the GLIM analyses that are the bases for this conclusion.

42. Küppers, N. Ulitzka, and P. Weingart, "Factors Determining the Award of Nobel Prizes in Physics and Chemistry 1901-1929," *Endeavor, New Series* 7, no. 4 (1983): 203-204, especially p. 203. The authors refer to an unpublished quantitative survey of the nominating process but say no more about the analysis of these data. Since a candidate is often nominated by more than one nominator, it would be necessary to know how multiple nominators are treated in the calculation before concluding that the laureates' candidates were successful twice as often as the candidates of other nominators.

43. Paul A. Samuelson, "Economics in a Golden Age: A Personal Memoir," in Gerald Holton, ed., *The Twentieth Century Sciences: Studies in the Biography of Ideas* (New York: W. W. Norton, 1972), 155-70, especially p. 155.

44. Crawford, *Nationalism and Internationalism in Science,* p. 136.

45. See Crawford, Heilbron, and Ullrich, *The Nobel Population.*

46. The dominance of United States scientists among prize-winners did not begin until after World War II.

47. It is, of course, possible that campaigns were mounted by the laureate-masters among their fellow laureates and that this accounts for the considerable number of laureate nominators registered here. Evidence for such campaigns will not, of course, be found in the archives of the Foundation but only in the private papers of the laureates.

48. Svedberg was repeatedly nominated for the prize mainly by his Swedish colleagues. See Milton Kerker, "The Svedberg and Molecular Reality," *ISIS* 67 (1976): 190-216.

49. Another instance of rule-governed constraints on the prize is the requirement that it be given for a particular discovery rather than the totality of a scientist's contributions. This has led, on occasion, to eccentric choices of prizeworthy work for a particular prize-winner. The most famous instance, of course, is Einstein's prize being given for the photoelectric effect rather than the special theory of relativity.

50. *Scientific Elite* takes up the most notorious of these cases, the award in 1923 for the discovery of insulin to the Canadians, Frederick G. Banting and J. J. R. Macleod. However, that discussion appeared before Michael Bliss' persuasive study of the discovery (*The Discovery of Insulin,* Chicago: University of Chicago Press, 1982). Before Bliss' book, it was widely believed that the insulin prize involved not one but three errors: that the committee excluded *two* important contributors, Claude Best and N. Paulesco, and that it included a marginal one, J. J. R. Macleod, the laboratory head who is said not even to have been in Toronto when the discovery was made. Banting, Best's co-worker, was "furious" when Macleod got the prize and drew attention to the committee's failure by sharing his prize money with Best. Macleod then shared his prize money with the fourth member of the group, J. B. Collip. The episode became all the more complicated when it was learned that Paulesco, a Rumanian, had anticipated the Banting-Best discovery by six months. The merits of Paulesco's claim to the prize have been recognized twice over by the Nobel establishment, by Arne Tiselius, one-time chairman of the chemistry committee and president of the Foundation, and by Rolf Luft, one-time chairman of the medicine committee. It was Luft who revealed that the committee in medicine thought it might have "a problem" with the insulin award and dispatched "its secretary to Toronto to get the truth. And he got it...from Macleod" (*Science* 211, no. 27 [March 1981]: 1404).

Bliss, a historian at the University of Toronto, records a different story. He concludes that Macleod was far more central to the insulin research than earlier accounts had it; that much of the time, Banting was anything but generous in recognizing Best's contribution; that Collip's biochemical contribution was central; that Banting and Best's having given Paulesco's work short shrift derived not from willful neglect, but from their poor command of French (the language in which Paulesco's paper was published); that, in any case, Paulesco's contribution was crude and nowhere near as well-developed as the Canadians'; and that the Nobel committee, drawing as it did on the firsthand report from August Krogh, a laureate in biomedicine, who visited the Toronto laboratory and talked with both Macleod and Banting, had grounds for its decision. Krogh wrote the committee that the prize should go to Banting and Macleod both for the isolation of insulin *and* for the clinical and physiological investigation of its use. As it turned out, the prize was given to them both but only for the discovery of insulin, no mention being given to its uses. Plainly, others looking at the historical record may disagree with Bliss' assessment. Yet it does make the

Nobel committee's decision more understandable than earlier accounts and suggests that imputed errors of judgment are themselves questionable.

51. Possibly in the same genre is the exclusion of Emil von Behring's Japanese collaborator, Shibasaburo Kitasato, from the first prize awarded in medicine in 1901 for the discovery of tetanus antitoxin. One significant paper reporting the discovery was published by von Behring alone and another jointly with Kitasato. Long after the fact, immunologists are of mixed minds on the matter of credit. Von Behring and Kitasato are cited for the discovery jointly in six of twelve major textbooks in immunology and in all three histories of the field. Sachi Sri Kantha, "A Centennial Review; the 1890 Tetanus Antitoxin Paper of von Behring and Kitasato and the Related Developments," *Keio Journal of Medicine* 40, no. 1 (1991): 35–39.

52. In all, ten out of 442 prizes given between 1901 and 1995 have gone to women.

53. Robert K. Merton, "The Matthew Effect in Science," [1968] in *The Sociology of Science* (Chicago: University of Chicago Press, 1973), chapter 20, p. 446. Merton's analysis is based in part on my interviews with the Nobel laureates.

54. Margaret Rossiter, "The Matthew Matilda Effect in Science," *Social Studies of Science* 23 (May 1993): 325–41, especially p. 335.

55. Otto Hahn, *My Life: The Autobiography of a Scientist*, translated by Ernst Kaiser and Eithre Wilkins (New York: Herder and Herder, 1970), 199. One suspects that had Rosalind Franklin lived and not been included in the 1962 prize to Watson and Crick (in place of Wilkins), she, too, would have had an "unhappy" conversation, in this instance, with Watson. Crick has been more generous in recognizing her contribution, claiming that Franklin would have solved the DNA problem in a matter of months, had she not been anticipated. Anne Sayre, *Rosalind Franklin and DNA* (New York: W. W. Norton, 1975), 214, fn 21.

56. The record of nominations shows that Hahn and Meitner were almost always nominated jointly by the same nominator suggesting that nominators believed they were both responsible for the research.

57. This is the view of William L. Duax of the Hauptman-Woodward Medical Research Foundation where Hauptman is President. The Karle daughter is Louise Karle Hanson of Brookhaven National Laboratory *(New York Times,* 17 October 1985, p. B12).

58. Dominique Stehelin, who had been a postdoctoral fellow in Michael Bishop's and Harold Varmus' laboratory, claims that he should have shared their 1989 award in medicine. In contrast, Bell has never publicly claimed her right to the award.

59. As William Lipscomb, the laureate in chemistry for 1976, put it, "I tell them [my students], suppose you join a successful archaeologist as a student assistant, and he tells you where to dig…you dig up a marvelous discovery. Now I ask you, who should get the credit: the director or the digger?" quoted in M. Browne, "Nobel Fever: The Price of Rivalry," *The New York Times,* 17 October 1989, p. C1.

60. William J. Broad, "Riddle of the Nobel Debate," *Science* 207 (4 January 1980): 37–38, especially p. 37.

61. It was also assumed that some consideration had been given to the then on-going patent suit involving Housfield's employer, EMI Ltd., and several U. S. firms. William Broad, *Science* 207 (4 January 1980): 207.

62. Bettyann Kevles, *Naked to the Bone* (New Brunswick: Rutgers University Press, in press).

63. Alison Bass, "Nobels are Distorting Science, Critics Say," *The Boston Globe,* 17 October 1988, p. 1.

64. Quoted in T. Rothman, "'Rabi' and 'Alvarez': Tedious, Vain Portraits," *The Scientist* (13 July 1987): 20.

65. Peter Medawar, *Advice to a Young Scientist* (New York: Harper and Row, 1979), 80.

66. Richard P. Feynman, *"Surely You're Joking, Mr. Feynman!" Adventures of a Curious Character* (New York: W. W. Norton, 1985), 311.

67. Most paradoxical of all is Richard Feynman's account of his speech at the Nobel dinner, "I couldn't just say thank you very much, blah-blah-blah-blah-blah; it would have been so easy to do that, but no, I have to make it honest. And the truth was, I didn't really want this Prize, so how do I say thank you when I don't want it?" R. P. Feynman, *"Surely You're Joking Mr. Feynman!",* 308. As noted in *Scientific Elite,* other laureates have been less than comfortable about accepting the prize.

68. *On Humor* (Oxford and New York: Basil Blackwell, 1988), 159–62.

69. See T. D. Lee's restrained and poignant account of the dissolution of his long and productive collaboration with C. N. Yang, "Broken Parity," in T. D. Lee, *Selected Papers,* vol. 3 (Boston: Birkouser, 1986), 487–512.

70. J. D. Watson, *The Double Helix* (New York: Atheneum, 1968).

71. Nicholas Wade, *The Nobel Duel: Two Scientists' 21-Year Race to Win the World's Most Coveted Research Prize* (Garden City: Anchor Press-Doubleday, 1981).

72. Ibid., 274–75.

73. Nicholas Wade, "Guillemin and Schally: A Race Spurred by Rivalry," *Science* 200 (5 May 1978): 510–13, especially p. 513.

74. Quoted in W. Broad, *Science* 207 (4 January 1980): 207.
75. Some scientists point also to the "inspirational" effects of laureates who, on more than one occasion, have been described as "role models." This is emphasized in a paper on Nobel Prize-winning surgeons. Its authors declare, "Yet we should not forget that the purpose of reviewing the accomplishments of our predecessors is not to make us surgical historians but to make us better surgeons.... Their pioneering vision is an inspiring example of the greatness of which surgery and surgeons are capable." Jon B. Morris and William J. Schirmer, "The 'Right Stuff': Five Nobel Prize-Winning Surgeons," *British Journal of Surgery*, 77 (1990): 944–52, especially p. 952.
76. Unpublished essay and letter from George J. Stigler, 3 April 1978, p. 2
77. Another faunal metaphor turned up in an interview with a laureate in physics who described high energy physicists as "hitting important problems like schools of piranhas."
78. Harriet Zuckerman and Jonathan R Cole, "Research Strategies in Science: A Preliminary Inquiry," *Creativity Research Journal* 7, nos. 3-4 (1994): 391–405.
79. Unpublished essay and letter from George Stigler, 3 April 1978.
80. Andrew Sullivan, *New York Times Book Review*, 22 March 1992, p. 1
81. *New York Times*, 28 July 1995, p. B9.
82. Columbia's president, Michael Sovern, was fond of saying that he would not worry as long as the university had more Nobel laureates than it had lost consecutive football games.
83. *New York Times*, 6 August 1995, p. 21
84. *New York Times*, 11 May 1979, p. A 15.
85. Roy Tate, *The Astrology of Genius: A Study of Nobel Prize Winners* (Miami, Fla.: Evolutionary Publications, 1989). According to the author, a chi-square test shows that the odds of the patterns he identified being the result of chance are less than 1/500. The volume is described as an exploration of "the most completely developed group of human beings on Spaceship Earth" and advertisements for the current edition note that a facsimile of the "classic 1975" out-of-print book edition is still available.
86. Harriet Zuckerman, "The Proliferation of Prizes: Nobel Complements and Nobel Surrogates in the Reward System of Science," *Theoretical Medicine* 13 (1992): 217-31.
87. *New York Times*, 4 March 1980, p. B10.
88. *New York Times*, 1 February 1991, p. 10.
89. The founders "wanted to help the areas of science not covered by the Nobels." M. Davis, "Filling the Nobel Gap," *Discover* (January 1981): 55.
90. G. Byrne, "NAE Creates New Prize," *Science* 242 (1988): 665.
91. *Manchester Guardian Weekly* 138, no. 20 (Week ending 15 May 1988): 17.
92. Crawford, *Nationalism and Internationalism in Science*, pp. 142 and 145.
93. Elisabeth Crawford, "Scientific Elite Revisited: American Candidates for the Nobel Prizes in Physics and Chemistry 1901-1938," in Stanley Goldberg and Roger Stuewer, eds., *The Age of Michelson in American Science* (New York: American Institute of Physics, 1988), 258–71.
94. Full citation in fn 39.
95. Crawford, *Nationalism and Internationalism in Science*, p. 128.
96. Ibid., p. 142.
97. Despite the attention devoted to occupants of the 41st chair throughout *Scientific Elite*, Crawford mentions them only once—in a footnote—and then only to say that a list can be found in an appendix to *Scientific Elite*. Ibid., p. 128.
98. Ibid., p. 143. Oddly enough, this opinion is being set forth by a scholar engaged in studying Nobel laureates and candidates for the prize.
99. Ibid., p. 144.
100. Ibid., p. 128.
101. Ibid., pp. 127 and 145. Crawford also manages to impute the same strange opinion to my colleagues, Robert Merton, Jonathan Cole, and Stephen Cole.
102. Barbara Rosenblum, *American Journal of Sociology* 85 (1979): 672–76.
103. Congressional Research Service, "The Nobel-Prize Awards in Science as a Measure of National Strength in Science." Report Transmitted to the Task Force on Science Policy, Committee on Science and Technology, U. S. House of Representatives. Washington, U. S. G. P. O.: September 1986. Science Policy Study, Background Report No. 3, p. xii for early data.
104. Roger Geiger and Irwin Feller, "The Dispersion of Academic Research in the 1980s," *Journal of Higher Education* 66 (1995): 336-60. They observe, however, that dispersion started as early as the 1950s. The concentration of funds in the ten top "performers" dropped from 20.2 percent in 1979–80 to 17.9 percent in 1989–90—not a substantial reduction. The largest losers, however, in share of funding over the last decade were the highest rated research universities. See pp. 342–43.

# PREFACE AND ACKNOWLEDGMENTS

First awarded in 1901, the Nobel prize is a comparative latecomer in the long history of honors bestowed for scientific accomplishment. During its first three-quarters of a century, the prize has become the supreme symbol of excellence in science, and to be a Nobel laureate is, for better or for worse, to be firmly placed in the scientific elite.

An inquiry into the careers and research accomplishments of Nobel laureates should therefore tell us something about the workings of the stratification system of science in relation to the development of scientific knowledge. As we follow the laureates from their social origins through their formal education and apprenticeships in the craft of science to their major contributions at the research front, we can catch a glimpse of the part played at each phase in their conspicuously upward mobility by the quality of their scientific work, on the one hand, and by such socially defined attributes as age, sex, religion, and ethnicity, on the other. It should come as no surprise—except, perhaps, to the polarized ideologues of purist science or of social discrimination run rampant—that both cognitive and social processes affected the actual course of these notable scientific lives.

When we examine the careers of the laureates in detail and systematically compare them with the careers of others in the scientific elite and of the rank-and-file, we find that the elite were successively advantaged—not least, in obtaining access to the human and material resources that greatly facilitate the doing of first-class research. This process of cumulative advantage helps to account for the growing disparities between the elite and other scientists in the extent and importance of their research contributions over the course of their careers. At a time when the great debate over the pros and cons of meritocracy continues to gather force, it may be of interest to observe the complex interaction of merit and privilege in a social institution, such as science, that is normatively committed to universalistic standards for judging and rewarding individual achievement.

The study draws upon various sorts of data: interviews with four-fifths of American Nobel laureates, historical and contemporary documents, and participant-observation of scientists at work. It brings together both quantitative and qualitative materials, in a style closely associated with the sociological research of the late Professor Paul F. Lazarsfeld and of Professor Robert K. Merton. Even if I had not happened to be trained in the research tradition associated with Lazarsfeld and Merton, close-up acquaintance with the evidence would have persuaded me that neither qualitative nor quantitative analysis alone could provide anything resembling a sound understanding of the subject.

As a sociologist, I was not, of course, prepared to venture knowledgeably and confidently into the vast terrain of modern physics, chemistry, and the biological sciences. Though greatly aided by experts in these fields, I have no doubt managed to commit various scientific blunders and gaucheries in my effort to convey the character of research carried forward by the Nobel laureates. But I retain the hope that these errors are comparatively few.

Fragments of the book have appeared before, in substantially different form. I began work on the subject for my doctoral dissertation a decade ago, and perhaps a sixth of what appears here is based upon that manuscript. Other minor sections of the book have been published in several papers (1967a, 1967b, 1968, 1972). And to a limited extent, I have drawn upon joint research with Robert K. Merton, who has kindly agreed to its incorporation in this book. A first draft of the book, written while I was a Visiting Scholar at the Russell Sage Foundation in 1971–72, was put aside for further work on problems that came to light in that draft. On the basis of this further research, the greater part of the final draft was written at the Center for Advanced Study in the Behavioral Sciences, where I was a Fellow during the year 1973–74.

During that year at the Center and intermittently since, I have worked on related problems in the sociology of science with the geneticist Joshua Lederberg, of Stanford University, who has a deep interest in the social as well as the cognitive organization of the sciences. I have also collaborated with the physiologist André F. Cournand, of Columbia University, on the normative code of science. Although both men are Nobel laureates, in neither case was our research closely related to the reward system of science in general or to the Nobel Prize in particular. Still, it was of grent value to me to have each of them patiently put up with my periodic barrages of questions.

My colleagues and friends, Professors Jonathan R. Cole of Columbia University and Stephen Cole of the State University of New York at Stony Brook, have always been ready to talk over general themes that run through our joint work, down to the smallest details of analysis. Our work sessions have been invariably stimulating and often noisy. In ways less easily described, though known to each of them, two other friends—Dr. Linda Schoeman, and Professor Yehuda Elkana of the Hebrew University—made it possible for me to finish this book. I thank them all.

Over the years, a succession of able assistants—Richard Lewis, Matthew Rosen, Lawrence Stern, and Jay Schechter—have tracked down research materials as needed. And nearly undecipherable drafts have been transformed into pristine manuscript by a series of secretaries —Sally Vernon, Francine Gleason, Joan Warmbrunn, and Priscilla Barton. It would have been difficult to get along without them. Thomas Gieryn and Jay Schechter helped greatly with the reading of proofs.

To Gladys Topkis of The Free Press I owe much for her fine editorial hand and to Charles E. Smith, for getting the book into print with dispatch. I am greatly indebted to the National Science Foundation for supporting my work through grants to the Columbia University Program in the Sociology of Science.

Most of all, I am indebted to Robert K. Merton, whose influence can be detected throughout the book. As others have had occasion to know, he represents, as scientist, critic, and teacher, the scholarly tradition at its best.

H. Z.

Columbia University
September 1976

# Chapter 1

# NOBEL LAUREATES AND SCIENTIFIC ELITES

When Alfred Nobel died in 1896, he left what was then a princely estate: more than thirty-three million kroner, or about nine million American dollars.[1] It was a legacy destined to become one of the most famous of its kind in modern history. His will specified that the bulk of the estate be put aside in a fund, its annual income to be divided among five prizes: three in science, one in literature, and one to advance the cause of world peace. Nobel could not, of course, have foreseen that his prizes in the sciences would become the ultimate symbol of excellence for scientists and laymen alike and that those in literature and peace, although more controversial than the others, would also carry their share of international prestige.

---

1. A nine-million-dollar estate these days seems skimpy for Nobel's purpose. In terms of the economists' standard 1967 dollars, however, his fortune comes to the far from negligible sum of thirty-seven million dollars and, in terms of the inflated currency of 1975, it amounts to sixty-one million dollars. See Bergengren (1962), the "official" biography of Nobel, and other biographies by Evlanoff and Fluor (1969), Halasz (1959), and Shaplen (1958).

Nobel, the ingenious inventor of dynamite, blasting gelatin, and smokeless powder, was paradoxically a pacifist and a self-styled idealist. It was therefore in character that he provide in his will for prizes that would honor "the most important discoveries or inventions" in physics, chemistry, and the composite field of physiology or medicine, "the most outstanding work in literature of an idealistic tendency," and "the best work for fraternity among nations" (Nobelstiftelsen, 1972, p. 10). He also specified that the prizes in science should be distributed by the Royal Swedish Academy of Sciences and the Royal Caroline Medico-Surgical Institute, in literature by the Swedish Academy at Stockholm (the literary counterpart of the Academy of Sciences), and in peace by the Norwegian Storting (parliament). The complex details involved in establishing the Nobelstiftelsen (Nobel Foundation) and the other institutional arrangements for distributing the prizes took no less than four years to settle. (See chapter 2 for more extended discussion of the statutes and customs governing the allocation of prizes.) Not until December 10, 1901, the anniversary of Alfred Nobel's death, were the first prizes in science and literature awarded, in a splendid ceremony held at Stockholm, one that has been repeated on that date every year since except for wartime interruptions.

## BIOGRAPHICAL NOTES ON THE STUDY

My research on the scientific elite began with my doctoral dissertation in 1963. I had decided to work in the thinly populated sociology of science, for it seemed to me then, as it does now, that science was a major social institution of our time and one largely neglected by sociologists. (See Merton, 1973, pp. 210–20, for some reasons for this neglect.) The investigation started with several specific questions that were central then but were destined to become subsidiary as the work developed. Is it true, as many observers of science have claimed, that important contributions to contemporary science are more often made by individual investigators than by teams? (See, for example, Eaton, 1951; Whyte, 1957; Jewkes, 1958; Jewkes, Sawers and Stillerman, 1959.) Are the outstanding scientists typically "lone wolves" who resist working in collaboration? And, if so, what were the effects on the advancement of scientific knowledge of the strong trend toward collaborative research in all the sciences and most notably in those exhibiting most rapid growth? I planned to interview a stratified sample of scientists about their research practices to find out whether the outstanding scientists tended to engage in solo research and whether this was especially true of their best work.

Before moving ahead with this program of research, I plainly needed to decide upon appropriate criteria for stratifying the sample of interviewees. It seemed sensible to assume that Nobel prize-winners constituted a small sample of the most accomplished American scientists making up the scientific elite. They could be compared with a larger sample of scientists from universities and research laboratories representing various strata in the community of science.

That spring, the detailed interview guide was pretested on a number of surprisingly uncomplaining members of the Columbia science faculty. (See appendix A for detailed analysis of the interview procedure.) It soon became evident that the value of the interviews increased when I reviewed the scientist's work and biography beforehand. This preliminary work became increasingly elaborate as the study progressed. Ultimately it amounted to reading everything accessible to a layman that each scientist had published as well as a highly selective sample of his scientific papers. Where the public record was fairly complete, preparation for each interview required scores of hours. By way of further groundwork, I read review papers and more popular accounts of the fields I had to cover, histories of science, and scientific memoirs. In the process, I learned a fair amount of science and a great deal of scientific terminology.

It was summer when I set out for California for the first round of interviews. Four laureates and a dozen or so other scientists in the San Francisco Bay area agreed to see me on that trip. The initial interviews turned out to be decisive. By and large, those with the laureates were richer and more instructive than most of the others. Evidently, the laureates were accustomed to talking about themselves and their work to visiting outsiders. Since more information about them and their research was publicly available, my questions could be more specific. The laureates also tended to be more reflective about the training and fostering of scientific talent and the organization of scientific work than were the others.

By the time I returned home from that first field trip, I had decided to confine the interviews to Nobelists and to try to talk to all those at work in the United States. My sense of the research problem had also changed by then. Although the agenda for my dissertation remained much the same, I began to plan a more comprehensive investigation of stratification in science: how scientists became members of the elite, why the gap in accomplishment between the elite and the rank and file is as great as it appears to be, and how the stratification system in science and the development of scientific knowledge are interrelated.

The interviews stretched on for more than a year. In the end, I had traveled to California twice, to the Midwest once, and up and down the eastern seaboard. I interviewed forty-one of the fifty-six laureates then at work in the United States and tape-recorded all but one of them. (Transcripts of interviews were deposited in the Oral History Collection of Columbia University with the proviso that they would be kept confidential for a period of years after the laureates' deaths. I still have no way of knowing whether my promise to keep the laureates' remarks confidential was decisive in securing the interviews. But those were the terms to which they agreed, and that is why the sources of so many of the quotations from the interviews in the pages that follow are unattributed.)

In 1965 I finished the dissertation, which dealt primarily with the somewhat modified questions about collaborative research in science (Zuckerman, 1965), and moved on to other work on stratification in science, the cultural structure of science, and processes of evaluation of scientific contributions.

Since then, I have gotten to know some of the forty-one Nobelists I interviewed a great deal better, and the same is true for a number of the more recent prize-winners. I have also worked jointly with two Nobelists, André F. Cournand and Joshua Lederberg, on investigations quite unrelated to this one (Cournand and Zuckerman, 1970; Lederberg, Zuckerman, and Merton, forthcoming). My encounter with Arne Tiselius was important in another way for the development of the investigation. As president of the Nobel Foundation, laureate in chemistry, and chairman of the Nobel Chemistry Committee, which awards prizes in that field, Tiselius embodied the Nobel establishment. He had read some of my preliminary reports on Nobel laureates and asked to see me when he was in New York. After a series of discussions, he invited me to the ceremonies at Stockholm and gave me the chance to talk off the record with many officials of the Nobel Foundation probably because rather than in spite of my papers having critically dealt with the unanticipated consequences of the prize. That trip renewed my interest in developing my study of the scientific elite further.

As I reviewed my work on elites, it became clear that more data on the careers of laureates and other scientists would be required if I was to begin to understand the larger question of elite formation in science and its system of stratification.[2] It seemed wise to enlarge the scope of the inquiry to include all American Nobel laureates from the first, Albert A. Michelson, who won the award in physics in 1907, to those who had won prizes by 1972. There was also a clear need for new data on the more extended scientific elite, members of the National Academy

<hr>

2. My Columbia colleagues Stephen and Jonathan Cole were also working on stratification in science and exploring ideas that our joint mentor, Robert Merton, had been developing for years. (See Merton, 1973, for his collected papers in the sociology of science, and Cole and Cole, 1973.)

of Sciences and rank-and-file scientists as well. Thus, what was originally a study of collaboration and individual research in science was transformed over the years into a broader investigation of the American scientific elite—how they are educated, recruited, sustained, and what contribution they have made to the advancement of science, which amounts to collective biography or what historians now call prosopography (Stone, 1971, pp. 48–57).

Although the study focused on scientific elites, I found myself drawing upon varying intellectual traditions, including contemporary studies in political theory and in social stratification, most particularly on the reward system of science. I also turned to "classical" treatments of elites by Saint-Simon, Mannheim, Michels, Mosca, and, of course, the Italian economist and social theorist, Vilfredo Pareto.

## PARETO AND ELITES

The history of social thought has its own variety of paradoxes. So it is that while Vilfredo Pareto neither originated the basic distinction between social classes and social elites nor did much to develop the distinction in his massive *Treatise on General Sociology* (1916), that work has nonetheless done more than any other to stimulate reflection and research on elites in the half century since it appeared. Pareto's informing idea held that the gradation of capacity and performance exhibited by people in every department of culture and social life resulted in socially identifiable hierarchies, and that the higher strata within each of these hierarchies could conveniently be described as "elites."

He put this idea graphically in the language that, as an acerbic engineer and economist, he found most congenial:

> Let us assume that in every branch of human activity each individual is given an index which stands as a sign of his capacity, very much the way grades are given in the various subjects in school examinations. The highest type of lawyer, for example, will be given 10. The man who does not get a client will be given 1—reserving zero for the man who is an out-and-out idiot. To the man who has made his millions—honestly or dishonestly as the case may be—we will give 10. To the man who has earned his thousands we will give 6; to such as just manage to keep out of the poor-house, 1, keeping zero for those who get in. . . . To a poet like Carducci we shall give 8 or 9 according to our tastes; to a scribbler who puts people to rout with his sonnets we shall give zero. For chess-players we can get very precise indices,

noting what matches, and how many, they have won. And so on
for all the branches of human activity. . . . So let us make a
class who have the highest indices in their branch of activity,
and to that class give the name of *elite*.[3]

Pareto's conception of the elite harbors a fundamental ambiguity,
as Coser (1971, p. 397), among others, has noted. And since this is an
ambiguity that often persists in current discussions of elites, leading to
an untenable doctrine of biological and social elitism, it needs to be
examined at the outset of any study of the scientific elite.

In the passage I have quoted, Pareto makes the strong double assumption that the members of any elite are there by virtue of their outstanding performance, and that this, in turn, is a "sign of [their]
capacity." In short, the various elites are here thought of as the product
of a meritocracy, with their members achieving lofty status through an
efficient process of social selection among the unequally endowed.
Pareto leaves unexamined the question whether these assumed differences in capacity are altogether biological or represent differences in
trained capacity. Yet elsewhere (1935, vol. 3, p. 1424) he takes
another tack and allows for errors in the process of social selection such
that membership in this or that elite need not invariably testify to great
capacity or outstanding performance. Thus,

in the concrete, there are no examinations whereby each person
is assigned to his proper place in these various [elites]. That de-
ficiency is made up for by other means, by various sorts of labels
that serve the purpose after a fashion. Such labels are the rule
even where there are examinations. The label "lawyer" is affixed
to a man who is supposed to know something about the law and
often does, though sometimes again he is an ignoramus. So, the
governing *elite* contains individuals who wear labels appropriate
to political offices of a certain altitude—ministers, Senators,
Deputies, chief justices, generals, colonels, and so on—making
the apposite exceptions for those who have found their way into
that exalted company without possessing qualities corresponding
to the labels they wear. Such exceptions are much more numerous than the exceptions among lawyers, physicians, engineers, millionaires (who have made their own money), artists
of distinction, and so on; for the reason, among others, that in
these latter departments of human activity the labels are won

---

3. Pareto (1935, vol. 3, p. 1423). First published in 1916 as *Trattato di sociologia
generale,* Pareto's work was translated into English and appeared in 1935 under the
title *The Mind and Society.* As the *locus classicus,* parts of this passage are much
quoted by students of elites; for example, Bottomore (1964, pp. 1–2) and Coser (1971,
pp. 395–97). The most comprehensive recent development of the theory of elites in
conjunction with newly available evidence (Keller, 1963) draws extensively upon Pareto
but somehow resists direct quotation of this passage.

directly by the individual, whereas in the elite some of the labels—the label of wealth, for instance—are hereditary.

Here, Pareto comes to recognize that elites are not all of a kind: that the relationship of capacity, performance, and status may differ among varieties of human activity. In effect, he thus raises a question about differences in the efficacy of processes of social selection in the various functional spheres of complex society. Like much else in the comparative study of elites, this issue has not yet received the sustained attention it requires.[4]

The amount of scholarly attention paid to elites in the major institutional spheres has varied greatly. Political elites have received the greatest attention, and there is also a growing literature on economic and religious elites. But scientific elites have been systematically investigated hardly at all—reflecting the lack of interest in science as a social institution among sociologists until recently. To be sure, much of the recent work in the developing sociology of science deals with those scientists to whom Pareto would give high marks in his scheme of things, but this focus is largely by inadvertence rather than by design since it is the more productive scientists who disproportionately turn up in the readily available rosters of scientists. With the growing interest in the stratification of science,[5] there may develop a basis for the

---

4. This does not overlook pioneering investigations of elites begun by Harold D. Lasswell and his associates in the 1930s and continued ever since (1952, 1961, 1965). Lasswell's early theoretical contribution identified elites in what he describes as skill, class, personality, and attitude groups (1936, pt. 1). Empirical studies deal primarily with political and ideological elites in various times and places rather than with comparisons of elites in different social spheres. Bottomore (1964) and Keller (1963), much more comprehensively, examine this latter kind of comparison of elites but, as they indicate, without having enough systematic comparative studies to draw upon.
5. Recent studies in the sociology of science have focused on the relationship between the quality of scientific achievement and the allocation of rewards. Merton (1957) analyzes the main components of the reward system in science, the composite of institutionalized procedures, positions, and organizations through which rewards are distributed. In "The Matthew Effect in Science" (1968f), he examines sources of differential distribution of recognition. Hagstrom (1965) and Storer (1966) emphasize the role of collegial recognition in scientists' motivations for research. Cole and Cole, together and separately (1970, 1972, 1973), have shown that the quality of role performance in physics is more closely associated with rewards than is quantity and have demonstrated how performance and rewards are temporally linked. Glaser's (1964) studies of industrial scientists also deal with the interconnections between rewards and continuing performance and focus on the differential meanings assigned to recognition in different sectors of science. More recently, Hagstrom (1974) has focused on competition for recognition in science and differences between fields in this regard. Other studies examine the effects of social origins and institutional affiliations on rewards and status attainment in American and British science (Crane, 1964, 1969, 1970; Gaston, 1973), the relations between gender and the distribution of recognition (Zuckerman and Cole, 1974; Reskin, 1976), communications behavior of various strata in science (Amick, 1973), and variations in recognition over the course of scientists' careers (Allison and Stewart, 1974). For a systematic review of much of this line of investigation and its significance for the stratification of esteem in science, see Zuckerman (1970).

sociological comparison of scientific elites not only with the rank and file of scientists but also with elites in other fields of activity. This would allow us to gauge what elites have in common by way of social origins, patterns of recruitment, training, career mobility, and linkage with other parts of the social structure just as it would allow us to identify what is distinctive of each elite. These are precisely the problems that Pareto edged up to but did not clarify or, of course, investigate empirically.

## SCIENTISTS AS AN ELITE AND ELITES WITHIN SCIENCE

In one sense, all scientists constitute an elite in complex industrial societies. Compared with other occupational groups, they rank high by any of the criteria ordinarily used to stratify populations socially. In the United States they are, on the average, in the top fifth of the income distribution (U.S. Bureau of the Census, 1974, pp. 390, 540). They are also accorded great social prestige. What is more, studies of occupational standing in this country show a dramatic rise after World War II in the prestige assigned to scientific professions. For example, the most highly rated scientific occupation, nuclear physicist, rose from eighteenth place among a cross-section of ninety occupations in 1947 to third place in 1963, putting nuclear physicists on a par in public esteem with cabinet members, governors, and higher diplomats (Hodge, Siegel, and Rossi, 1964). And nuclear physicists are not entirely atypical: the aggregate of physical and biological scientific occupations rose seven places in the same period. More recent but truncated studies show that scientists have held their own in occupational prestige; they continue to be ranked only second to physicians among the professions (National Science Board, 1975, p. 146). Moreover, the much discussed resentment toward science and technology—a phenomenon greatly disturbing to scientists but not yet systematically investigated—seems not to be reflected in public sentiment that science has changed life for the better (National Science Board, 1975, p. 145).

But if the aggregate of scientists seems to outsiders to be a more or less homogeneous elite, whose members possess esoteric knowledge and sport an impressive array of academic degrees, this is far from being so for insiders. The community of scientists is sharply stratified. Scientists incessantly engage in assessing one another's work and capacities. The process of evaluation helps to generate the system of stratification within science and to locate individual scientists within that system. Although some scientists seem to deny the existence of stratification by affirming that all scientists are peers, the everyday idiom suggests that the opposite is true. The physicist and biologist Leo Szilard, who never doubted the stratification of capacity and performance, liked to measure himself by the quality of his opponents and was ready to announce that he never argued "with third-rate scientists, ... only with the first-rate" ones (Gilman, 1965, p. 143). Luis Alvarez, a physicist and Nobel laureate given to plain talk, put the general point this way: "There is no democracy in physics. We can't say that some second-rate guy has as much right to [an] opinion as Fermi" (Greenberg, 1968, p. 43).

Prestige in the scientific community is largely graded in terms of the extent to which scientists are held to have contributed to the advancement of knowledge in their fields and is far less influenced by other kinds of role performance, such as teaching, involvement in the politics of science, or in organizing research. Even great influence in national politics and science policy, for example, does not earn a scientist the same kind of esteem as scientific contributions judged to be truly important.

With more objective indicators of prestige, we can adopt the logic of Pareto's procedure for identifying elites to sketch the broad-based pyramid of stratification of the American scientific community. Recent data indicate that there were approximately

493,000 men and women in the United States who described themselves as "scientists" when asked their occupation in the national census (U.S. Bureau of the Census, 1974, p. 57), a self-definition that does not of course always correspond to institutionally legitimated definitions. To take another figure, there were about

313,000 scientists as estimated in the biennial survey making up the National Register of Scientific and Technical Personnel prepared by the National Science Foundation (1970), once the most reliable and thorough source on scientific man- and womanpower. There were about

184,000 scientists by the somewhat more exacting criteria adopted by *American Men and Women of Science* for listing in its pages, which are: achievement equivalent to that associated with holding a doctorate or continuing research activity or holding a position of responsibility requiring scientific training.[6] There were about

---

6. Social change has caught up with *American Men of Science,* which in its twelfth edition has enlarged its title and presumably its scope to *American Men and Women of Science.* The criteria for inclusion have remained much the same since its inception except for the much greater coverage of the social and behavioral sciences since the

175,000 scientists, if these are confined to people holding a Ph.D. in one or another science or mathematics, a criterion that roughly testifies to advanced scientific training (National Science Foundation, Doctoral Roster for 1973, in U.S. Bureau of the Census, 1975, p. 555. The National Research Council reports about 13,000 fewer scientists with doctorates for the same year [U.S. Bureau of the Census, 1974, p. 540]—a commentary on the status of official records on U.S. scientific personnel.) Among these, in turn, there are about

950 who have been elected to the National Academy of Sciences, a distinction widely acknowledged within the American scientific community as signifying superior scientific work on the whole, if not unimpeachably in every individual case. (The number increased to 1133 in 1975.) Finally, there were

72 Nobel laureates living in the United States, almost universally judged as having made scientific contributions of the first importance. (They numbered seventy-seven in 1976.)

Putting this arithmetic of stratification somewhat more compactly, we note that for each Nobel laureate in the United States, there are about

6,800 self-defined scientists;

4,300 scientists identified in the *National Register;*

2,600 scientists of sufficient stature to be included in *American Men and Women of Science;* and

2,400 scientists holding a Ph.D.; and, to move into a higher stratum, about

13 members of the National Academy of Sciences.

If the membership of the National Academy constitutes an extended elite in the domain of science, then the Nobel laureates, located at the apex of the hierarchy of prestige and esteem and fewer in number than the Academicians, can be fairly described as the ultra-elite.

---

ninth edition in 1956. The criteria are put in these terms: "(1) Achievement, by reason of experience and training, of a stature in scientific work equivalent to that associated with the doctorate degree, coupled with presently continued activity in such work; or (2) research activity of high quality in science, as evidenced by publication in reputable scientific journals; or, for those whose work cannot be published because of governmental or industrial security, research activity of high quality in science as evidenced by the judgment of the individual's peers; or (3) attainment of a position of substantial responsibility requiring scientific training and experience to the extent described for (1) and (2)." *American Men and Women of Science* (1973, p. vii).

## The Ultra-Elite of Science

As the *ne plus ultra* of honors in science, the Nobel prize elevates its recipients not merely to the scientific elite but to the uppermost rank of the scientific ultra-elite, the thin layer of those at the top of the stratification hierarchy of elites who exhibit especially great influence, authority, or power and who generally have the highest prestige within what is a prestigious collectivity to begin with. But no more than other social arrangements can the institution of the Nobel prizes in science be assumed to operate with total efficiency. As we shall see, Nobel laureates are not all equally outstanding pathbreakers and pacesetters in the science of their time. There are some whose contributions, in the judgment of many qualified scientists, have not advanced the frontiers of their fields as much as others who do not wear the Nobel laurels. These uncrowned laureates either have not yet or will not win Nobel prizes. Nevertheless, in the aggregate, the Nobelists, together with the publicly less visible scientists of comparable stature, take their place in the highest stratum of the loosely formalized hierarchy of science. They are worthy of our attention not merely because they have prestige and influence in science but because their collective contributions have made a difference in the advancement of scientific knowledge. L. Pearce Williams, a historian of science, goes perhaps too far in rhetorically asking "Is what *most* scientists do really relevant to what science in the long run is? Do we weight the opinion of, say, Peter Debye [a Nobel laureate in chemistry] equally with that of a man who measures nuclear cross-sections?" (1970, p. 49). But plainly there are cognitive and historical reasons as well as political and sociological ones for focusing on the elite stratum of scientists.

Seen in historical perspective, a Nobel prize obviously does not in itself ensure immortality to its recipients. Historical immortality for scientists derives from the enduring quality of their scientific work, whether or not this happens to be initially confirmed by formal awards. When the measure of achievement far transcends what is required to qualify for an award, it is the recipient who confers honor upon the award rather than vice versa. Had the Nobel prizes existed in the seventeenth century, for example, they would almost surely have been awarded to Newton, Hooke, Boyle, and Galileo (at least before the Inquisitorial summons to Rome). Just as surely, their acceptance would have honored and heightened the prestige of the prizes and would have given added pride, pleasure, and standing to others invited to join the company of these giants. But, as the record of this century testifies, if a Nobel prize does not confer immortality, it does confer contemporary preeminence upon scientists and significantly affects their future scientific work and other roles in society.

In effect, Nobelists are life peers in the world community of scientists. Having been recruited into the aristocracy of science on the basis of their own achievements, they can only pass on to their children an enlarged access to opportunity. They cannot of course ensure that their own offspring will enter the ultra-elite of science. In short, as we shall see in chapters 3 and 4, this limited elite is replenished and renewed by processes that are more visible and formalized but functionally not much different from the processes of cumulative advantage that renew the larger scientific elite.

Elites in every department of social life are plainly numerical minorities, as Pareto and Simmel have reminded us at length. (See Simmel, 1950, pp. 90–91, on "Aristocracies," and Pareto, 1935, pp. 1423–25.) The part of the scientific elite represented by Nobel laureates has always been exceedingly small, in absolute or relative terms. (See Simmel, 1950, pp. 97–98, and Merton, 1968b, p. 228; 1968e, 366–68, 465.) From their beginning in 1901 to 1976, Nobel prizes have been awarded to only 313 scientists of the world total of well over a million. Moreover, the age distribution of recipients has been such as to ensure a relatively high mortality among laureates so that a considerably smaller number than this are alive and an even smaller number are actively engaged in scientific work at any given time. All this means that there is not only a small *absolute* number of laureates but, in light of the exponential increase in the number of scientists generally and the comparatively unchanging rate of additions to the ranks of laureates, a decreasing *relative* number. This may serve to retain or increase the scarcity value of the Nobel prize and its rate of return in terms of prestige and derived influence for its recipients, but it also means that other criteria, informal as well as formal, are used to identify the growing numbers of scientists who have contributed as much to science as the laureates but are not Nobel prize-winners.

By and large, the Nobel laureates and their uncrowned peers are only peripheral to Pareto's "governing elite" (1935, pp. 1423–25) or Mills's "power elite" (1956). Rather, they are members of what Keller (1963) describes as a "segmental elite," their influence being largely confined to the one domain in which they have made their mark. They have little power but exercise considerable and possibly growing influence in the larger society following the tendency for social concentrations of great prestige in one sphere of society to spread into other spheres. Those who are so inclined can convert the prestige of being a laureate into influence on consequential decisions within the domain of science, and, in some cases, they convert this influence into power, as they seek or are drawn into higher political circles such as the once active President's Science Advisory Committee (PSAC), which was charged with formulating science policy and bringing scientific knowledge to bear on other policy issues.

Thus, not a few of them become members of the "strategic elite," claiming or being assigned influence on significant decisions in the society as a whole. There is the prospect that this tendency, hitherto identifiable to only a limited extent, will become more marked with the growing interaction between science and government, on the one side, and the more diffuse pressures for the politicization of science, on the other. (See Rose and Rose, 1969; Ravetz, 1971; Salomon, 1970, 1972, 1973.) Even now, Nobel laureates are increasingly called upon to symbolize and to act in terms of the growing role assigned to science in contemporary society, in both popular imagery and actual fact.

Still, as Daniel Greenberg has documented in his anything but reverent account of "the politics of pure science," all this does not add up to there being a monolithic "American Scientific Establishment." He therefore finds it necessary (1968, pp. 3–4) to reject "elitist theories [and the notions of] compact tables of organization, or directions to a central command post." That is not the way power, authority, and influence work in the domain of science.

The ultra-elite of science, comprised of Nobel laureates and their appropriate peers in the elite, is therefore an aggregate rather than a social group. Being numerically scarce and being highly esteemed by their fellow scientists, they exercise considerable influence and some power in scientific affairs, but they typically do so as individuals rather than as members of an organized group. The influence and power they wield are not, however, the prime focus here. Instead, the operations of the evaluation and reward systems of science that made possible their entry into the ultra-elite, and the consequences of their having arrived there are the principal subjects of inquiry.

## PLAN OF THE BOOK

This study centers on the ninety-two Nobel laureates selected between 1901 and 1972 who worked in the United States on the research that was to bring them their prizes. It draws upon a variety of data on their careers and scientific work: the detailed interviews with forty-one of the laureates that were mentioned earlier and less intensive interviews with other scientists; observations of laureates in their laboratories; autobiographical and biographical accounts of scientists; and an array of comparative data on the productivity and career patterns of laureates, members of the National Academy of Sciences (the larger elite), and a sample of scientists matched with the laureates in age, field of specialization, and type of organizational affiliation. By focusing on differences between laureates and other scientists and on attributes laureates have shared over time, my study neglects the impact of major historical developments on science such as the Depression, World War II, and the ensuing Cold War between the United States and the Soviet Union. By focusing on American laureates, the study also neglects the distinctive features of stratification in other national scientific communities, but since the Nobel prizes and the careers of the laureates have been conspicuously international, the boundaries of the inquiry occasionally stretch beyond those provided by U.S. geography.

Drawing upon these diverse kinds of qualitative and quantitative data, the book deals first with the workings of the reward system of science as illuminated by the Nobel prizes. Then the life cycle of the laureates is examined, beginning with their social origins and education and moving on to the time after they were elevated into the ultra-elite of science.

Chapter 2 examines the evolution of the Nobel prizes as cardinal symbols of scientific excellence in the twentieth century and as an institution that soon acquired a degree of functional autonomy. (A list of all Nobel prize-winners in the sciences from 1901 to 1976 can be found in appendix B.) This chapter traces the ramifying symbolic uses of the prizes in academic and research organizations as well as in other sectors of society and the tendency toward goal displacement, in which acquiring the prize almost becomes an end in itself. Here I also examine the fact that a number of scientists have contributed as much to scientific knowledge as the laureates but, for a variety of reasons, have not won Nobel prizes. (Appendix D provides a list of biological scientists whose work the Nobel Committee considered important enough to merit an award but who lost out to successful candidates.) Chapter 3 deals with processes of accumulation of advantage. These processes create and maintain the structure of stratification in science and result in "the rich" becoming richer at a rate that makes "the poor" progressively poorer. The social and educational origins of future Nobel laureates, members of the National Academy of Sciences, and a sample of rank-and-file scientists are examined in light of accumulation of advantage. In particular, I try to show how ascribed statuses of future scientists affect their differential access to various kinds of educational institutions and to first-class scientific training.

Chapter 4 treats the critical period in the training of scientists— the time of graduate study and immediately afterward—and centers on the character of master-apprentice relations for scientists who will later become members of the ultra-elite. Here I examine the implications of the pronounced tendency for Nobel laureates to have been trained by older Nobel laureates. Chapter 5 traces the careers of the future laureates: early signs of excellence in their research performance, their upward mobility from the time of their first jobs, and how they fared in comparison with future members of the larger elite of the National Academy and the sample of other scientists. (The laureates' age-specific rates of published productivity are compared to the matched sample of scientists' in appendix E.) In the process, I note that future laureates benefited from the *noblesse oblige* of their eminent sponsors, a pattern they soon begin to exhibit vis-à-vis their own students. The conception of reenactment of roles identified in this study of scientific elites may have general significance beyond the special case.

Chapter 6 deals with the personal and social contexts of the research that would later be honored by Nobel prizes: the age at which the laureates did that work (I examine the widespread belief that the pathbreaking scientific contributions are made primarily by the young); the extent of collaborative and individual work; and the role of "evocative environments" in that work. (The changing foci of research selected for a prize are shown in appendix C.) Chapter 6 also deals with the extent and kinds of recognition elicited by prizewinning research and suggests that many future laureates are already members of the ultra-elite in science before they actually receive their prizes.

Chapter 7 centers on the experiences and attitudes of laureates after they receive their prizes: their ambivalence toward the award and the ramified, sometimes ironic consequences the prize had for their scientific work, their relations with other scientists, and their standing and roles outside the scientific community.

The final chapter interweaves these strands of observation and analysis about scientific elites, stratification in science, the operation of the reward system, and the efficacy of the Nobel prize as the ultimate scientific award. It is not entirely disappointing that this attempt raises questions more than it provides definite answers to them.

# Chapter 2

# THE SOCIOLOGY OF THE NOBEL PRIZE

Since 1901, when the first awards were given, 313 scientists have won Nobel prizes in physics, chemistry, and medicine and physiology.[1] Almost from the beginning, the prize has been regarded by laymen and scientists alike as the acme among symbols of scientific achievement. It is probably the only award known by name to a sizable fraction of the public at large, and it appears to be the only one that is universally known in the community of scientists. In a sample of some 1300 American physicists, for example, a Nobel prize was the only one of 100 scientific awards, a good many far from obscure, that could be

---

1. In 1969 the Swedish Central Bank established the Alfred Nobel Memorial Prize in Economic Science. Although Nobel made provision for the prizes to be given annually, wartime interruptions in scientific activity have occasionally broken the customary pattern. This has happened six times in the case of the physics award, eight times in the case of the chemistry award, and nine times in the case of the award in medicine-physiology (Nobelstiftelsen, 1972, pp. 155–59, 638–45).

identified by all, and it was ranked in prestige far ahead of the rest (Cole and Cole, 1973, p. 270).

Throughout, this study will examine the characteristics, behavior, and structure of the ultra-elite of Nobel prize-winners and their role in the development of science. Here I examine the sociology of the Nobel prize itself: how the laureates are selected, the processes through which the prize has come to generate an ultra-elite in science, the diverse uses to which it has been put within the social system of science and the larger society, and the growing constraints on the prize as a valid measure of scientific achievement.

## SELECTING NOBEL PRIZE-WINNERS

That the Nobel prize now serves as the prime symbol of achievement in science and outside is consistent with Alfred Nobel's intent, although he could scarcely have foreseen how far social process would outrun that intent. He wanted to promote human welfare, to honor great scientific accomplishment, and, above all, to foster research by providing "such complete economic independence for those who by their previous work had given promise to further achievement that they could ever afterwards devote themselves entirely to research" (Nobelstiftelsen, 1962, p. 143).

These announced purposes were in large part a function of the particular time and place in the development of science. In the late nineteenth century, there was no substantial support for research, and comparatively few scientists were employed as researchers. As an inventor and scientist of sorts, Nobel knew from his own experience that the costs of most research were modest. It was not unrealistic for him to believe that he could accomplish the double purpose of underwriting the work of a handful of first-class scientists and providing for their support as well. As late as 1920, for example, the research grant for the entire Cavendish Laboratory was only 2,000 pounds, or about ten thousand dollars (Holton, 1974, p. 164). But when the Nobel prizes were inaugurated in 1901, the awards came to $42,000 each. By way of comparison, this was seventy times as large as the honorarium attached to the Royal Society's Rumford Medal, one of the major prizes in science available at the time. For the general public, the Nobel's princely sum constituted a symbolic message asserting in a way that could be understood by the informed and uninformed alike that science and scientists really mattered. For many of the scientists themselves, it provided both symbolic and public recognition within their own ranks of major contributions to scientific knowledge.

When Nobel set down the requirements for his prizes in science, he stipulated that they be given in just three fields: physics, chemistry, and physiology or medicine. Mathematicians like to explain the absence of a Nobel prize in mathematics by telling the story of the rivalry between Alfred Nobel and the Swedish mathematician, Gosta Mittag-Leffler, for the hand of an unidentified lady. Nobel, the story goes, was the unsuccessful suitor. By excluding mathematics from the fields eligible for prizes, Nobel retaliated and made sure that Mittag-Leffler would never get one of his prizes. Appealing as the story is, it seems to be more fantasy than fact. It appears in none of the standard histories of mathematics nor can a number of historically minded mathematicians say just where they encountered it or even that it is consistent with what they know of Mittag-Leffler's private life. One might better conclude, along with Stig Ramel of the Nobel Foundation, that Nobel excluded mathematics from the fields to be covered by the prizes because he wanted to benefit mankind in a concrete, rather than abstract, way.

Each prize was to be given for "discoveries," "inventions," or "improvements" (the language varies for each science) made "during the preceding year," and nationality was not to be taken into account in the process of selection. (See Nobelstiftelsen, 1962, pp. x–xi for a facsimile and translation of Nobel's will.) He left it up to his executors, the Royal Swedish Academy of Sciences and the Royal Caroline Institute, to draw up the detailed rules that would govern the actual selection of recipients. It took five years for the interested parties (including the Nobel relatives) to agree on procedures and the wording of the official statutes and to make the first selections. (See Nobelstiftelsen, 1972, pp. 49–72, for an account of the negotiations and of the establishment of the Nobel Foundation, and pp. 619–34 for the official statutes.) Since then, elaborate customs have evolved about the selection of candidates and the conferring of the awards by the king of Sweden in Stockholm, but the statutes themselves have not changed.

Although the statutes governing selection of prize-winners differ somewhat for the awards given by the Academy and by the Caroline Institute, in their essentials they are the same. Most important, the statutes provide for the establishment of three committees, each comprised of five qualified scientists who have primary responsibility for inviting nominations, investigating candidates, and selecting the winners. The names of those selected are presented for formal approval to the membership of the Academy and the faculty of the Caroline Institute on specified dates. Committee members serve for renewable terms of three to five years and are elected by the Academy for physics and chemistry and by the Institute for medicine and physiology. The physics and chemistry committees always include the chiefs of the physics and chemistry sections of the Nobel Institute, and the committee for physiology and medicine includes the rector of the Caroline Institute. Since the committees determine who can nominate in a given year and since they control the evaluation process, their power to select and reject potential laureates should be evident.

Nominations are solicited from two sets of proposers: those with permanent rights to nominate (members of the Academy of Sciences, the faculties of the Caroline Institute and of the eight Scandinavian universities in the appropriate sciences, and past laureates) and others who are invited to do so year by year. In 1900, when the first nominations were solicited, the committees supervising the prizes in physics and chemistry each sent out 300 invitations (Nobelstiftelsen, 1972, p. 282). Now, upward of a thousand scientists the world over are asked to nominate for each of the prizes (Schimanski, 1974, p. 10), although far fewer actually do so. The committees hold that they systematically solicit nominations from scientists at minor as well as major research centers.

Nominators may propose several candidates, but the prizes may not be divided into more than three shares, presumably so as to maintain the honor and the honorarium at the level Nobel intended. Thus far, this has meant that laureates are laureates whether or not they have shared their prizes. In effect, no laureate is a fractional Nobel prize-winner. Beyond this, all nominees must be alive not only at the time they are proposed but also at the time the award is made. Prizes in the sciences are never given posthumously.

The application of these rather simple procedures for nomination and selection has produced a distinguished roster of Nobel laureates in science. As we shall see, however, these same procedures leave room for complaint and skepticism about the preeminent award in science.

## PREEMINENCE OF THE NOBEL PRIZE

The Nobel prize has become a metaphor for supreme achievement. When a movie actor, Walter Matthau, is described as so profligate a gambler as to "surely have won him the Nobel prize for masochism" (Meehan, 1971, p. 6), we know that he loses money in truly extraordinary style. When psychoanalysts speak of the "Nobel Prize Complex," we also know that they have tagged a syndrome involving the pursuit of the most ambitious goals (Tartakoff, 1966; Sperber, 1974–75). And when scientists read the ominous announcement that appropriations for research have been so reduced that the National Science Foundation "cannot assume support even for the best people—Nobel laureates," they quickly sense their own diminished chances for research funds (President's Task Force on Science Policy, 1970, p. 25).

No one attribute of the Nobel prize accounts for its extraordinary visibility and prestige. Rather, these appear to derive from its uniformly high ranking on a composite of interacting attributes, although it may be matched or outranked on any particular one such as its comparative venerability, wealth, or the standing of the awarding body. In this respect, it appears that the unplanned and tacit contest for prestige among awards is something like the planned and explicit contest of the decathlon in sports, with the Nobel emerging as champion through its high ranking in a variety of attributes making for prestige.

To begin with, other prizes in science boast much longer histories than the Nobel prizes. Established in 1901, the Nobel prize is something of a latecomer when compared, say, with the Copley and Rumford medals established by the Royal Society of London, the one in 1736 and the other in 1800. Still, its seventy-five years are enough to give the Nobel the patina of an established institutional symbol. Moreover, unlike *nouveaux riches* in other fields, the Nobel was linked from the beginning with institutions of comparatively ancient and distinguished lineage. Yet the Royal Swedish Academy of Sciences and the Royal Caroline Institute, which were charged with the task of selecting recipients for the prizes, can scarcely be said to outrank such other scientific academies as the Royal Society of London, the Académie des Sciences, or the National Academy of Sciences in the United States.

There are also other prizes in science as rich as or richer than the Nobel prizes. The John and Alice Tyler Ecology Award now heads the list with an honorarium in 1975 of $175,000. The Robert A. Welch award in chemistry carries an honorarium of $100,000, and the Prix Balzan one of $52,000, while the Nobel prizes have varied over the years from a low of $30,000 to a peak of $160,000 in 1976. Moreover, as noted, the Nobel prizes are frequently divided into two or three portions, making the awards to individual scientists smaller than those provided by other prizes. And, although the cash value of the prizes has been rising in recent years, as figure 2–1 shows, the purchase value (adjusted for inflation) still lags behind the originally substantial sum given in 1901 and no longer provides anything like the liberating endowment that Nobel intended. Nevertheless, the original sum was magnanimous enough to achieve Nobel's announced purpose and immediately established the prizes in the public mind as unique (*New*

**Figure 2-1   Purchase Value and Cash Value of Nobel Prizes (1901-75)**

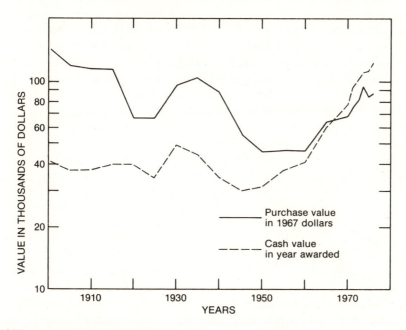

NOTE: Purchase values are calculated from the Consumer Price Index base 1967, prepared by the Bureau of Labor Statistics (1969, p. 282; 1975, p. 316). Data for the cash value are based upon Nobelstiftelsen (1951; 1972) and the *New York Times* when they are not available from official Nobel sources.

*York Times,* 13 August 1901, p. 13), and over the years the Nobel awards have been generous enough to attract considerable attention. No other science prizes of comparable value would become available for another fifty years. In the meantime, the Nobel prizes acquired other attributes that, in combination, put them in a symbolic class by themselves.

The Nobel prizes are again like a variety of other awards with respect to their being awarded without regard to nationality. (There has always been discussion about the Nobel committees having favored scientists of one or another country, but an international set of prizes was plainly specified in Nobel's will.) The Fields Medals, for example, are generally regarded as ranking above all other prizes in mathematics and are awarded to outstanding mathematicians of whatever nationality—appropriately enough, by the International Mathematical Union. But the Nobel prize is in effect the most visibly international award in science. The fact that scientists of twenty-six nations have received a Nobel prize not only attests to its international character but greatly extends its visibility. In turn, the prestige of the prize contributes to the worldwide publicity that annually renews the public awareness of its internationality. And its visibility and prestige in turn make for intense international competition that further reinforces the visibility and prestige of the prize as the mass media (and not infrequently scientists themselves) anxiously await the annual announcements and then promptly recompute the slowly changing national boxscore of the prize-winners. Plainly, the mutually reinforcing interaction of its attributes accentuates the great prestige and high visibility of the Nobel prize as an ecumenical symbol of scientific accomplishment.

The comparative multiplicity of the Nobel prizes also enhances their visibility. That the awards have been given since the beginning in the three sciences (and in economics since 1969) as well as in literature and peace reiterates their public identity. The number of recipients is considerably larger than the number of prizes. In the sciences and economics, there can be as many as twelve recipients in any given year, since prizes are granted to as many as three investigators in each field. There is of course no magic number that will maximize the visibility and prestige of the prizes. But, intuitively, the range of from four to twelve recipients a year seems wide enough to provide for diversity of social impact and narrow enough to keep both the cash and the symbolic value of the awards from becoming excessively fragmented.

Other scientific societies and institutes also make awards in several fields. But some of these—as, for example, the Franklin Institute of Philadelphia, with its high-ranking awards in physics and chemistry among others—do not achieve a public organizational identity. As the Coles have found (1973, pp. 53–57), relatively few scientists are likely to know, for example, that the Cresson Medal in chemistry, established in 1848, and the Franklin Medal in physics, established in 1914, are both awarded by the Franklin Institute. But the Nobel prizes in the several fields of science are unmistakably identified as being of a kind and thus maintain a worldwide public presence.

The annual focus of attention on the Nobel prizes also marks them off from a number of other important awards in the sciences that are not given annually, such as the Fields Medals, given every four years, and the Barnard Medal, every five. But there are a great number of annual awards in science and, once again, through the conjunction of its other attributes, the Nobel prize has greater visibility and prestige than any of them. Contributing to this result may also be the pomp and circumstance that has the king of Sweden taking a major role each December 10—the anniversary of the death of Alfred Nobel—in the actual awarding of the prizes at ceremonies that today seem a bit *fin de siècle*. (For a lively account of the week-long celebration by a laureate, see Hench, 1951.)

What is really distinctive about the Nobel prizes, however, is the extraordinary roster of recipients. Here the prizes in science differ most significantly from those in literature and peace. There are many who believe that the literary and peace awards have gone to candidates of questionable merit far more often than those in the sciences and that selection of laureates in literature and peace has often been affected by political or other functionally irrelevant considerations. These differences cannot be taken for granted as inevitable correlates of the more "objective" evaluation process in science. There are, after all, other awards for scientific work, such as the Prix Balzan, that are said to consider the religious and political suitability of scientist candidates.

## SYMBOLIC USES OF THE NOBEL PRIZE

As the financial significance of the prize has contracted—with the current honorarium serving as a convenient nest egg for recipients rather than as the lifelong support Nobel intended—its symbolic significance has been enhanced for the recipients, their scientific specialties, their universities, and their nations.

As it turns out, the prizes and the prestige they confer have been enlisted in the service of a variety of ends, some in keeping with Nobel's intentions and many not. The prestige of laureates and of the prize has been used to confer legitimacy on all manner of ventures—ideological, political, commercial, and military. Thus, when forty-four American laureates asked the president to bring the war in Vietnam to a rapid conclusion, the antiwar movement took on new authority (*Science,* vol. 168, 1970, p. 1325). This does not mean, of course, that the public opinions of laureates have been universally accepted: some scientists described the laureates' attack on the war as an irresponsible political act, particularly since they had offered the president no concrete suggestions as to how he might disengage troops from Southeast Asia (*Science,* vol. 169, 1970, p. 927). But many favorable public references to the consolidated action of the laureates indicated their legitimizing function for the antiwar position. Quite clearly, scientists and others have come to believe that the names of Nobel laureates, especially in large numbers, carry weight in issues of science and public policy. On its publications, the Federation of American Scientists, guardians of the collective conscience of American science and its sometime lobby, unambiguously identifies the prize-winners among its sponsors (at last count, there were twenty-nine) and in the process capitalizes on their prestige and scientific authority.

American laureates are not unique in using the prize to achieve political ends. In response to the physicist Andrei Sakharov's first public condemnation of Soviet political repression, all five Soviet Nobelists and thirty-five other academicians jointly signed a letter to *Pravda* denouncing Sakharov as "an instrument of hostile propaganda against the Soviet Union" and his action as "radically alien to Soviet scientists" (*Science,* vol. 182, 26 October 1973, p. 334). As if mobilized to answer them directly, eighty-nine Nobelists from ten countries in their turn published in the *New York Times* a letter to President Podgorny supporting Sakharov and expressing their distress at "the resumption of ... the campaign of harassment against [our colleagues] in the Soviet Union" (7 December 1973, p. 6E). (Since then, Nobelists Frank, Kantorovich, and Tamm have refrained from joining the collective attacks on Sakharov by Soviet academicians (*Science,* vol. 190, 19 December 1975, pp. 1152–54). And when François Jacob, André Lwoff, and Jacques Monod, the first French laureates in science in thirty years, vigorously denounced national policies on education soon after receiving their prizes, even President Charles de Gaulle took notice (McElheny, 1965, pp. 1013–15). Soon afterward, the same three laureates succeeded in bringing about the resignation of the Board of the Pasteur Institute and the long-sought reform of its archaic organizational structure (*New York Times,* 24 January 1966, p. 9).

The individual and collective authority of laureates appears most obvious in scientific matters. Laureates have banded together to denounce astrology as unscientific (*Science,* vol. 189, 19 September 1975, p. 979). They also acted in 1974 when religious conservatives tried to insert the biblical explanation of creation into science textbooks used throughout California. All nineteen Nobel laureates then at work in the state rallied in support of Darwinian theory and pronounced the proposed alternative unscientific. But collective scientific authority is one thing and political power is another—in this case, as in others. The laureates' efforts to exclude theories of divine creation from public school curricula failed, and Darwinian theory was demoted from the status of purported fact to that of untested speculation.

The prestige of the laureates has been borrowed for a wide range of other activities. When publishing houses proudly announce that their authors include laureates or industrial corporations place them on their boards of directors, they unblushingly take this to attest to the soundness of their judgment. This would-be transfer of prestige can be endlessly extended. Here, for example, is McGeorge Bundy legitimating research support by the Department of Defense in the case of "the great von Békèsy ... we [Harvard] let the Navy pay the whole bill. His work on the human ear was pure science, and eventually he won the Nobel

Prize; so much for the notion that all defense-supported research is bad" (1970, p. 566).

Closer to Nobel's original intent is use of the prize as a measure of scientific merit, for example, as a gauge of the fruitfulness of new lines of research. Max Born appraises fellow Nobelist Otto Stern and his work:

> Stern became a great physicist, as I had predicted. The method of molecular radiation which he introduced into atomic physics has become one of the main instruments of present day research; his teaching has spread all over the world, and has produced numerous discoveries of the first rank as well as a significant number of Nobel prize winners [1971, p. 54n].

It is one thing for scientific contributions to be appraised by their association with the prize and its recipients, and quite another for Nobel prizes to be further elaborated into second-order indicators of intellectual importance. Thus it is said that Karl Popper "is regarded by many as the greatest living philosopher of science. Nobel Prize winners who have publicly acknowledged his influence on their work include Sir Peter Medawar, Jacques Monod, and Sir John Eccles" (Magee, 1973, p. 1). This testimony is apparently cited as validation of the direct importance of Popper's work as well as of its indirect multiplier influence. Such appraisals by exacting scientific minds are taken to be direct testimony to the significance of another scholar's ideas for their own work.

The proliferating metaphor of the Nobel prize as a symbol of ultimate achievement also means that association with laureates is used in competition for prestige among all kinds of social entities: nations and universities, business corporations and research institutes. For some, the recruitment of laureates has become an end in itself; greater attention is paid to hiring them than to providing a research environment that helps to produce first-rate scientific work. As such displacement of goals occurs with the widening symbolic use of the Nobel prize, the symbol and the underlying reality become dissociated. In these cases, especially, the prize has been victimized by its own success.

## Counting and Claiming Nobel Laureates

Almost from the beginning, the Nobel prize has been adopted as a universal measuring rod for the scientific standing of nations and organizations. Hitler's impact on German science has been gauged by the number of Nobelists he forced into exile (Needham, 1941, p. 32), and Germany's fall from scientific preeminence by the downward trend in Nobel prizes won by its nationals (Gray, 1961; Cantacuzene, 1969). Similarly, the ascendance of the United States as a major producer of scientific knowledge has also been reckoned by the number of Nobel prizes its citizens have taken home (Lukasiewicz, 1966; Verguèse, 1975; National Science Board, 1975). Before World War II, German scientists dominated the competition. The thirty-five Germans who were named laureates (by 1939) constituted more than a fourth of all winners, with their nearest rivals, the British, able to claim only twenty-two.[2] After the war, the United States took a commanding lead, with ninety-one laureates between 1943 and 1976—that is, ninety-one whose prize-winning research was done in the United States.

The growing accomplishments of American science are taken to become immediately evident by comparing this record of ninety-one postwar awards to the total of fourteen won in the forty years before World War II. The figures are put in various ways: fifty-one percent of all laureates named since the war have been Americans as compared to only 11 percent before, and by 1976, "American" scientists [3] had accumulated more Nobel prizes than nationals of any other country.[4] Following behind their total of 105 are Great Britain with 58, Germany with 50, and France with 21.[5] No other country has had as many as

2. Since the prizes can be divided among several scientists, it is possible to count prize-winners and prizes separately. In this study, which focuses on the careers of laureates, the individual winner rather than the prize is always treated as the unit of analysis. The latter, of course, provides a more exacting measure of the "productivity" of nations, universities, or research laboratories since it is not inflated by divided awards. But counting prizes rather than people also involves the use of fractions of awards and lends a specious air of precision to an altogether imprecise indicator.

3. The influx of emigré scientists into the United States has contributed significantly to the excellence of "American science" and to the number of prizes "Americans" have received. Altogether, 31 of the 105 who did their prize-winning research in the United States and had won awards by 1976 were immigrants, but only 11 arrived between 1930 and 1941 as refugees from Hitler's Europe.

4. John Bardeen is counted twice among American prize-winners in science since he won the prize on two separate occasions. Linus Pauling, the other American double winner, holds awards in chemistry and in peace.

5. In reporting laureates' nationalities, the Nobel Foundation uses country of citizenship as its criterion. Although citizenship is unambiguous for the most part, it gives few cues to the milieus in which significant scientific training and research are done. Laureates often were educated and did their prize-winning research in countries other than those of which they were citizens. Thus, whenever I allocate Nobelists to one country or another, it is the country in which they did their prize-winning research that is enumerated unless otherwise specified.

By way of example, Max Born is officially credited to Great Britain, but since he was born in Germany, educated there, and did his prize-winning research in Göttingen ten years before he emigrated, for our purposes he is credited to Germany. Among the "American" laureates, official enumerations according to citizenship credit Max Theiler, a resident of the United States from 1922 to his death in 1972, to South Africa, where he was born, or to Switzerland, since he held dual citizenship. Similarly, the physicists T. D. Lee and C. N. Yang are officially credited to China although both were educated and did their prize-winning research in the United States. Albert Claude, born in

ten, a fact of some concern to Russians who set considerable store by the science awards, whatever their opinions of the prizes in literature and peace. As one Soviet scientific bureaucrat observed, Nobel prize statistics indicate "an acute deficit of researchers of the highest qualifications" in the USSR, officialese for the fact that only nine Russians have won Nobels, two of them before the Revolution (Astrachan, 1973, p. 117).

Outside the Soviet Union, these enumerations also have political implications. National totals of Nobels serve for some as indicators of success or failure of policies for science. It has been observed that the rate of American acquisition of prizes "correlates well with the generosity of federal funding. The percentage of prizes brought home to this country . . . [rose] to 60 to 70 percent in the most recent four or five years." But, as Gerard Piel also suggests, the luster of these statistics is somewhat dulled by an observation of Lord Rothschild. Recently he was able to cheer his countrymen with this rhetorical question: "Have we not got 4.6 Nobel Prize–winners per ten million of our people, in comparison with America's 3.3?" (Piel, 1973, p. 229). Switzerland and the Netherlands have both produced more prize-winners per capita than Great Britain (National Science Board, 1975, p. 162).

Even if association with Nobel laureates is almost universally taken as an indication of national or organizational excellence, in Great Britain at least the economic crisis has made some Britons (including Lord Rothschild) skeptical. The desirability of supporting Nobel laureates and providing for their kind of fundamental research is no longer self-evident. Noel Annan of the University of London recently observed that people are not sure any more that they need Nobel prize-winners more than highly competent applied scientists and, in a similar mood, the Duke of Edinburgh remarked that "a university should measure its success just as much by the number of millionaires as by the Nobel prize-winners it produces" (Quoted in Jevons, 1974, p. 143).

The diversity of symbolic meanings acquired by the Nobel prizes permits slight variations in their distribution to be taken as signs, indicators, or even iconometrics of the changing condition of science or one or another of its aspects. Thus, the physicist administrator in charge of

---

Luxembourg and holder of a professorship at the Catholic University of Louvain in Belgium, did his prize-winning research in the United States as did Renato Dulbecco, an Italian by birth and now a resident of Great Britain. On the other hand, Ben Mottelson, born and educated in the United States, did his prize-winning research primarily in Copenhagen, has since become a Danish citizen and thus is credited to Denmark in national comparisons. Otto Stern, having been educated at Breslau and having done his Nobel research while at Frankfurt, Rostock, and Hamburg, also is not included among the 105 Americans whose careers are analyzed here although he is officially credited to the United States by virtue of citizenship.

research and development at Xerox sees multinational corporations as taking on an enlarged share of support of basic science as support by governments diminishes; the evidence offered is the fact that in 1973 Esaki and Giaever, two of the three recipients of the Nobel Prize in Physics, had their work supported by industry (*New Scientist,* vol. 60, 1 November 1973, p. 354).

Criteria used in counting and claiming prize-winners vary considerably and tell almost as much about those seeking to identify themselves with laureates as about the actual distributions. The grand total of American Nobel prize-winners in the sciences, as of 1976, can be counted in various ways. If it is taken to include those who immigrated to this country after they won the prize, as Enrico Fermi and Albert Einstein did, then the number is 114. If one restricts the count to those who were in this country when they received the award, the total is 111; if to those who were here when they did their prize-winning work, the total is 105; and if to those credited to the United States by virtue of citizenship, then it is 104. An even stricter measure, which also indicates the performance of a nation's universities in identifying and nurturing scientific talent, is the number who got their doctoral training there; that number for American universities is eighty-five.

As the United States has moved up to first place in the cumulative total of prizes bestowed, the claiming and counting of prize-winners has come to serve as a mode of competition for prestige among the country's universities, colleges, and research organizations, including even the most prestigious of them. Although such competition seems trivial, it is serious business. All manner of organizations need to appraise their performance. Business enterprises seize on percentage of profit and growth of sales and assets. Academic and scientific organizations have no comparably objective and seemingly precise measures of accomplishment. When it comes to the training of scientists and the support of their work, the Nobel prizes are assumed to be one relatively objective measure of how well organizations are doing. But what does the counting of Nobel prizes in fact tell about the performance of the institutions of higher learning that claim their winners? Drawing on a large number of self-appraising publications of these institutions and on biographical data on Nobel laureates, we can go some distance toward answering this question.

The American universities and research institutes with which seventy-seven Nobel prize-winners were affiliated as of the academic year 1975–76 derive honor from these men. (There is not one living American woman laureate; the Nobelist Gerty Cori died in 1957, and the Nobelist Maria Goeppert Mayer in 1972.) The University of California at Berkeley takes pride in the presence of nine laureates on its faculty (including professors emiriti). Harvard and Stanford each have eight and Rockefeller University has seven.

The spread of glory does not stop, however, at present affiliation. Every institution with which the prize-winners have been affiliated at any time and in any way can and does take pride in the fact. The official documents of these institutions suggest that at least ten kinds of affiliation serve as grounds for public claim to a Nobel prize-winner. He can be claimed by the institution at which he did his undergraduate work and by the one from which he got his doctoral degree. Some institutions claim him if he studied there without getting a degree. A postdoctoral year or two may also suffice for a claim. Membership on a faculty in any rank, from research assistant to full professor, seems to establish a legitimate connection, whether the scientist held his appointment before, during, or after he did his prize-winning work. Presence at an institution when the award is made, or after that, also justifies a claim. At the end of the prize-winner's career, when he takes a visiting professorship after his official retirement, he brings honor to his new employer. The prestige of a Nobel prize is such that 80 percent of the institutions that can lay claim to a prize-winner on the basis of any one of these ties do so.[6]

Reckoning of this kind has several intriguing consequences. In the first place, it multiplies the presence of the prize-winners, bringing honor to a number of institutions much larger than the number of laureates. The approximately 500 positions occupied at one time or another by the 111 scientists who were in the United States at the time they won the Nobel prize (the basis on which national claims are often grounded) allow them to be claimed by some 170 institutions. Academic mobility thus expands the pool of institutional prestige. At one extreme, the biochemist Fritz Lipmann can be claimed by eleven institutions and the chemist Harold C. Urey by ten. On the other hand, there are three prize-winners who can be counted only once because each spent his entire career at a single university: the physicists P. W. Bridgman (Harvard) and C. D. Anderson (California Institute of Technology) and the physical chemist W. F. Giauque (University of California at Berkeley).

No attempt has been made to determine whether the claims laid to

---

6. The business of claiming laureates has become sufficiently routinized to be satirized at the University of Chicago:

> We have agreed that those to be counted are: (1) those who have studied at an institution, (2) those who have been on the faculty payroll either before, after, or at the time of receiving the prize. It is not considered fair to count those who have delivered a single lecture, attended a conference, or visited a friend on campus as being representative of the University.... As a general rule, if a recipient of a Nobel Prize lists your University in *Who's Who* or *American Men of Science*, it is fair to lay claim to him. In fact, it would be churlish not to do so [Larsen, 1967, p. 19].

prize-winners in all the diverse capacities mentioned carry equal weight with the audiences to which they are directed, or even whether close ties to a single laureate throughout his career bring greater advantage to an institution than more tenuous ties to a large number of laureates associated for shorter periods. The claims themselves, however, tell something about the relative standing of institutions. In general, the more distinguished the institution, the more restrained the claim.

For the most part, institutions are remarkably precise in characterizing a prize-winner's affiliation with them. Washington University counts the biochemist E. A. Doisy in its roster of prize-winners with the qualification that he "did most of his work at St. Louis University [but] served for a time on the Washington Medical School faculty." The University of California at Berkeley shows the same care in claiming the biochemist J. H. Northrop: "[He] has been a member of the Rockefeller Institute since 1916, contributing the major part of his numerous research accomplishments at the Institute's former laboratory in Princeton. He came to the University of California in 1949, but has continued to serve as a member of the Rockefeller Institute."

By such precision, institutions concede the legitimacy of competing claims and avoid embarrassing accusations of encroachment by others. By the same token, they are able to extend their claim to scientists whose association may have been slight, to say the least; implicit in the careful specification of claims is the assumption that any claim is justified as long as it is spelled out. As a result, the geneticist H. J. Muller appears on the Cornell University list on the ground that he spent a year there as a student; the physicist Ernest O. Lawrence, on the University of Chicago list because he began his doctoral work there under the Nobel prize-winner R. A. Millikan; and the Swedish chemist Arne Tiselius, on the Princeton University list because he spent a postdoctoral year at its Frick Chemical Laboratory. The Princeton list also includes the physicists P. A. M. Dirac and Wolfgang Pauli on the basis of visiting professorships. The short careers of Rudolf L. Mössbauer at Caltech and of Hideki Yukawa at Columbia qualify them for inclusion on the lists of those universities because they happened to be there at the time the award was made. Chicago stakes a claim to Eugene P. Wigner and Glenn T. Seaborg on the basis of their wartime work in the Metallurgical Laboratory of the Manhattan Project, which was housed on the Chicago campus.

These claims are the principal but not the only means used by institutions to assert publicly their association with Nobel laureates. The Case Institute of Technology, for example, has established the Michelson Prize to honor its first professor of physics and America's first Nobel laureate in any science. And the College of the City of New York has given a prize to Arthur Kornberg, its first alumnus to win the

Nobel—perhaps the first time someone has won an award for having won an award.

A few institutions observe self-denying ordinances that limit their claims to laureates. Harvard does not list its laureate alumni, nor does it count those who were on its faculty at some time or another before or after winning the Nobel prize. It claims only those who did their prize-winning research at Harvard and those who were on its faculty when the award was made.[7] The University of California at Berkeley observes most of the same restraints but counts its alumni as well. The Rockefeller University, merely listing the names of laureates in its catalogue, is most restrained in adopting the inverse strategy of making no published claims at all. But public restraint does not mean that Rockefeller's laureates are ignored by some university powers that be. When one of its members suggested that the university might be "coasting", David Rockefeller, a member of the Board, retorted that an institution that "has nine scientists win the Nobel Prize in the last five years doesn't sound to me like an institution that's coasting" (*New York Times,* 26 September 1976, p. 51). Even the Rockefeller University needs to guage how well it is doing although David Rockefeller's estimate of the number of its Nobel prizes is somewhat inflated.

The kind of stratification observed in the style with which institutions wrap themselves in the Nobel mantle is roughly paralleled in the way they invoke other indicators of their standing and performance. This involves a seeming paradox. Although the range of measures of performance available to organizations varies directly with the organization's rank, the number of measures actually used tends to be more restricted. At each rank, as though a Guttman Scale were being invoked, the most exacting available measure is the one put on display. Thus, highest-ranking universities generally confine themselves to mentioning winners of major scientific awards; those at the second and third levels of prestige do not neglect what award winners they have but will also indicate, for example, which faculty members hold office in scientific societies. Those at the lowest level are largely confined to using such indicators as the number of papers published by their faculty members, as the best measure they can muster. What is of immediate interest here is that the top-ranking universities, which can be selective in what they mention, make use of the Nobel awards to exhibit or enhance their prestige.

---

7. Harvard has been said by some of its faculty to use the number of laureates as a measure of quality of its own departments. One Harvard scientist who works in a field not covered by the Nobel prizes observes with some feeling that "those branches of science that are not eligible have a hard time to prove that they are, perhaps, just as good or even nearly as good [as the departments that have laureates]." As I have noted, in the eyes of the public also, the value of branches of science is often mistakenly measured in terms of the *absent* Nobel prizes.

Universities are not the only institutions to seek in the counting of Nobel prizes an objective measurement and affirmation of their performance. The National Institutes of Health also uses the Nobel prize as a universal marker of excellence for its multiple audiences. A few years ago, it announced proudly in its annual report that "unprecedented recognition came to NIH during the year when Dr. Marshall W. Nirenberg, Chief of the Laboratory of Biochemical Genetics of the National Heart Institute, was named a co-winner of the Nobel Prize." The report went on to point out that "Dr. Nirenberg, whose entire career as a scientist was spent at NIH, shared the award with two other scientists whose work has long been supported by NIH grants, Drs. H. Gobind Khorana . . . and Robert W. Holley. . . . Dr. Nirenberg is the first federally employed scientist ever to win a Nobel Prize, although another American winner, Dr. Arthur Kornberg . . . had previously been employed by NIH. Of the remaining American winners of the Nobel Prize in Medicine or Physiology, 40 had received support from NIH for their research" (National Institutes of Health, 1969, p. 2). Since then three more Nobel prizes went to NIH scientists, Christian Anfinsen, Julius Axelrod, and Carleton Gajdusek, thus providing the Institutes with fresh evidence that by this criterion it was doing rather well in its double role as research facilitator and research laboratory.

Organizations which themselves bestow scientific awards gauge their success by the number of laureates they honored before the Nobel Foundation did so.[8] The Guggenheim Foundation, with its modest stipends covering all fields of learning and the arts, takes understandable pride in the finding that by 1975 it had supported the work of twenty-seven Nobel prize-winners in the physical and life sciences and economics before they won a prize and ten more after they did so. Similarly, the Research Corporation reported in 1971 that it had honored fifteen future laureates out of thirty-eight awardees, and the Rockefeller Foundation, with its extensive programs of support for science, can claim to have sponsored nearly one hundred laureates "almost always before [they] received their prizes"[9] (Bernon, 1964, p. 86). The

8. Similarly, the frequency of prizes has been used to gauge the imagination and scientific taste of Warren Weaver, who worked to get strong support for molecular biology at the Rockefeller Foundation. George Beadle, himself a laureate, writes that seventeen of the eighteen prizes going to scientists in what could be broadly defined as molecular biology were for work supported by the Rockefeller Foundation. And, he continues, it is especially important to note that, on the average, these awards were made eighteen years before the Nobel prizes. This, for Beadle, is the ultimate test of Weaver's foresight and judgment (1967, p. xi).

9. If the prize givers are solemn about their Nobel prescience, not all the recipients are. When one recent winner of the Lasker award was told that the Lasker had anticipated the Nobel prizes at least twenty times, he retorted, "Not at all, that's all wrong. The Nobel people have been good enough to select twenty Laskers."

fact that some scientific awards have a history of anticipating Nobel awards sets the context for unfulfilled expectations on the part of their recipients, who might not have previously considered themselves potential Nobelists. These premonitory prizes not only honor scientific achievement but also define membership in the pool of potential laureates.

It has been said of *Scientific American* that, in its coverage of a much greater diversity of science than is recognized by the Nobel prizes, it published the work of forty-two prize-winners before that work won them their prizes. Even the National Academy of Sciences must stand scrutiny on this standard of performance. The Academy itself has not, of course, publicly assessed its own prescience, but D. S. Greenberg, writing in 1967, reported that the Academy had previously elected thirty-six out of the forty-five United States scientists who had won the Nobel prize since 1950 (1967, p. 362; by 1976, the corresponding numbers were fifty-eight out of seventy-seven).

The Nobel Foundation itself has taken what is perhaps the ultimate step in adopting the Nobel prize as the most exacting measure of prime achievement in science. The Foundation has in effect gauged its own performance by noting the impact of Nobel prize-winning research on later Nobel prize-winning research. Thus the Foundation's account of the Wilson Cloud Chamber:

> Some of the most important achievements using the Wilson chamber were: the demonstration of the existence of Compton recoil electrons, thus establishing beyond doubt the reality of the Compton effect (Compton shared the Nobel Prize with Wilson in 1927); the discovery of the positron by Anderson (who was awarded the Nobel Prize for 1936 for this feat); the visual demonstration of the processes of "pair creation" and "annihilation" of electrons and positrons by Blackett and Occhialini.... [Blackett] in 1948 received the Nobel Prize on account of his further development of the cloud chamber and his discoveries made therewith...; and that of the transmutation of atomic nuclei carried out by Cockcroft and Walton [and honored by a 1951 Nobel prize] [Nobelstiftelsen, 1965, p. 216].

This practice of gauging success by the extent of anticipation of Nobel selections is only one way in which the prizes stand as a brooding presence over all scientific awards; they are, in fact, even used as the unit of measure for gauging the significance of other awards. The Arches of Science Medal, for example, is described as the "American Nobel Prize" (*Science Yearbook,* 1967, p. 411), the Fields Medal of the International Mathematics Union as "the Nobel Prize for younger mathematicians," and Harvard's Ledlie Prize as its "in-house Nobel."

Even though the Nobel prizes are accorded great esteem, having one does not always make a laureate an attractive candidate for other honors. Some organizations actively resist the tendency to have their evaluations in effect preempted by the academies in Stockholm. Thus, a member of a university committee on honorary degrees remarked to me that his colleagues refused to follow along after the Nobel. In some circles, Nobel prizes are not even considered sufficient evidence of achievement. A member of the American Philosophical Society, generally considered to be the most venerable of learned societies in the United States, recently observed, "We keep saying we're going to get younger people. But when someone proposes one of those bright young Chinese who gets Nobels at 32, they say he has not proved himself" (Catherine Drinker Bowen, quoted in *New York Times,* 14 November 1972, p. 93). When resistance to preempting authoritative evaluation is combined with ambivalence toward early rewards, younger laureates, in particular, may be subject to something like an anti–Matthew Effect [10]—that is, to them that hath a Nobel no more shall be given, while to them that hath *not* a Nobel an abundance shall be given.

So the counting and claiming goes on and not surprisingly some become skeptical of what the *New York Times* calls "the public numbers game of Nobel Laureates" (2 January 1968, p. 36).[11] This skepticism, born of the belief that enumerations come to be an end in themselves rather than a true gauge of performance, is only partly justified.[12]

---

10. The Matthew Effect in science, according to Robert Merton, "consists of the accruing of greater increments of recognition for particular scientific contributions to scientists of considerable repute and the withholding of such recognition from scientists who have not yet made their mark" (1968f in 1973, p. 446). Merton draws upon the stately language of the Gospel According to St. Matthew:

> For unto every one that hath shall be given and he shall have abundance;
> but from him that hath not shall be taken away even that which he hath.

I shall have more to say about the operation of the Matthew Effect in the careers of Nobel laureates in later chapters.

11. Having said this in 1968, even the *Times* was not above using the Nobel prizes to make a much needed point several years later. Thus, in editorializing on the Nixon administration's reductions in research expenditures, it observes: "The brilliant record American scientists have amassed in winning Nobel Prizes . . . testifies to the quality of the national science research enterprise. But how long can such top quality survive in the face of the present economic stringency and the threat of even worse times ahead?" (5 January 1973). Again, in 1976, the *Times* commented on the United States' "clean sweep" of Nobel awards as an "impressive tribute to the quality of American research since World War II" (20 October 1976), implying, of course, that the coincidence of national origins of all five laureates in the physical and biological sciences (and, as it happens, in economics and literature as well) is especially significant as an indicator of national excellence.

12. Students' evaluations of professorial performance can, in principle, provide evidence on the laureates' effectiveness as teachers. Without putting much confidence in the particular data set, I note, as an illustration, that at Berkeley, all five of the laureates whose teaching was appraised received top grades from students as lecturers. Glenn Seaborg, for example, is described as "the best chemistry teacher they ever had" and

In the absence of better criteria for appraising the effectiveness of national research efforts, universities, laboratories, and even honorific awards themselves, the use of Nobel prizes as a touchstone will continue. At the same time, in some universities and research organizations, greater priority is often given to multiplying laureates than to fostering conditions that will help produce them. As a result, the prizes lose some of their luster and require reassessment even though the roster of prize-winners has always been distinguished.

## NOBEL LAUREATES AND THE NOBEL PRIZE

The Nobel prizes in science began with the great advantage of having a backlog of the still-living giants of nineteenth-century science to draw upon. Arne Westgren, an official historian of the awards, reports that the great problem in the early years was not to identify enough scientists who were deserving of a prize but to determine the order in which the prizes were to be awarded to them (Nobelstiftelsen, 1962, p. 354). Nobel laureates named in the first five years included scientists of the first class such as Roentgen, Lorentz, Zeeman, Becquerel, Rayleigh, and Pierre and Marie Curie among the physicists; van't Hoff, Emil Fischer, Arrhenius, and Ramsay among the chemists; von Behring, Pavlov, and Koch among the physiologists and physicians. These first lists of luminaries helped set the standards for later selections and served to legitimate the prizes by association with scientific distinction.

The scientific excellence of the early Nobelists also set in motion a social process of reciprocal transfer of esteem and prestige between the recipients and the prizes themselves. (See Davis, 1949, pp. 93–94, for the theoretical significance of prestige, an attribute of positions in the social structure, and esteem, an evaluation of individual role performance.) At the outset, before the prizes had acquired their present standing, the Nobel Foundation honored itself in the course of honoring scientists of great stature. In effect, the Nobel prizes received annual infusions of prestige borrowed from the esteem long accorded the eminent scientists who agreed to accept a prize. This process of prestige accumulation by the Nobels meant that later, when awards were occasionally made to scientists of considerably lower standing, they would acquire a worldwide eminence largely derived from their having been

---

Melvin Calvin as "an excellent and concise lecturer" in the *Primer,* a student-edited course review. But the laureates do not get uniformly high grades on their accessibility or their patience (*San Francisco Sunday Examiner and Chronicle,* 7 December 1975, p. 9).

named Nobel laureates. This component in the process that established the preeminence of the prize began in the early years and was maintained by the annual additions of laureates whose work was truly outstanding.

The interplay between the esteem accorded to scientists and scholars and the prestige of organizations with which they are associated provides a link between the separate systems of stratification in which individuals and organizations are ranked. In the academic world, for example, scholars of great repute contribute to the prestige of the universities with which they are affiliated, and, varying inversely with their repute, affiliation with a great university will contribute to their own standing. In the case of the Nobel prize, its prestige is augmented by selections of the most eminent scientists—Einstein, Rutherford, Bohr, or Koch, for example—and it in turn confers prestige on laureates of considerably less distinction.

In the long run, the standing of the Nobel prizes for scientists and ultimately for the public hinges on the esteem in which the prize-winners are held—that is, the belief among knowledgeable scientists that the laureates are the most deserving among potential candidates. Were the quality of scientific contributions honored by the prizes and the esteem accorded the laureates to decline for a time, the prestige of the prizes would also decline, although not without some lag. This has not occurred, although there have been "strong" prizes and "weak" ones, some controversial prizes, and two or three outright and acknowledged "errors" in selection. Johannes Fibiger, for example, is regarded after the fact as one of the least meritorious of laureates because his work on the propagation of malignant tumors was altogether mistaken (Nobelstiftelsen, 1972, pp. 188–89). The Nobel Committee for Medicine was so embarrassed by this episode that, drawing a dubious inference, it declined to give a prize for cancer research for almost forty years. The selection of J. J. R. MacLeod was another "error." As director of the laboratory in which Banting and Best isolated insulin and studied its therapeutic use in human diabetes, he may have facilitated their work, but he was not even present when the experiments on insulin were done (Nobelstiftelsen, 1972, p. 225). He is, according to the historian of science Donald Fleming, "the only palpably undistinguished investigator in the whole list of laureates in science" (1966, p. 55).[13]

From the outset until now, the community of scientists has been more likely to charge the Nobel committees with sins of omission—than

---

13. Fibiger and MacLeod represent cases of outright error in selection for Nobel awards but there are still others who received prizes for research that was known to be incorrect when the awards were made. This was so, as we shall see, for Svedberg and Fermi.

with sins of commission. Why this is so seems, naturally enough, to be the outcome of the great scientific distinction of most laureates and their research.

## Nobel Laureates: Indicators of Excellence

The significance of the laureates' scientific contributions and of their standing in the scientific community can be roughly gauged by four indicators: the impact of their research before receiving the prize; their overall versatility as investigators, some having made not one but several contributions considered to be prizeworthy; the extent of consensus among nominators for the prizes; and the continuing influence of the laureate's work after it was honored by the prize.

*Impact of Research by Future Laureates.* In the aggregate, future Nobel laureates have greatly influenced ongoing research in their fields, as can be seen in part from an analysis of citations to their published work. With all their limitations, citation counts have been found to be a useful though crude indicator of the impact of research on subsequent scientific development (Cole and Cole, 1971). In terms of this measure, laureates-to-be, scientists who we know by hindsight will win Nobel prizes, have been among the most influential of contemporary scientists. Their work is heavily cited—in fact, almost forty times as often as the average author whose research has been cited at all and who therefore is listed in the Science Citation Index. Laureates named between 1965 and 1969, for example, averaged 232 citations in the 1965 index as compared with an average of six citations to other scientific authors listed there. This rate of citation placed 85 percent of the laureates-to-be among the top 0.2 percent of authors cited in the scientific literature that year. Not surprisingly, laureates-to-be also frequently turn up on the Science Citation Index list of the fifty most cited authors. Four on the 1967 list—Gell-Mann, Herzberg, von Euler, and Barton— were to win Nobel prizes in the next five years (Garfield, 1970a, p. 199).

As measured by citations, the impact of research by future laureates plainly derives from their considerable overall scientific productivity, but they also rank high among authors of the individual papers most frequently cited in the science literature. Of the fifty papers most often cited in 1967—the first year for which such data are available—five are by laureates-to-be: Gell-Mann, Bardeen, Cooper, Schrieffer, Moore, and Stein (Garfield, 1970a, p. 190). Since papers reporting new methods or research instruments tend to be cited more often than theoretical or experimental contributions (Garfield and Malin, 1968,

p. 4), the fact that future laureates' papers appear on this list at all testifies to the substantial impact of their research. Some papers by prospective laureates also become "landmarks" in the sense of being much used and long lived. They are frequently cited for years and do not exhibit the usual pattern of declining citation soon after publication. The paper Nirenberg and Matthaei published on the genetic code in 1961, for example, was cited more than a hundred times between 1964 and 1968, when Nirenberg got his prize. Gell-Mann's paper on symmetries of baryons and mesons, published in 1962, was cited about 150 times a year in the same period. All this suggests that the scientific work of laureates has been enormously influential *before* they received their prizes.[14] So much for these quantitatively crude estimates of the impact of research by future laureates.

*Multiple Prize-worthy Contributions.* Although some laureates have focused on a single line of investigation, most have moved from problem to problem, making diverse contributions sometimes to quite distinct fields of inquiry. When scientists do several pieces of fundamental work, their candidacy for a Nobel prize is validated. The fact that the same investigator has done several important things reduces disagreement about selecting him if not about the merits of the particular contribution for which he received an award. There is, for instance, the case of Otto Warburg, whose research on the Pasteur reaction survived the rigorous selection process and was judged by the committee for the prize in medicine to be "prize-worthy" in 1927. Such judgments are kept confidential since, by statute, the committees deliberate in secret. But the rule of silence was once breached by Göran Liljestrand, official historian of the prize in medicine, in his published review of awards in that field. It was there that he named sixty-nine scientists, among them Warburg, whose work had been officially, if secretly, judged to be prize-worthy. (See Nobelstiftelsen, 1962, and appendix B for the list of those whose research was considered to be of prize caliber.) Ironically, Warburg lost out that year to Johannes Fibiger, one of the most conspicuous Nobel "errors." But, by 1931, Warburg had produced another contribution on the respiratory enzyme that was also judged to be prize-worthy and for which he won his award. In 1944, he was nominated for a Nobel prize once again, this time for his studies of the role of enzymes in intermediary metabolism, but the Committee decided not to make a second award (Nobelstiftelsen, 1962, pp. 248, 291–93). Warburg, then, could have been named a laureate on any (or all) of three occa-

---

14. This is not to say that citation counts would be useful in selecting Nobel prize-winners or that the committees are at all influenced by the frequency with which candidates' publications are cited.

sions, and this multiplication of eligible contributions confirmed his standing as a laureate.

The most consequential scientists are of course those who have made several fundamental contributions. To take an obvious instance, Einstein won an award in 1921 for his discovery of the photoelectric effect. His monumental work in 1905 on the special theory of relativity and the theory of Brownian motion and in 1916 on the general theory of relativity were not mentioned in the prize citation, but they would have surely gained a prize for him later had his work on the photoelectric effect been considered ineligible. Or take the case of Edward C. Kendall, whose research on thyroxin (between 1914 and 1926) was proposed for the prize in medicine and judged prize-worthy (Nobelstiftelsen, 1962, p. 227). Kendall, however, did not win his prize for the thyroxin investigations; they were, for unknown reasons, passed by. He had to wait until 1950 to become a laureate, when his studies of the biochemistry of cortisone and its use in treating chronic rheumatoid arthritis finally brought him a publicly visible award. The list of multiple near laureates, scientists who qualified several times for the Nobel prize, also includes the biologists Robert Koch and Paul Ehrlich, the biochemists Emil Fischer and Adolph Windaus, and the chemist J. H. van't Hoff. But the career of Frederick Soddy provides the clearest case of one laureate's multiple prize-worthy contributions.

As a close, indispensable collaborator of William Ramsay, Soddy nonetheless did not share in the chemistry prize that Ramsay won in 1904. Soddy missed out again four years later when the prize went to Ernest Rutherford, even though Rutherford emphatically stated that Soddy had been a full collaborator on the research cited for his award. Soddy finally received his own prize in 1921 for his investigations of isotopes, thus demonstrating his persistence and putting to rest questions that had been raised about the wisdom of the Chemistry Committee's earlier decisions. The ultimate cases of validation of Nobel selections are of course Marie Curie and John Bardeen, the only researchers to have won two Nobel prizes in science.

*Consensus on Nominations.* As gauged by the clustering of nominations, there has been marked consensus on the merits of many scientists nominated for a prize. This includes multiple nominations in the same year for one or more candidates as well as their repeated nomination year after year. Although there is evidence that campaigns have been mounted for one or another candidate (a matter to be discussed later), there is no reason to suppose that nominators did not believe in the merits of the candidates for whom campaigns had been waged. Arne Tiselius, once chairman of the Chemistry Committee, president of the

Foundation, and laureate in chemistry, has remarked from his distinctive vantage point that "despite [the] system of circulation of invitations for nomination of candidates, certain names appear year after year, and not so few of them. . . . As a whole our experience would indicate that there is an international opinion about who represents the elite in certain fields" (1967, p. 3).

Since the Nobel committees remain silent about the names proposed for awards, systematic data on the extent of agreement among nominators are unavailable. But the public record does occasionally contain gross statistics on nominations for particular years and, now and then, the names of scientists nominated many times. From these scattered clues, we can piece together some information on the extent of agreement on nominations.

In the first years of the prizes, there was considerable agreement. Roentgen, the first laureate in physics, received seventeen out of twenty-nine nominations (Nobelstiftelsen, 1962, p. 449) and van't Hoff, the first winner in chemistry, eleven out of twenty. In 1904, Ramsay won twenty-two out of thirty-two nominations, and in the following two years Adolph von Baeyer and Henri Moissan also received a majority of nominations for the prize in chemistry (Nobelstiftelsen, 1962, pp. 354–55). Once the backlog of distinguished scientists was exhausted the extent of agreement declined (Nobelstiftelsen, 1962, p. 349) although a measure of consensus about certain candidates recurs. Thus, to take an extreme case, the neurophysiologist C. S. Sherrington received a cumulative total of 134 nominations from thirteen countries over a span of thirty years, until he finally got the prize in medicine in 1932 (Nobelstiftelsen, 1962, p. 310). The question remains, of course, why the Committee waited so long to recognize Sherrington's work.

Evidently scientists who have been passed over in a given year are not "lost causes." In the case of the prize in medicine, at least two thirds of the nominees each year have been proposed before (computed from Nobelstiftelsen, 1962, pp. 158–65, tables 1 and 2). This is less often the case in chemistry, where about a third of the candidates are renominations (Litell, 1967, p. 54). Gauged by the frequency of multiple nominations and of renominations, consensus apparently obtains for a significant fraction of candidates. Although it appears that many scientists chosen for a prize have been candidates for some time, the committees insist that they are not influenced by the number of times a candidate has been nominated.

***Enduring Influence of Laureates' Research.*** Finally, the quality of laureates' contributions can be assessed in terms of their continuing scientific impact. Such "after-the-fact" validation of decisions by the

Nobel committees does not result merely from the halo effect of the prize, as we shall see. Most laureates who received prizes between 1901 and 1964 still appear often in the Science Citation Index. Their mean of ninety-seven citations each in 1965 places them within the top 1 percent of all scientists cited in the index (but significantly below the laureates-to-be who were selected later on and were cited an average of 232 times that year, as indicated earlier). Although the work of prospective laureates should break new ground, according to the conventional wisdom about the mortality of research, the work of past laureates should be thoroughly outdated. But on the whole they are not conventional scientists, and conventional wisdom fails to meet their case. At least some of their research continues to be used. Among the fifty most cited authors listed in the Science Citation Index from its inception in 1961 to 1972 are 13 Nobel prize-winners, including some "old timers" such as Max Born, whose important work on the statistical interpretation of the wave function dates back to 1926 (Garfield, 1973, p. 7).

Another indication of the significance of laureates' research is the frequency with which they are singled out by fellow scientists as having made important contributions. Stephen Cole (1971) found for a sample of some 300 American biochemists, physicists, and chemists that laureates rank consistently high in lists of the "five scientists who have contributed most to the field in the last fifteen years." Although the laureates comprised only 23 percent of the names given in response to Cole's question, they received 63 percent of the mentions. That is, each of the laureates listed received an average of 10.8 mentions as compared with 1.9 each for all other scientists. The top-ranking scientist in each of the three fields is a laureate, as were all ten of the top-ranking biochemists, nine of the ten top-ranking physicists (Brian Josephson, the tenth, having since won his prize), and, for possibly interesting though unknown reasons, "only" half of the ten ranking chemists.

Like the other indicators, Cole's data testify that the contributions of laureates have continuing relevance for the development of scientific knowledge. This has further ramified consequences. For one thing, the merit of their scientific work and the consequent esteem accorded them contribute to the prestige of the Nobel prize that, in turn, confers enhanced prestige upon the later recipients. For another, the Nobel prize itself has come to symbolize great scientific achievement for scientists as well as laymen. Often enough, scientists can be overheard remarking that a colleague "should have the prize" or "will surely win the prize." Such remarks are easily misunderstood by outsiders, who conclude that the prize itself has paramount importance. In fact, these remarks indicate that the prize is used as a shorthand for a certain level of accomplishment. At the same time, there is evidence that scientists are increasingly skeptical about the Nobel, in no small measure because they know that excellent work is not confined to the ranks of the laureates.

## THE FORTY-FIRST CHAIR IN SCIENCE

Every year, more scientists are eligible for Nobel prizes than can win them. This means that there has always been an accumulation of "uncrowned" laureates who are the peers of prize-winners in every sense except that of having the award. These scientists, like the "immortals" who happened not to have been included among the cohorts of forty in the French Academy, may be said to occupy the "forty-first chair" in science (Houssaye, 1886; Merton, 1968f). Scientists of the first rank who never won the Nobel prize include such giants as Mendeleev, whose Periodic Law and table of elements are known to every schoolchild, and Josiah Willard Gibbs, America's greatest scientist of the nineteenth century, who provided the foundations of modern chemical thermodynamics and statistical mechanics. They also include the bacteriologist Oswald T. Avery, who laid the groundwork for explosive advances in modern molecular biology, as well as all the mathematicians, astronomers, and earth and marine scientists of the first class who work in fields statutorily excluded from consideration for Nobel prizes.

In the strict sense, occupants of the forty-first chair at a given time are scientists who are known to have been serious contenders for Nobel prizes—that is, those especially worthy candidates selected by the committees for painstaking "special investigation" that precedes the final decision (see Nobelstiftelsen, 1962, pp. 157–59) but who have not won Nobel prizes—and their peers in fields not included within the purview of the prizes.[15] But since the Nobel committees almost always abide by the rule of secrecy, occupants of the forty-first chair must be identified less rigorously as those who are widely esteemed in the scientific community for having contributed as much to the advancement of science as some laureates. (See appendix D for a listing of officially designated occupants of the forty-first chair.) In the hyperbole of Arne Tiselius, who was something of an insider's insider when it came to the Nobel prizes, "The world is full of people who should get the Nobel Prize but haven't got it and won't get it" (*Minneapolis Star,* 4 May 1963, p. 6a).

One archetypal occupant of the forty-first chair is the American

15. Laureates-to-be whose work is widely known and esteemed before the awards are therefore temporary occupants of the forty-first chair.

anatomist and embryologist Herbert M. Evans. Evans was responsible for at least three major contributions. He identified both the growth hormone and the oestrus cycle in the rat, which made endocrinology come of age as a discipline. He also discovered the antisterility vitamin (Vitamin E) and determined its structure.[16] His work on Vitamin E, according to the official history of the Nobel awards, "has been held to deserve a Nobel Prize, even if it has not been victorious in the competition with others" (Nobelstiftelsen, 1962, p. 242). Evans's work was not ignored. He was made a full professor at the University of California at the then early age of 33, was elected to the National Academy of Sciences, and received nearly a dozen honorary degrees. Still, although each of these investigations was of Nobel prize caliber, he was never summoned to Stockholm. Occupants of the forty-first chair, like Evans, continue to be esteemed by their fellow scientists, and their standing is unimpaired by the Nobel oversight. The Evanses of the world of science provide evidence of the imperfect operation of the reward system at the topmost level.

Indications from Stockholm that certain scientists could have gotten the prize but did not is one kind of evidence for the existence of occupants of the forty-first chair. Another is the widespread impression that certain scientists have won the prize when in fact they have not. Consider the case of Samuel Goudsmit and George Uhlenbeck, the physicists who discovered the spinning electron. Goudsmit writes:

> There are many colleagues who believe that we received the Nobel Prize for introducing electron spin. In fact, [Lee] DuBridge recently introduced me as an early Nobel Prize winner. I have also seen it in print. This is all very flattering but it does not supplement my TIAA pension [Goudsmit, 1976, p. 42].

It does, however, suggest that Goudsmit and Uhlenbeck are unofficial laureates.

But the best evidence for the existence of occupants of the forty-first chair is the fact that some of them eventually get the prize. The laureate in chemistry, Lars Onsager, is an extreme but instructive example. Onsager's work on reciprocal relations between voltage and heat was published in 1931 and eventually came to be known as the "fourth law of thermodynamics." But its significance began to be recognized in the 1940s. Scientific contribution is not, after all, an event but a process in which implications are revealed as the cognitive context of the contribution changes. As it turns out, Onsager was another temporary

---

16. I. D. Raacke, one of Evans's students, kindly provided me with this appraisal of his scientific contributions.

occupant of the forty-first chair, finally becoming a laureate in 1968. Long before the award, however, there was little doubt among knowledgeable scientists that his research was of Nobel caliber.

On occasion, imperfections in the awarding procedures are repaired by eulogizing great scientists who should have received the Nobel prize but have not and who thus stand out as distinguished omissions. Some believe this is so for O. T. Avery, whose place in the history of science is now doubly ensured by the importance of his demonstration, with McCarty and MacLeod, that DNA is the hereditary material responsible for transformation and by the fact that he is one of the very small number of scientists whom Nobel officials publicly regret having excluded from the Nobel roster (Nobelstiftelsen, 1972, p. 201).

As it happens, these sins of omission occur both because of the sheer scarcity of top-level awards and because of shortcomings in the selection procedures for the prizes. The scarcity of awards must of course be reckoned relative to the number of truly innovative and distinguished scientists at work at a given time. It now appears that the numbers occupying the forty-first chair in science have been increasing over the seventy-five years that the Nobel prizes have been given. The number of prizes awarded has, as we shall see, grown very little, but the population of scientists is estimated to have multiplied about thirty times (Price, 1963, p. 7). Even if the proliferation of scientists has not produced correspondingly great advances in scientific knowledge, as Weiss (1971, p. 135) and Ziman (1969, p. 361) have argued, the number of uncrowned laureates must have increased considerably since the prizes were established.

This increase can be roughly gauged by the marked secular increase in the number of scientists nominated for the prize in medicine or physiology, the only field for which data of this kind are available (figure 2–2). But there are also great fluctuations in both the number of candidates and the number subjected to "special investigation" (that is, serious contenders who reach the last round of consideration) within the overall pattern of secular increase. The same figure also shows that the abrupt downturn in the number of candidates nominated during the two world wars is a temporary perturbation, since the figures resume their upward course after war's end.

Arne Tiselius (1967) and R. W. Hodge (1966, pp. 11–12) have separately suggested that changes in the numbers of candidates in the sciences covered by the prizes also reflect changes in the sciences themselves, particularly in their pace of development. Tiselius adds that scientists working in certain specialties such as nuclear physics and molecular biology have dominated the nominations when those specialties were most active.

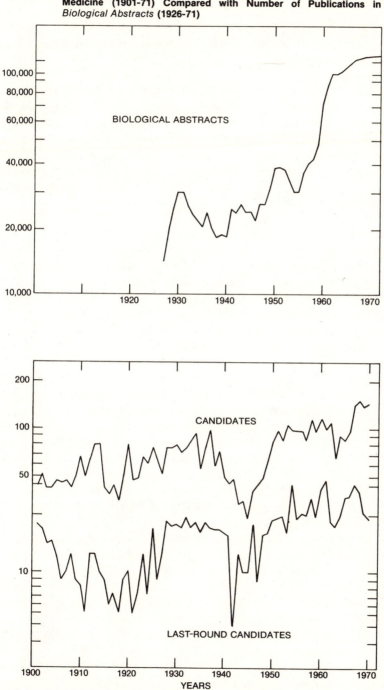

**Figure 2-2  Total Number and Last-Round Candidates for Nobel Prize in Physiology/ Medicine (1901-71) Compared with Number of Publications in** *Biological Abstracts* **(1926-71)**

NOTE: Data are reproduced from Nobelstiftelsen (1972, p. 152). Detailed data for the period 1901-61 are available in Nobelstiftelsen (1962, pp. 158-59).

That variations in the number of nominees reflect changes in the extent of research activity is also suggested by a comparison with the number of papers published at the same time. Thus, as figure 2–2 shows, the annual number of nominees for the prize in the biological sciences ("physiology or medicine") tends to follow much the same course as the number of publications registered in *Biological Abstracts* which, since its establishment in 1926, has covered the world literature in the biological sciences.

The increasing number of candidates subjected to special investigation obviously indicates more intense competition for the Nobel prizes as time has passed. What is less clear is the possible association between the intensity of competition and the caliber and kinds of scientific work chosen for the prizes. Are the times of most intense competition the times of "strong" prizes? Do the nominees who lose out in these demanding times appear again as candidates in subsequent years? Do these exceptional though unsuccessful competitors turn up as prize-winners in less competitive years? If so, then the contributions chosen when competition is least intense should be older on the average and perhaps more representative of whole lines of work rather than being single, easily identified, dramatic discoveries. But this is not so.

There is no apparent relationship between the extent of competition for the Nobel and the age and type of research cited for the prize. So, for example, the prize for 1951, a year of apparently intense competition, went to Max Theiler for his research on yellow fever and the vaccine he developed to combat it. This research was done in the 1930s, and its significance was not established for years, until the effectiveness of the vaccine was confirmed. Conversely, a few years earlier, when **nominations were at a low ebb, the prize was given to C.P. Henrik Dam and E.A. Doisy for the discovery and investigation of the effects of Vitamin K, the vitamin that affects blood coagulation. These investigations produced dramatic results and were recognized by the prize a scant seven years after they were published.**

Although the intensity of competition apparently bears no direct relation to the kinds of scientific work selected for a prize, years of rich harvest in this or that science tend to produce uncrowned laureates whose work is widely recognized as more significant than that of laureates who have been crowned during the lean years. This circumstance results from the constraints placed upon the committees of selection by Nobel's terms of grant requiring them to make their awards principally for "recent" discoveries.[17] Thus, front-running researches that lost out

---

17. Nobel's stipulation that the prize-winning research be done in the year preceding the award proved unworkable. The committees have interpreted the rule to mean that the significance of the work must have become evident in the years immediately preceding the award.

in years of severe competition become outmoded and cannot remain in contention.

The criterion of recency therefore has the unintended consequence of relegating some deserving candidates to the forty-first chair. This was the case, one gathers, for Mendeleev's work. He was nominated for the prize in chemistry in 1905 and 1906. His work was rejected because, as one committee member put it, it "had been lectured on from all the chairs of chemistry in the world and treated in all textbooks as something which, in spite of its imperfections, had firm foundations in nature herself." In other words, it was too old. In 1906 Mendeleev lost out to Moissan by a vote of five to four, and he died in 1907 (Nobelstiftelsen, 1962, p. 368).

Such constraints upon the selective process are widely sensed by the knowledgeable strata of scientists and provide one reason why occupants of the forty-first chair are sometimes more highly esteemed than laureates in the same field who gained their prize when the going was comparatively easy.

Another circumstance deriving from the selection process produces occupants of the forty-first chair: the comparative neglect of certain fields of investigation that, although included in the terms of reference, are not likely to be honored by a prize. Investigations bearing upon cancer were, as I noted, passed over for years after the traumatic selection of Fibiger's ultimately mistaken work. Contributors of important research to this field of inquiry, as to some others, were long-time occupants of the forty-first chair, a position Peyton Rous held for fifty-six years after his discovery in 1911 that a malignant tumor (*Rous sarcoma*) was produced by a virus.

Confronted with uncertainties in appraising the validity of such work, the Nobel committees have preferred the risks of conservatism to the risks of daring. Thus, Sten Friberg of the Nobel Committee on Medicine observed that the validity of the Rous work was still in contention as late as 1959, and another member of the Caroline Institute claimed that had Rous been awarded a prize in that year "the world would have said we were crazy" (Litell, 1967, p. 53). Within this context of uncertainty, Rous had to survive to his eighty-fifth year to get his prize; other investigators of cancer have been passed over entirely.

In practice, then, the Nobel committees of selection have tended to adopt a particular decision rule: one designed to minimize errors of commission rather than of omission. Of course, these appointed groups of judges appraising the significant scientific work of their time want to avoid errors of judgment altogether. But, recognizing that they cannot be infallible, they prefer the risk of making one kind of error to the other. When the validity of a scientific contribution is at all in doubt, the judges reject it, preferring seemingly less consequential but more thoroughly attested contributions.[18] Since some of the chancier candidates turn out to have done truly consequential work, they have had at least a temporary term as occupants of the forty-first chair and, when mortality acts before the Nobel committee does, a permanent one. This was the fate of O. T. Avery, who had the bad fortune to have done the research on DNA when he was 67 years old and did not live long enough for the committee to conclude that the work was valid and prize-worthy.

But perhaps the most copious source of occupants of the forty-first chair is the almost unchanged constraint upon the range of scientific work that comes within the boundaries covered by the Nobel prizes. The statutory limitation of prizes to physics, chemistry, physiology, or medicine, and, recently, economics obviously means that even the most important work in mathematics and all the other sciences is excluded from consideration. Yet the blurred public image of the Nobel awards is often taken to mean that their recipients have done the most significant work in the entire spectrum of "science." The prime contributors to the statutorily excluded fields thus become institutionally predetermined occupants of the forty-first chair.

All this contributes to a sense of uneasiness about the prize. Beyond that is a degree of skepticism about the prizes being allocated on strictly universalistic criteria, that is, in terms of their assessed contribution to the advancement of scientific knowledge and without reference to such particularistic criteria as friendship, nationality, religion or politics.

The question of whether universalism or particularism governs the allocation of the prizes is badly put. Each may be applied or both, first one, then the other. The model of decision making for the Nobel prize may be thought of in terms of successive phases of selection from smaller and smaller pools of candidates. In the first phase, significance of scientific contribution should take precedence in the sorting process since laureates, on the whole, are generally considered to have made major advances in their fields. Only rarely, however, does the first cut generate a small number of decisively superior contributors who stand out above the rest, as is likely to have been the case, for example, when the physicists Lee and Yang were selected for having demonstrated theoretically that parity (then a universal rule in physics) was not conserved in weak

---

18. An analogy may be drawn between the selection of Nobel laureates and the canonizing of saints. John Noonan, a specialist in church law, has observed that the procedures for canonization have been formalized and restricted so as to rule out the possibility of erroneous selections. It is more important that there be no false saints than that there be deserving candidates who have not been canonized (private communication). The Church and the Nobel committees apparently observe the same decision rule.

interactions. But generally, the first cut produces a cadre of candidates who on a first approximation seem pretty much on a par. Since some additional bases for selection are required if a choice is to be made, secondary criteria are called into play: some of these remain functionally relevant to the advancement of scientific knowledge but others are particularistic. Secondary but still universalistic criteria might include scientific specialty in so far as the choice of one specialty is thought to move science ahead more rapidly or the youthfulness of contributors if the committees judge younger scientists to have longer and more fruitful careers ahead of them and thus to have a better chance than older ones to advance scientific research. In any given year, however, such secondary but universalistic criteria may not differentiate a few candidates from the rest.

Particularistic criteria that have nothing to do with the advancement of science may also and perhaps are most often applied when increasingly fine-grained universalistic judgments fail to generate a small enough number to receive the prize. Such particularistic criteria might include the candidates' nationality, politics, or even their affability, rectitude, or the tidiness of their domestic lives. And one laureate in physics is said to have concluded that his prize was delayed until "the Nobel people finally saw that he had settled down . . . and that he himself had become the model of a family man" (Wallace, 1968, p. 174). A co-worker of H. M. Evans has observed that his arrogance probably alienated his scientist-colleagues sufficiently to affect his prospects for the award.

According to this model of successive approximations in the application of diverse criteria for selection, the question of whether selections for the prize are universalistic or particularistic misses the mark. In a given year, when there are a number of candidates with approximately equal claims to having advanced scientific knowledge, the award may go to the candidate from this or that underrepresented country or the one with close ties to the Swedish scientific community. But the crucial first cut is generally made on universalistic grounds even though, in the eyes of some scientists, committee judgments even on these grounds are occasionally mistaken.

The possibilities for particularism to enter the selection process also makes politicking for or against worthy candidates a potentially effective strategy. Now that the papers of Nobel laureates such as Rutherford and Millikan have become available to historians, it is evident that these scientists actively supported a series of successful candidates for prizes. Only members of the Nobel committees can say whether campaigns for or against particular candidates were influential, since the choice of this or that candidate is no proof that politicking had an effect. It may be that some laureates have an exaggerated sense of their influence on the selection of subsequent prize-winners; their nominations may have only coincided with but not determined committee choices.

Members of the Nobel committees are not naive about intrinsic difficulties in selecting laureates. As Arne Tiselius observed:

> The Nobel Committees are perfectly aware of the fact that it is impossible to discover who is best; for the simple reason that one cannot define what is the best. All one can do is to try to find a particularly worthy candidate. Even if one does one's best, accusations of negligence or injustice are inevitable. Accordingly, one simply applies the appropriate principles of evaluation [translated from the French, Nobelstiftelsen, 1964, p. 17].

But awareness of difficulties plainly does not prevent them from being consequential, and the existence of a cadre of occupants of the forty-first chair raises questions about the validity of the Nobel prize as the preeminent symbol of accomplishment in science. Its value has been eroded further as it has been appropriated for use in other parts of the social system. To a degree, the Nobel prize has become a victim of its own success.

## NEW PERSPECTIVES ON THE NOBEL PRIZES

Never free of criticism, the Nobel prizes are now increasingly subject to attack. In part, their cumulating significance as *the* prime symbol of scientific excellence invites criticism. But the criticism also results from the fact that the procedures for allocating Nobel awards have scarcely changed in seventy-five years, a time when the cognitive substance and social organization of science have been transformed.

The most evident anachronistic aspects of the awards are the limited fields in which they are given, the stipulation that they must be made for a "discovery," and the customary restriction of the number of recipients to no more than three in one field.

### Fields of Science Eligible for the Awards

Until the past few years, the Nobel committees have narrowly interpreted their mandate to name winners in the three spheres of science under their jurisdiction. From the beginning they elected to exclude work in astronomy and astrophysics from the contest for the prize in physics. W. W. Campbell of the Lick Observatory was nominated for the first prize in physics but the committee decided that, physicist or not, the astronomically oriented Campbell was ineligible for an award (Nobel-stiftelsen, 1972, p. 296). This has meant that astronomers of the first class such as E. P. Hubble, Jacobus Kapteyn, and Harlow Shapley could at best assume a place in the select but outer circle made up of occupants of the forty-first chair. All this may change now that the Physics Committee has broken a longstanding taboo and awarded the 1974 prize to the astronomers Anthony Hewish and Martin Ryle.

The field of mathematics, however, is now and always has been excluded, and so Henri Poincaré, George Birkhoff, and John von Neumann, among many others, could not appear on the roll of Nobelists. So, too, are the earth and marine sciences. Exhibiting no small amount of *Tendenzwitz,* the form of humor that makes expressions of hostility socially permissible, a select group of oceanographers have designated themselves "Laureates of the Albatross." In this way, they manage to twit the Nobel committees for the arbitrariness of the boundaries of the prizes without declaring that oceanography *ought* to be included in official considerations. A recent recipient of the Albatross, Roger Revelle, at once parodied the traditional humility of Nobelists: "It hasn't changed my life-style at all. . . . I'm really being quite modest about it" (*Science,* vol. 181, 7 September 1973, p. 926).

In biology, the awards were long concentrated in the specialties having direct implications for medicine or relating directly to human development, with no regard for work in evolutionary biology of the kinds developed by George Gaylord Simpson or Ernst Mayr. But, again, the committees have recently shown some inclination to enlarge their perspectives. In 1973 the award in medicine went to the pioneer ethologists Karl von Frisch, Konrad Lorenz, and Nikolaas Tinbergen, and thus opened the door for consideration of other kinds of biological research. Still, it is the case that scientific grounds for such restriction have been increasingly called into question by changes in the scope and texture of knowledge.

That the boundaries of the sciences and their internal geography have greatly changed since 1901 is a commonplace. Scientific specialties now exist that were not dreamt of in Nobel's time. As a consequence, a larger proportion of significant contributions to knowledge than ever before fall outside the boundaries of eligibility and it should be increasingly difficult to allocate candidates on the edges of the "established" fields to competent committees for appraisal. Moreover, there are intimations that, confronted with enlarged numbers of contributions of Nobel caliber, the committees of selection prefer to make awards that clearly conform to the statutes and elect to pass over major work in fields at the margins of eligibility.

The restriction of the prizes to the fields of physics, chemistry, and the biological sciences raises questions of inclusion as well as of omission. It is claimed that once active but now quiescent specialties continue to receive Nobel awards even though the work that is honored is unexciting. One critic observed, "Some chemists speak most disparagingly of some of the recent awards. It seems that great breakthroughs in chemistry are rare; the making of an unusual new chemical compound [cited for several recent awards] may be a technical achievement but rarely is it an intellectual achievement." A second observes that "physics is in the doldrums these days. The selections are increasingly dubious." No doubt there are many who would disagree with these views, but they focus attention on the fact that the richness and pace of scientific development vary over time, and the terms of the bequest may not be sufficiently flexible to take these changes into account. They also implicitly address themselves to a difficult epistemological question that must continue to plague the Nobel committees: Are discoveries in different disciplines and specialties commensurable, and if so on what grounds? Nobel might have guessed that the contours of the sciences would change but he evidently did not anticipate the speed or extent of the transformation. The great changes that have occurred contribute to the apparently growing conviction among scientists that the prizes have become increasingly parochial and governed by a set of rules no longer adequately meshed with the realities of modern science.

## Eligible Contributions

The statutory requirement that prizes must be awarded for "a discovery," "improvement," or "invention" has its own share of ambiguity. In practice, it has meant that statements of scientific principles or organizing conceptions have often been ruled out of consideration. Yet, as many observers have noted, it is these principles, not particular discoveries, that make science coherent and consequential. The historian Donald Fleming argues that

> anybody solely dependent on following the Nobel citations would be imbibing a narrowly positivistic conception of science as an accumulation of many hard little pellets of empirical knowledge to be shaken free of any conceptual matrix in which they are unaccountably embedded. It is a peculiarly end-of-the-nineteenth-century view, comprehensible in a man of Nobel's generation

and outlook but now hopelessly antiquated as a way of looking at science and the dynamics of scientific progress [1966, pp. 58–59].

Since the Nobel awards are made for conceptual contributions only when these are validated by the discovery of "new facts," ideas that unify large bodies of data without directly giving rise to a new discovery—of the sort represented by Darwin's principles of evolution and Cannon's concept of homeostasis—do not qualify.

Arne Tiselius of the Nobel Committee for Chemistry defends this position in terms of hardline positivism, though the latter part of his formulation rather resembles the position advocated by the philosopher of science Imre Lakatos (1973, pp. 118–19).

> Concepts are extremely useful to the human mind . . . but concepts change and change very fast. What remains are the facts, the experimental facts . . . concepts are instruments in scientific research. They help you make new discoveries. If they can't, they have no justification. And the only way you can prove the justification of a concept . . . is to see if it not only explains already-known facts but also leads to a new and unexpected discovery. That proves there is something new in the concept [quoted in Litell, 1967, p. 49].

The general rule protects the committees from honoring contributions that do not prove out. But this decision rule means that prizes are more likely to be awarded for rather conventional empirical contributions than for consequential theoretical ones. This penchant for the "sure thing" was evidently expressed in the decision to award a prize to Einstein for his work on the photoelectric effect rather than for his theory of relativity. It also accounts, as we shall see, for the dissatisfaction of some laureates with the choice of the work cited as the basis for their prizes. They maintain that the prize has brought attention, not to their best work, but merely to good work that clearly conformed to the statutory requirements.

The Nobel committees have also opted for a conception of science that emphasizes the culmination of complex research efforts rather than their original sources. Often, prizes honor those who harvest rather than those who sow; the successors rather than the originators. The physicist John Ziman observes, "The experts themselves will tell you how many superb scientists have been passed over because their discoveries were not 'important' to the view of the day, or because they were only stepping stones in a very long investigation by many different research workers." He maintains, for example, that Crick and Watson, who have been widely acclaimed as having opened up a new era of biological research on DNA replication, were in fact like sergeants who, "after a mighty assault, finally planted the flag upon the summit of the citadel. By the time they had ventured into battle, victory was certain; it was largely chance that put the symbol of it into their hands" (n.d., p. 9). To judge from Watson's account of the discovery of the structure of DNA (1968), Ziman's metaphor is apt. Qualified observers who emphasize the immense value of the Crick-Watson discovery also go on to note that it involved neither the profound foresight required for explorations into largely unknown territory nor the deep skepticism needed to recast old ideas in fundamentally new ways. Crick and Watson no doubt found it odd that they were made Nobel laureates before Max Delbrück, Alfred Hershey, and Salvador Luria, the older pioneers who did so much to create the perspectives of molecular biology. In this symbolic anticipation, they were hardly alone. Before the prize finally came to the three founding fathers in 1969, it had gone to fifteen molecular biologists and biochemists for investigations built on foundations the three pioneers had laid down.

## Constraints on the Number of Eligibles

Although the stipulation that the prizes be limited to no more than three recipients in each field helps to maintain their honorific and cash value, it is incompatible with important facts of contemporary scientific life. It will become evident that these are not difficulties attributable to population growth and consequent increases in the numbers of scientists eligible for prizes. They derive, rather, from changes in the social organization of scientific work since Nobel established his awards.

Not only are the vast majority of all scientific investigations now collaborative efforts but the "prime movers" in collaboration sometimes number more than three. Thus it is often difficult to identify the principal contributors among a number of co-workers. Moreover, significant scientific contributions are often the outcome of distinct but coalescing investigations by many researchers or involve multiple, although independent, discoveries by more than three investigators.

The steady increase in the average number of laureates named annually in the three fields of science from 3.6 in the first decade of the award to 6.2 in the most recent one may reflect the growing significance of collaborative research and the partial or piecemeal character of scientific contributions. But this change does not appear to be an equitable or efficient mode of adaptation to social change in the organization of scientific inquiry.

The difficulties involved in identifying the prime contributors to collaborative work have also acquired new importance with the growing frequency and scale of joint research. Although the official chronicle of the Foundation holds that "it has been possible to find adequate grounds for limiting the award to the person or persons who have had the decisive share in the discovery" (Nobelstiftelsen, 1962, p. 333), this declaration seems anachronistic in light of the marked division of labor in many investigations. Even Arne Tiselius, who was ardently committed to limiting the number of laureates in chemistry not to three but to one, was well aware of the pitfalls in trying to identify *the* key investigator in joint research. "This often calls for much 'inside information' which is difficult to obtain . . . there is always the danger of injustice" (1967, pp. 4–5).

Tiselius agreed that an injustice was done in the 1923 award for the discovery of insulin, which was given to Banting and MacLeod (who was not directly responsible for any phase of the research) but excluded Banting's collaborator, **Claude Best,** and it turns out that Best was not the only scientist excluded from the controversial award. The Romanian, N. Paulesco, came to many of the same conclusions as Banting and Best and did so six months earlier. He published his work in *Archives International de Physiologie,* a journal that was not obscure. Like Best, Paulesco was never even proposed for the prize given for insulin and thus was excluded from the Nobel roster. The merits of the Paulesco claim were unofficially recognized by the Nobel establishment belatedly and apologetically. In a letter to the Romanian Academy, Tiselius wrote:

> I have thoroughly studied the documents you have sent me and I have also discussed the case with colleagues, especially with Prof. Ulf von Euler, President of the Nobel Foundation and, as you know, himself a physiologist and endocrinologist of the highest reputation. As you know well, the Nobel Prize to Banting and MacLeod has been criticized by many, especially the fact that Best was not included. In my opinion, Paulesco was equally worth the award. As far as I know, Paulesco was not formally proposed, but naturally the Nobel Committee could have waited another year. . . . Unfortunately there is no mechanism by which the Nobel Committee could do anything now in this or similar cases. Personally, I can only express the hope that in an eventual celebration of the 50th anniversary of the discovery of insulin, due regard is paid to the pioneer work of Paulesco [Quoted in Pavel, Bonaparte, and Sdrobici, 1972, p. 489].

Others believe that injustices were done when co-workers of some laureates were excluded from awards. In the case of laureate Selman

Waksman, a court of law had determined two years before he won the prize that Albert Shatz was a full collaborator and was entitled to a share of the royalties deriving from the discovery of streptomycin.[19] (See Waksman, 1954, for his version of the episode.) The Nobel Committee has indicated that it was not aware of the circumstances surrounding the case (it made front-page news in the American press) and, more important, that Shatz was never nominated for the award although a number of scientists had nominated Waksman alone (private communication). Since Waksman was a major figure in the development of antibiotics generally and Shatz's contribution appears to have been limited to the research on streptomycin, the prize judges may well have concluded that Waksman alone deserved the award. But, since the merits of leading candidates are presumably subjected to painstaking study, it is surprising that Shatz's role in the investigations did not come to light before the prize.

Recent controversy about the role of Jocelyn Bell Burnell in Anthony Hewish's discovery of pulsars raises similar questions about distinguishing between full-fledged scientific collaboration and supervised research assistance, between irreplaceable and replaceable scientific contributions to prize-winning research. Burnell's own account squares well with Hewish's version of her contribution to the pulsar discovery, and she has been the model of restraint in the face of great temptation to claim a share in the Nobel for herself (see *Science,* vol. 189, 1 August 1975, pp. 358–64). But, as in the Shatz case, it is not evident how lines can be drawn appropriately between collaborators, especially when the research cited for the prize is described as a particulate discovery rather than a line of research work and hence presumably can be attributed to specific investigators even if they have had numerous collaborators.

The practice of setting limits on the number of recipients of a prize has odd consequences for its announced purpose of honoring fundamental contributions to science. For the requirement that full recognition be given to all major participants in scientific investigations of unquestioned importance may mean that achievements representing multiple contributions are simply ruled out of consideration. By their own

19. An apparently similar suit brought against Owen Chamberlain and Emilio Segrè by Oreste Piccioni now seems to be an altogether different matter. In 1972, Piccioni claimed that the two laureates had not acknowledged his contribution to the discovery of the antiproton (the research that brought them the prize in 1959) even though they mentioned him and his suggestions in their original publication and in their Nobel addresses (Shapley, 1972; *New York Times,* 17 June 1972). Unlike both Shatz and Burnell, whose cases both had active supporters, scientists in a position to know Piccioni's contribution have not pressed his case. A year after the suit was filed, a judgment of dismissal was rendered but the case was not settled for another year and a half and a total of 65 separate legal actions (Piccioni v Segrè and Chamberlain, Alameda Co., #425000).

THE SOCIOLOGY OF THE NOBEL PRIZE

account, committees of selection have done just that in several cases. To take just a few examples, they have passed over significant work on Vitamin D (the antirachitic provitamin), on the biological significance of sex hormones, on local anesthesia, and on the treatment of otosclerosis by fenestration of the ear because each case involved "too many" scientists (see Nobelstiftelsen, 1962, pp. 235, 241, 328, 331). In much the same way, they could not bring themselves to identify distinctive roles in the basic work on the binding and transport of oxygen and carbon dioxide in blood, thus consigning J. G. Priestley, J. B. S. Haldane, and L. J. Henderson to the forty-first chair. Since 1962, the Committee on Physiology and Medicine has shown signs of recognizing and trying to cope with this sort of problem. Although they had reported that "the number of contributors make it difficult to make a selection for the Prize" for research on bacteriophages (Nobelstiftelsen, 1962, p. 211), they have hurdled that difficulty to make several awards in the field.

All this does not, however, suggest that the Nobel committees of selection may in the future follow the example set by the American Academy of Arts and Sciences, which awarded the Rumford Premium to three sets of scientists, not as individuals but as groups. This may violate the heroic conception of scientific development in which no good can come of large-scale collaborative inquiry, but there are evidently cases testifying otherwise. "The Academy's presentation of the [Rumford] award to teams rather than to individuals was made in an effort to reflect the way in which much scientific research is conducted today" (*Physics Today,* June 1971, p. 69).

Changes in the organization of science have come to exert pressure for change in the awarding of the Nobel prize and have put the almost immutable rules governing eligibility for it under great stress. In some quarters the long-continued failure to modify the rules and policies has eroded the prestige of the prize. Because it has become a measure of ultimate accomplishment in science, the prize is especially subject to the process of displacement of goals. For some scientists acquiring the prize has become an end in itself just as acquiring laureates has become an end in itself for universities and other organizations. In scientifically knowledgeable circles especially, ambivalence about the prize is reinforced as its symbolic uses multiply. Every reward system, however appropriate when first instituted, acquires dysfunctional features when the context of its operations changes while it does not; as these dysfunctions accumulate, they press for change. In the case of the Nobel prize, there is a perceptibly growing sense that its almost universal acceptance as the symbol of supreme scientific achievement is beyond reason. There is, in effect, a rising awareness of the increasing numbers of scientists who are uncrowned laureates in every respect the peers of Nobelmen. The sense that too much has been made of the prizes and their laureates may be reflected in the remark of the laureate Max Delbrück: "By some random selection procedure, you pick out a person, and you make him an object of a personality cult. After all, what does it amount to?" (quoted in *Time,* 3 July 1972, p. 53). Still, for the present, the Nobel prizes continue to serve as the touchstone of scientific excellence.

# Chapter 3

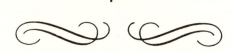

# THE SOCIAL ORIGINS
# OF LAUREATES

     In science, rank is primarily achieved, not ascribed. But even when upward mobility is based on meritocratic principles, it involves a process of accumulation of advantage [1] that helps to shape, maintain, and modify the structure of stratification in science, as it does in other merit-oriented systems. To learn something about cumulative effects in the careers of Nobel laureates and other members of the scientific elite, we turn first to more general aspects of this self-reinforcing pattern in science.

## ACCUMULATION OF ADVANTAGE IN SCIENCE

     Advantage in science, as in other occupational spheres, accumulates when certain individuals or groups repeatedly receive resources and

---

1. Having identified the "principle of cumulative advantage" in his 1942 paper on the normative structure of science, Merton (1973, p. 273) sketches out some of its individual and institutional consequences in the domain of science in his analysis of the Matthew Effect (1968f reprinted in 1973, pp. 416, 457–58; 1975). The formulation is developed further in S. Cole (1970), Zuckerman (1971b, p. 245), Cole and Cole (1973, passim, see especially pp. 237–47), Allison and Stewart (1974) and Faia (1975).

rewards that enrich the recipients at an accelerating rate and conversely impoverish (relatively) the nonrecipients. Whatever the criteria for allocating resources and rewards, whether ascribed or meritocratic, the process contributes to elite formation and ultimately produces sharply graded systems of stratification.

Advantage can accumulate in two ways: by addition or by multiplication. In the additive model, people who begin their careers with certain ascribed advantages continue to benefit, to receive resources and rewards on grounds that are "functionally irrelevant"—that is, irrespective of their occupational role performance.[2] In the second model, people judged on functionally relevant criteria as the most likely to make effective use of resources are also the most likely to receive them.[3] Recipients are advantaged in the sense of being more able to begin with, of getting more of what is needed to perform their roles, and of consequently achieving more. The resulting gap in attainment between the advantaged and the others is far greater than under the conditions of the additive model, in which the ability to use resources for further achievements is randomly distributed among recipients and nonrecipients.[4] To the extent that resources are allocated to the same individuals on functionally relevant criteria over the course of their careers, discrepancies in attainment tend to grow ever larger between the initially advantaged and everyone else.[5]

Since the application of functionally relevant criteria in allocating resources to individuals results in ever larger differences in their attain-

---

2. The concepts of "functionally relevant and irrelevant criteria" for status that have been incorporated into these two models were developed by Merton in lectures at Columbia University (1955–71). For pertinent uses of these conceptions in the dynamics of status sets, particularly with regard to functionally irrelevant statuses, see Epstein (1970, esp. chapter 3).
3. The same process involving privileged access to resources sometimes takes the form of exponential rather than merely multiplicative increases in advantage. Comparable processes are of course much better understood and analyzed in the domain of economics.
4. Differentials in occupational attainment among men and women, in the United States for example, can also be construed as the result of additive processes in which resources for mobility tend to be differentially allocated to men on functionally irrelevant criteria involving their culturally defined location in the occupational and educational systems. See Zuckerman and Cole (1975) and Reskin (1976).
5. Discrepancies in attainment can be further increased by "practice effects." That is, the ability to use resources efficiently may grow with the number of opportunities for use, leading those who are consistently advantaged to develop differentially greater skill in utilizing resources put at their command.
   Allison and Stewart (1974) and Faia (1975) present evidence of growing differentials in scientific productivity with chronological age for a synthetic cohort of university scientists. There is reason to suppose that the differentials they observe might be even greater if based on productivity data from a single cohort as it aged. Less productive scientists who initially began in university departments are more likely to drop out and take jobs in colleges or industry than more productive ones. As a consequence, the older scientists in the Allison-Stewart sample may be more highly selected for continuing productivity than the younger ones. See Shockley (1957, pp. 285–86) for other interpretations of productivity differences.

ments, the resulting distribution of attainments tends to be sharply stratified, with those judged most competent at the top of the hierarchy. This, then, is one process that produces elites of achievement. It also enables elites to develop self-serving justifications for their position: since recipients of resources are more likely to achieve, it is argued that the system of allocation is effective and legitimate.

These general observations on the accumulation of advantage have special meaning in the domain of science. The normative structure of science, with its abiding emphasis on extending certified knowledge, calls for the application of universalistic or meritocratic criteria in distributing resources for scientific investigation and rewards for contributions to science. On these grounds, facilities "should" be allocated in accord with scientific merit, and varying degrees of recognition "should" accord with contributions to knowledge. The scientific ethos thus provides a normative basis for establishing a multiplicative process of accumulation of advantage along lines that tend toward meritocracy.

It follows from these observations that the earlier individuals are identified as meritorious—in science, this means being judged as a "comer"—the sooner they gain privileged access to resources, the earlier they begin to develop, and the greater their head start over other young people judged less capable or not noticed at all. Since the accomplishments and the promise of young scientists are generally gauged in comparison with those of their age peers, a head start is particularly advantageous in such age-graded systems of evaluation. Thus, in examining the social and educational origins of Nobel laureates, we will want to look for evidence of their being advantaged early in their careers in ways their ultimately less accomplished rank-and-file colleagues were not.

The accumulation of advantage in science involves both getting ahead initially and moving farther and farther out in front. As we have noted, enlarged access to resources enables talented individuals to perform more effectively and, in accord with the merit principle in science, to be rewarded more copiously.[6] Since collegial recognition and esteem are the prime rewards for scientific achievement,[7] honored standing must be converted into other assets more directly applicable to further occupational achievement: including assets such as influence in allocative

---

6. Norman Storer suggests that two kinds of facilities enter into this process. The first kind—money, for example—has little effect on later achievement once it is "spent." The second—education, for example—is an internalized resource and, once acquired, continues to enter into the process of accumulating advantage. Those facilities that contribute to the scientist's capability for doing further research provide an advantage that accumulates beyond the initial one of access to resources.
7. For the emphasis on collegial recognition in the reward system of science, see Merton (1957 in 1973, pp. 286–324), Hagstrom (1965), and Storer (1966).

decisions, access to gate-keeping positions (such as editorships and slots on panels allocating research funds and awards), and, most important for the accumulation of advantage, new facilities for work. These resources can be used to improve subsequent role performance, thus renewing the cycle.[8] This formulation of career mobility suggests that Nobelists and other members of the elite are comparatively more productive than other scientists early in their careers, are rewarded promptly, and are reasonably successful in transforming recognition into resources for further work. Thus the reward system affects the distribution of achievement in science and makes for continuing interplay between honor and access to the means of scientific production.

While early identification of talent generally leads to its support and reinforcement, recognized talent and capability are no substitute for role performance of the first class. Great expectations must ultimately be buttressed by scientific production if scientists are to continue to receive scarce resources for research. Cases of unfulfilled promise in science (those that fizzled early and those that never really got started) are numerous and well known. How long such scientists continue to be well regarded and to receive support for their work without tangible results probably varies among the sciences and over the course of the scientific career. Still, the merit principle in science does not permit the accumulation of advantage to be automatic for very long.

The same merit principle also provides for recognition of contributions by scientists not initially identified as talented but who do good work nonetheless. But there is obviously no way of gauging the frequency with which unknown talents are completely overlooked or not encouraged to develop at all.

The accumulation of advantage in science is reinforced still further by allocative processes that are not strictly based on functionally relevant criteria even though they are related to past achievement. The Matthew Effect describes the process by which "greater increments of recognition . . . [go] to scientists of considerable repute and . . . such recognition [is withheld] from scientists who have not yet made their mark" (Merton, 1973, p. 446). Those who have done good work and earned the esteem of their colleagues get more recognition for new work and get it faster than they would were it not for their previously

8. Various studies have identified the interaction between recognition and subsequent high-level performance in science. Barney Glaser (1964, p. 23) describes the process by which "recognition" elicits motivation, which in turn leads scientists to spend more time at work, which then makes for high-level performance. Cole and Cole come to similar conclusons in their study of physicists: "When a scientist's work is used by his colleagues [receives recognition] he is encouraged to continue doing research." They observe that physicists' productivity is substantially influenced by recognition received for early work (1973, p. 113, and chapter 4, *passim*).

achieved status. The Matthew Effect also operates in the distribution of honorific awards: those who already have them are most likely to receive new ones.[9]

Much the same processes of accumulative advantage apply to organizations as to individuals. The most distinguished universities attract better applicants, on the average (Bayer and Folger, 1966; Blau, 1974). The students acquire a certain prestige from having gone to these universities and, in turn, lend prestige to their alma maters later in their careers. The same universities are more generously supported and thus have facilities that attract accomplished faculties (National Science Foundation, 1976, p. 13). The interaction of processes of accumulative advantage at the individual and organizational levels further reinforce the discrepancies in individual achievement already noted. In light of these observations on cumulative effects in the stratification system in science, we turn now to the social origins of laureates and other scientists and to the ways in which they influenced access to undergraduate and advanced education, noting that where scientists get their training generally has much to do with where they end up years later.

Ideally, estimates of the extent to which members of the scientific elite (or any other group) have benefited from accumulative advantage are based on repeated comparison of their achievements and rewards with those of others in the same age cohort. Since there are so few American Nobelists—by 1972, just seventy-one were reared in the United States and only ninety-two did their prize-winning research here—historical and cohort-specific analyses are not feasible. As a consequence, we will treat laureates as a single cohort and compare their careers with those of members of the National Academy of Sciences and the scientific rank and file having roughly the same age distribution.

## SOCIOECONOMIC ORIGINS OF ELITE SCIENTISTS

Even in societies with comparatively high rates of intergenerational mobility, elites in nearly all departments of social life come in disproportionate numbers from the middle and upper occupational strata. This has been found to be the case in the United States, now as in the past, for business leaders, admirals and generals, diplomats, senators,

---

9. The Coles observe, for example, that scientists located in the most distinguished university departments receive more recognition than would be the case if only their contributions were being taken into account (1973, p. 121).

and justices of the Supreme Court (Keller, 1963, p. 292ff).[10] It also
holds true for the seventy-one scientists who were raised in the United
States and later became Nobel laureates, as table 3–1 shows. Like
Supreme Court justices and admirals and generals, American Nobel
laureates are far more likely to be the sons of professionals than of
businessmen.

Whether this is also the case for Nobelists of other nationalities we
cannot say. The same pattern appears to hold for the twenty-one laureates who emigrated to the United States as adults (after completing their
undergraduate studies) but before doing their prize-winning research;
however, since the only study (Moulin, 1955) is confined to a third of the
prize-winners named before 1950, it provides no basis for assessing the
representativeness of the truncated sample. That study does conclude that
the origins of the laureates are "never humble." In the United States, at

**Table 3–1. Elite Origins of American-Reared Laureates (1901–72) and Other Elites**

| American Elite | | FATHER'S OCCUPATION | | |
| --- | --- | --- | --- | --- |
| | | Professional | Manager or Proprietor | Sum |
| Supreme Court Justices | | 56% | 34% | 90% |
| Nobel Laureates[a] | | 54 | 28 | 82 |
| Admirals and Generals[b] | | 45 | 29 | 74 |
| Business Leaders | | 15 | 57 | 72 |
| Diplomats | | 32 | 36 | 68 |
| Senators | | 24 | 35 | 59 |
| *Employed Males*[c] | 1900 | 3.4 | 6.8 | 10.2 |
| | 1910 | 3.5 | 7.7 | 11.2 |
| | 1920 | 3.8 | 7.8 | 11.6 |

a. These consist of the sixty-five American-born laureates and the six who emigrated
to the United States before going to college.
b. The data for the military, business leaders, diplomats, and senators are reported in
Keller (1963, p. 311), who draws upon the work of Janowitz (1960) and Matthews
(1960) as well as her own (1953). The data for justices of the Supreme Court, 1889–
1957, are from Schmidhauser (1959, p. 7, computed from table 1).
c. U.S. Bureau of the Census (1960, p. 74, Series 74, 75). These dates roughly correspond to the periods when fathers of these elites were in the work force.

10. The American Catholic hierarchy provides one exception to the rule of elites
originating in the upper levels of the stratification system. Donovan (1958) finds that
only 5 percent of cardinals, archbishops, and bishops had fathers who were professionals and only 10 percent had fathers in managerial positions. Similarly, saints of the
Roman Catholic Church, unlike secular elites of this world, have for several centuries
been drawn from the middle and lower classes (George and George, 1955, p. 87).

any rate, having a relatively well-placed family [11] seems to facilitate one's entrance into the ultra-elite of science just as much as into other elites that are recruited in larger numbers, and presumably on the basis of more ascriptive criteria.

This observation needs to be put in perspective. True, as can be seen in table 3–1, the family origins of American laureates were much higher in rank than those of the population at large. Eighty-two percent of them had fathers who were professionals, managers, or proprietors— about eight times the representation of these occupational groups in the male labor force in the years the fathers were at work. But closer examination of the composition of the professional stratum from which many of them came suggests that not all were from the more lucrative professions: the group included four secondary school teachers, five clergymen, six college professors, and eleven physicians (including two on medical school faculties), four engineers, two college presidents (one of them a clergyman as well), two lawyers, and four others: a dentist, diplomat, author, and artist.[12] There is a hint in these figures that it may have been the educational environment rather than opulence that mattered most, an impression confirmed by comparing the socioeconomic origins of the laureates with those of the run of scientists.

As in other occupations, so it is in science: attained rank is related to the occupational rank of family of origin (Duncan, Featherman, and Duncan, 1972). As table 3–2 shows, the laureates are almost twice as likely as scientists of roughly comparable age holding doctorates to come from professional families and somewhat more likely to have had fathers in business.[13] Toward the other end of the distribution, only about 15 percent of the laureates and a third of the other scientists come from blue-collar and white-collar families, in contrast to the proportions of employed males in these occupations at a comparable period.[14] All

---

11. There is substantial agreement among students of stratification that occupational rank is highly correlated with other measures of socioeconomic status and is its best single measure (Siegel and Hodge, 1969).

12. This illustrates the fact that the correlation between income and socioeconomic status, though strong in general, is far from perfect in particular instances (Hodge, Siegel, and Rossi, 1964).

13. Harmon's survey of doctoral recipients in the sciences also shows that physical scientists are somewhat more likely than biological scientists to have fathers in the professions or business: 52 percent compared with 42 percent among those receiving their degrees in the late 1930s (1965, p. 39). Among the laureates, 83 percent of the physicists and chemists and 69 percent of the biologists had fathers in the professions and business.

14. Studies of the socioeconomic origins of other samples of scientists of varying eminence report inconsistent findings. Scientists starred in *American Men of Science* (Visher, 1948) and the distinguished scientists studied by Anne Roe (1953) are re- ported to have come largely from the upper middle class while the scientists in the *Fortune* study (1948, pp. 106ff.) and those listed in *American Men of Science* (Knapp and Goodrich, 1952; 1967) are reported as having lower-middle-class origins. These

**Table 3–2.  Socioeconomic Origins of American-Reared Laureates (1901–72), Scientists Receiving Doctorates (1935–40), and Employed Males**

| Father's Occupation | Nobel Laureates[a] | Science Doctorates[b] | Employed Males[c] |
|---|---|---|---|
| Professionals | 53.5% | 29.1% | 3.5% |
| Managers and Proprietors | 28.2 | 18.7 | 7.7 |
| Farmers | 2.8 | 19.5 | 34.7 |
| Sales, Service, and Clerical Workers | 7.0 | 13.1 | 12.8 |
| Workers: Skilled and Unskilled | 8.5 | 18.0 | 41.3 |
| No Information | — | 1.5 | — |
| Total | (71) | (2695) | (29,847,000) |

a. These data are drawn from interviews, questionnaires, and biographical sources. The stated occupations of fathers do not uniformly refer to a fixed comparable age but to "principal" occupations.
b. Harmon (1965, p. 39).
c. Data are for males employed in 1910. U.S. Bureau of the Census (1960. p. 74, Series 89–105).

this suggests that professional families provide joint social and educational advantages. This impression is consistent with the fact that the professional fathers of laureates tended to be in science or in science-related occupations to a greater extent than is true for the general distribution of professionals: 55 percent were physicians, engineers, teachers of science, or full-fledged research scientists. And though there are no cases among American-reared laureates of a special kind of occupational inheritance—multigeneration Nobel prize-winners in the same family—there are seven cases of occupational inheritance in the strictly technical sense of the term—that is, seven laureates had fathers who were working scientists.

Although the substantial proportion of laureates with fathers in science-related professions immediately suggests that many had access to appropriate training and role models and were directed toward scientific careers early on, no evidence of their having had a head start can be detected. Laureates whose fathers were scientists, engineers, or physicians did not graduate earlier from college, did not get their doctorates sooner, and were not promoted more rapidly, nor were they more precocious in doing the research that brought them their Nobel prizes than laureates whose fathers were not in science-related occupations.

investigations deal with different samples and may register actual variations in the social origins of different sectors of the scientific community. It should be noted, however, that the Knapp and Goodrich indicator of "class origins"—the costs of attending particular colleges (1967, p. 292)—is crude at best and ignores variability in the social origins of students of the same college.

The absence of such differences does not, of course, mean that having a scientist father confers no advantages on the run of scientists or, as a matter of fact, that elite standing is altogether unrelated to having a parent in the *same* occupation.[15] It turns out that occupational inheritance of this kind is at least as frequent among laureates (29 percent had scientist or engineer fathers) as among any other elite studied (Keller, 1963, p. 324), and far more frequent than the rate observed for all men for much broader occupational categories (Blau and Duncan, 1967, p. 39). The nature of the connections between the occupations of elites and their parents is therefore yet to be determined.

Unlike the family origins of the other scientists holding doctorates, those of the laureates have not changed much through the years. The representation of professionals among fathers of the Nobelists has dropped slightly, from 56 percent for those born before 1900 to 50 percent for those born in the decades afterward. However, as the lower tier of table 3–1 indicates, the proportion of professionals among males in the labor force at large has been increasing. Thus the ratios of professionals among Nobelists' fathers relative to the proportion in the work force have declined. The same is true for other scientists, where the ratios have gone from 8:1 for those receiving their doctorates in the late 1930s to 7:1 in the late 1940s to 5:1 in the late 1950s (Harmon, 1965, pp. 38–40). Roughly comparable ratios for the laureates are 14:1 for those who took doctorates in the 1930s and 10:1 for those who did so in the 1940s. (So few laureates received degrees in the late 1950s as to make computation of ratios unreliable.) Harmon also observes a slight upward trend in the proportion of scientists coming from families of skilled and unskilled workers, a trend that may reflect the enlarged educational support made available to students of "limited means but high ability" (1965, p. 38). But there is no such trend in the social origins of laureates; instead they are increasingly drawn from families headed by businessmen.

This means that while inequalities in the socioeconomic origins of American scientists at large have been significantly reduced during the past half century, this has not been the case for the ultra-elite in science. Even in a system as meritocratic as American science, in which *identified* talent tends to be rewarded on the basis of performance rather than origin (Cole and Cole, 1973), the ultra-elite continue to come largely from the middle and upper middle strata. Whatever the ultimate explanation of this fact—the interplay between genetic and social components in the process is far from having been worked out—one aspect of the

15. There is ample evidence of anticipatory socialization among physicians' offspring (Merton, Reader, and Kendall, 1957).

fact itself is clear: the social origins of Nobel laureates remain highly concentrated in families that can provide their offspring with a head start in access to system-recognized opportunities.

## RELIGIOUS ORIGINS

The religious origins of Nobel laureates exhibit a pattern of subsidiary concentration, that is, overrepresentation of religious origins occurs not for the majority of laureates but only for a minority among them.

Protestants turn up among the American-reared laureates in slightly greater proportion to their numbers in the general population. Thus 72 percent of the seventy-one laureates but about two thirds of the American population were reared in one or another Protestant denomination—mostly Presbyterian, Episcopalian, or Lutheran rather than Baptist or Fundamentalist. However, only 1 percent of the laureates came from a Catholic background, one twenty-fifth the percentage of Americans counted as adherents to Roman Catholicism (U.S. Bureau of the Census, 1958). Jews, on the other hand, are overrepresented: comprising about 3 percent of the U.S. population, they make up 27 percent of the Nobelists who were brought up in the United States.[16] (This double pattern of a deficit of Catholics and an excess of Jews among scientists relative to their numbers has often been noted in studies of religious origins of scientists and shall be discussed presently. See Lipset and Ladd, 1971; Hardy, 1974.)

These figures, it should be emphasized, refer to the religious *origins* of this scientific ultra-elite, not to their own religious preferences. Whatever those origins, laureates often describe themselves as agnostics, without formal religious affiliation or commitment to a body of religious doctrine.

The large representation of Jews among laureates is by no means a uniquely American phenomenon. By rough estimate, Jews make up 19 percent of the 286 Nobelists of all nationalities named up to 1972, a percentage many times greater than that found in the population of the countries from which they came and one that may be related to the

---

16. The picture for the ninety-two laureates who did prize-winning research in the United States is not much different: about two thirds came from Protestant families, 28 percent from Jewish families, and just under 5 percent from Catholic families.

It should be understood, however, that the overrepresentation of Jews among American Nobel laureates tells nothing about the extent to which their religious origins may have facilitated or hampered their achievement of scientific eminence.

often documented proclivity of Jews for the learned professions in general and for science in particular (Singer, 1960; Weyl and Possony, 1963; Lipset and Ladd, 1971).

These data would begin to put in question the often expressed belief that the notable representation of Jews among "American" laureates resulted from the great migration of talented young scientists in the wake of Hitler's rise to power. It is true that many scientists did escape to the United States, where they significantly augmented the *numbers* of Jews among the scientific elite as well as among the rank and file. But the refugees did not materially increase the *proportion* of Jews among the future laureates. Nineteen of the 71 laureates raised in the United States (27 percent) and seven of the 21 raised abroad (33 percent) were of Jewish origin. The seven Jewish emigré laureates-to-be, though a substantial addition to the ultra-elite, raised the proportion of Jews among all ninety-two future laureates by only 1 percent. To put this in another way, two thirds of the twenty-one future laureates who emigrated as adults to the United States were not Jews—and all but two of these non-Jews came for reasons unconnected with the Nazi assumption of power. But, on closer inspection of both the religious origins of laureates-to-be and the dates on which they emigrated, we also find that seven of the eleven who came between 1930 and 1941 were Jews; thus every one of the Jewish emigrés was a refugee from Hitler. Not one of the six who emigrated before 1930 was Jewish, nor were any of the four who came after World War II.

## "Hitler's Gift to American Science"

Many of the emigré laureates-to-be coming just before World War II were already well known scientists, although they had not yet done the work for which they would win their prizes. In fact, as the historian of science Charles Weiner tartly observes, the situation for younger European scholars at that time could be characterized as "to have published or to have perished" (1969, p. 217). The most illustrious of all the emigrés of this period, of course, were the eight who had already received a Nobel prize: the physicists Niels Bohr, Albert Einstein, Enrico Fermi, James Franck, and Viktor Hess; the chemist Peter Debye, the pharmacologist Otto Loewi, and the biochemist Otto Meyerhof. Scarcely less illustrious were the three who were soon to receive it: the physicists Wolfgang Pauli and Otto Stern and the biochemist C. P. Henrik Dam. Although Bohr, Pauli, and Dam returned to Europe at the conclusion of the war, the others remained as permanent additions to the ultra-elite of American science.

Along with the Nobelists came many occupants of the forty-first chair such as the Hungarians Leo Szilard, Edward Teller, and John von Neumann; the Germans Heinz Fraenkel-Conrat and Curt Stern; the Austrians Erwin Chargaff, Kurt Gödel, Maurice Goldhaber, Victor Weisskopf, and Paul A. Weiss; and the Poles Richard Courant, Samuel Eilenberg, Stanislaw Ulam, and Mark Kac.

Contributions by emigré scientists to the war effort and particularly to the development of the atom bomb have by now become the conventional measure of their first impact on American science. But, as we shall see in chapter 4, their influence was more farreaching. Many made their mark not only by their own scientific work but also by training apprentices who would in turn make major scientific contributions. Thus, confining ourselves to emigré Nobelists alone (although the pattern also holds for occupants of the forty-first chair), we find that after coming to the United States they served as masters to six future laureates (Bardeen, Chamberlain, Kornberg, Lee, Watson, and Yang) and Bardeen, in his turn, has been master to two laureates, Cooper and Schrieffer. Thus, to gauge the true extent of "Hitler's gift" requires that we take into account not only the scientific work of the emigrés themselves (during the war and afterward) but also their multiplier effect as mentors to new generations of scientists, including a good many Nobelists.

We should pause for a moment to consider how the same events can be viewed from the complementary perspective of what the Nazi hegemony meant for science in Germany. As noted in chapter 2, Germany dominated the Nobel awards up to World War II. By 1933, when the Nazis came to power, the combined total of Nobels awarded to scientists who had done their prize-winning research in Germany came to thirty-one: 30 percent of the 103 prizes awarded since their founding in 1901. After 1934 and up to 1976, only nineteen who worked in Germany won prizes, or 9 percent of the total of 210 for the period. Part of this decline involves the drastic reduction in the number of Jewish laureates from nine to just two. Not even these two, Max Born and Otto Stern, chose to ride out the storm in Germany. Born settled in Edinburgh in 1936, ten years after publishing his statistical interpretation of the wave function. Stern accepted a chair at Carnegie Institute of Technology and emigrated in 1933, having already developed the molecular beam method and measured the magnetic moment of the proton. While officially credited to Great Britain and the United States, respectively, both should be counted as Germans since their research was done in Germany.

More telling perhaps than the virtual elimination of Jewish laureates from Germany after 1933 is the fact that the number of Gentile laureates also declined by 20 percent. The Nazi effect on German science cannot be attributed exclusively to the persecution of Jews.[17] The Nazis' dismantling of much of the scientific establishment and the impoverished conditions prevailing after the war help to account for the decline in the numbers of German laureates.

In reckoning the extent of the Nazi effect, we cannot indulge in conjectural history and suppose that the young Hitler emigrés who left Germany and later did prize-winning research would have done work of the same significance had they stayed. Indeed, as more than one said in the course of my interviews with them, having been forced to leave Germany turned out to be the best thing that could have happened to them. The United States provided an active and hospitable climate for their work, and for many ample resources as well. But if emigration was highly beneficial for some of the scientists individually, it was scarcely the best thing that could have happened to German science, and its effect on world science is still not clear.

## Selective Processes

Thanks to a comprehensive study of the religious origins of 60,000 American academics by Lipset and Ladd (1971), we can assemble a series of comparative figures that, by way of context, help to clarify if not to explain some of the selective processes resulting in the heavy representation of Jewish scientists among the American ultra-elite. (Lipset and Ladd also kindly provided unpublished data for this analysis.) As we shall now see, each phase of the selective process involves an increasing proportion of Jewish scientists appearing in given sectors of the scientific community and thus a decreasing *dis*proportion of Jewish laureates to their numbers in that sector.

A first selective process leading to the marked overrepresentation of Jews among the laureates apparently stems first from their tradition, which, as Veblen among many others has noted, sets great value upon the higher learning. Thus, a distinctively large proportion of Jewish youth go on to college—in the early 1970s, for example, 80 percent of American Jews of college age were in college as compared with 40 percent for the college-age population as a whole (Lipset and Ladd, 1971, p. 99). A large and increasing proportion have managed to enter the professoriate in spite of discriminatory barriers lowered only recently.

---

17. Weyl and Possony (1963, p. 144) report slightly different figures on the number of German-born Gentile Nobelists named before 1933 and afterward. Aside from several errors in their counts, they do not take into account the emigration of Gentile laureates after the Nazis came to power.

Considering only the academics in 1968 who came from the three principal religious backgrounds, 9.3 percent were Jewish, 70.7 percent were Protestant, and 20 percent Catholic [18] (recomputed from Lipset and Ladd, 1971, p. 92, table 1; see similar findings reported by Hardy, 1974). Thus, while the proportion of Jewish laureates is nine times the proportion of Jews in the general population, this early phase in the process of self-selection and social selection results in the Jewish laureates being "only" three times the proportion of Jewish faculty members.

But Nobel prize-winners, both Christian and Jewish, tend to come from the elite colleges and universities, first as students and then as faculty members. Once again the educational selective process brings a higher proportion of Jews into those academic populations from which the scientific elite tend to be recruited and placed. Restricting the analysis to the three major religious groups, it turns out that in the elite universities and colleges, Jewish faculty have more than doubled their share, from 9.3 percent among the American professoriate in general to 20.9 percent, while Protestants have not quite held their own in these elite institutions (64.3 percent vs. 70.7 percent) and the Catholic percentage has decreased from 20 percent to 14.8 percent (recomputed from Lipset and Ladd, 1971, p. 93, table 2; see also Steinberg, 1974, pp. 104, 107). In terms of this aspect of the academic selective process, the disproportion of Jewish Nobel laureates is reduced from three times that in the professoriate generally to less than 1.3 times that in the elite professoriate.

Obviously, if we are to trace the selective process involved in producing the religious composition of the American ultra-elite in science, we should consider the professoriate not in all fields but only in the physical and biological sciences, in which (along with the recent addition of economics) the Nobel prizes are awarded. In these broad fields, Jewish faculty comprise 9.6 percent of the total, thus returning the proportion of Jewish laureates to a multiple of almost three times their share of the professoriate in the sciences under the Nobel umbrella.

As Lipset and Ladd observe in another connection, such aggregate numbers mask significant religious differences within each broad domain of science. And some of these differences help us to understand the strong showing of Jewish scientists among the laureates. In the biological sciences, for example, the proportions of Jewish faculty are

18. In drawing extensively upon the valuable Lipset and Ladd study (1971) for context, I have recomputed their figures on religious backgrounds of American faculty members, omitting the category of "Other & None" to provide for closer comparison with my data on religious origins of the laureates. The new computations do not materially change the observed patterns.

comparatively small in the fields of botany (4.1 percent), zoology (3.8 percent), and agriculture (a negligible .8 percent). These happen to be the more traditional fields of biology which, it is often said, have recently developed at a slower pace than others. Equally to the point, they are fields that have been accorded few if any Nobel prizes. In strong contrast, Jewish faculty are most heavily represented in the dramatically advancing fields of biology, making up 15.6 percent of the faculty in bacteriology (taken by Lipset and Ladd to include molecular biology, virology, and microbiology) and as much as 22.8 percent in biochemistry. And, as we have seen, these are the fields in biology that have been singled out for recognition by the Nobel prize committees.

Bacteriology and biochemistry also happen to be related to medicine, a profession favored by Jewish tradition. As of the late 1960s, Jews comprised 23.8 percent of the medical faculty coming from the three major religious backgrounds in the United States. As Lipset and Ladd comment: "It is impossible to tell from the data how much of the attraction of these 'health'-linked fields has been a substitute for fulfilling the Jewish dream of becoming a 'doctor.' Probably, many Jews who were unable to attend medical school picked such subjects as a 'second choice' " (1971, p. 94). But this is clearly not the case for the Nobel laureates. Only one, Julius Axelrod, can be described as a physician *manqué* in the fairly precise sense of one who turned to science only after he had tried and failed to get to medical school (City University of New York, *Graduate Newsletter,* 7 April 1971, p. 1). As Axelrod himself has observed:

> The problem was getting into a medical school in 1933. At that time, I think about 90 percent of the City College graduates were Jewish and at that time, most medical schools, in fact all, had a sort of unwritten quota system that they will accept only a certain percentage of Jews in their class and this meant that you had to be extraordinarily good or the son of a doctor to get in and I just didn't get in because of that [American Association for the Advancement of Science, 1975, pp. 4–5].

But the common pattern among laureates in the biological sciences runs counter to the one hypothesized for scientists in general: actual or prospective M.D.s transform themselves into scientific investigators as their preferred choice rather than taking the opposite route. Through the year 1972, M.D.s had been acquired by twenty-three of the forty-two laureates in what the Nobel authorities continue to describe as the field of "Physiology or Medicine," although in fact it increasingly encompasses work in varied fields and subfields such as biochemistry, bacteriology, virology, genetics, molecular biology, and biophysics.

That most biologist-laureates were not would-be physicians but rather physicians and others who chose careers in biological research is suggested by another simple statistic involving M.D.s and Ph.D.s alike: the site of prize-winning research was the medical school for eighteen laureates; for seventeen others, it was a department of science elsewhere in a university; the remaining seven did their work in such research organizations as the Rockefeller Institute (later University), the National Institutes of Health, and the Mayo Clinic.

Two prototypal cases serve to bring out ways in which laureates turned from medicine to science. The pattern of a prospective M.D. turned biologist is perhaps best seen in the case of Joshua Lederberg. After spending two years at the College of Physicians and Surgeons of Columbia University, he took a leave of absence in 1946 to do research in genetics with E. L. Tatum at Yale. He never returned to his medical studies. During those years of research, while still in his early twenties, he discovered the mechanism of sexual recombination in bacteria, the discovery that, eleven years later, was to win him a Nobel prize.

Like Lederberg, the biochemist Arthur Kornberg was precocious. He took his bachelor's degree at 19 and his graduate degree (in his case an M.D., not, as in Lederberg's case, a Ph.D.) at 23. Kornberg actually practiced medicine for a year, as an intern. But immediately afterward he turned to scientific research, first, for a decade, at the National Institutes of Health and then at the Washington University School of Medicine, where he did his path-breaking and prize-winning work on DNA replication, and then at the Stanford University Medical School. Unlike Lederberg, Kornberg had for a time considered being a physician but his dissaffection with prevailing practices in medicine and his early interest in science converged to lead him into a scientific career. Evidently, for Lederberg and Kornberg, as for many other laureates, Jewish or otherwise, scientific investigation in biology was no second-best substitute for life as a practicing physician.

In any case, the cognitive and organizational connections between the biological sciences and medicine are enough to warrant comparing their religious composition separately and in combination among the professoriate and the laureates, as is done in the top half of table 3–3. Much the same pattern emerges in both comparisons. The Protestants have a slight deficit of laureates relative to their numbers in the professoriate (as indicated by ratios of 0.78 and 0.93) and the Catholics, greater deficits (0.43 and 0.65), while laureates of Jewish origin appear at 1.49 times their proportion among faculty in medicine and the biological sciences combined and 2.55 times their proportion in biology alone.

The "overrepresentation" of Jews among laureates in physiology or

**Table 3–3. Religious Origins of American Professoriate and Laureates (1901–72) in Fields of Science Eligible for Nobel Prizes [a]**

| BIOLOGICAL SCIENCES (without Medicine) [b] | | | | BIOLOGICAL SCIENCES AND MEDICINE | | |
|---|---|---|---|---|---|---|
| | Professoriate | Laureates | Ratio of Laureates to Professoriate | Professoriate | Laureates | Ratio of Laureates to Professoriate |
| Protestant | 68.2% | 53.6% | 0.78 | 65.4% | 60.9% | 0.93 |
| Catholic | 16.4 | 7.1 | 0.43 | 15.0 | 9.7 | 0.65 |
| Jews | 15.4 | 39.3 | 2.55 | 19.6 | 29.3 | 1.49 |
| Total Number | (2143) | (28) | | (4323) | (41) [c] | |

| CHEMISTRY | | | | PHYSICS | | |
|---|---|---|---|---|---|---|
| | Professoriate | Laureates | Ratio of Laureates to Professoriate | Professoriate | Laureates | Ratio of Laureates to Professoriate |
| Protestant | 74.1% | 84.2% | 1.14 | 68.2% | 58.6% | 0.86 |
| Catholic | 19.2 | 5.3 | 0.28 | 16.1 | — | — |
| Jews | 6.7 | 10.6 | 1.58 | 15.7 | 41.4 | 2.64 |
| Total Number | (1710) | (19) | | (1464) | (29) [d] | |

a. The figures for the American professoriate as of 1968 have been recomputed from the data reported by Lipset and Ladd (1971, p. 95, table 4) on a base of the three principal religious backgrounds.

b. The "biological sciences" in this table refer to fields traditionally covered by the Nobel prizes: bacteriology (in which Lipset and Ladd have included molecular biology, virology, and microbiology), biochemistry, and physiology. The next quadrant of this table combines these fields of biology and medicine.

c. One laureate is omitted from this tabulation because the religious categories are not applicable.

d. Two laureates are omitted from this tabulation because the religious categories are not applicable.

medicine is nevertheless not as great when gauged by the combined base of the biological and medical sciences as it is when gauged at earlier phases in the process of social selection and self-selection. The ratio of Jewish laureates to the proportion of Jewish faculty in the biological fields where a prize has been awarded (1.49:1) is nothing like their ninefold ratio to the proportion of Jews in the unselected population and is even somewhat lower than the threefold ratio to their proportion in the American professoriate generally.

In other words, as we examine the *selective flow* of scientists of differing religious backgrounds into the ultra-elite, the relatively large proportion of Jews is seen to relate to a prior proclivity for certain fields of science, as well as a tendency toward representation in the faculties of elite universities, and these, in turn, relate to a still earlier pattern of seeking higher education in the first place.

The pattern in the biological sciences also holds to a differing extent in the physical sciences, as can be seen from the lower half of table 3–3. As the Lipset and Ladd data show, a larger proportion of academic physicists than of academic chemists are of Jewish origin: 15.7 percent compared with 6.7 percent. The difference may reflect a longstanding inhospitality to Jews in chemistry. As late as the end of World War II, it was possible for the chemist Albert Sprague Coolidge of Harvard to explain to a legislative committee in Massachusetts, "We know perfectly well that names ending in 'berg' or 'stein' have to be skipped by the board of selection for students for scholarships in chemistry." And he explained this practice as stemming from the department's understanding that there were "no jobs for Jews in chemistry" (Lipset and Ladd, 1971, p. 91). In light of this observation, it is only fitting that one of the two Americans of Jewish origin to receive the Nobel Prize in Chemistry should have his monosyllabic name not only *end* in "stein" but also begin with it—and, in the bargain, that he should have taken his undergraduate degree and first graduate year at Harvard.[19]

As table 3–3 also shows, the religious composition of the professoriate in physics is practically the same as it is in those biological sciences

---

19. In view of the propensity of Jews for the biological and medical sciences, it is appropriate in yet another way that William Howard Stein should be one of the two Jewish laureates in chemistry. For, as he reports in an autobiographical fragment (1972, p. 125), after poor performance in his first year of graduate work, "I was almost ready to abandon a career in science when it was suggested to me that I might enjoy biochemistry much more than straight organic chemistry." The suggestion was not a bad one. He transferred to the Department of Biochemistry at the College of Physicians and Surgeons of Columbia University, where he took his degree, and then went to the Rockefeller Institute where, together with Stanford Moore, he did the work on chemical structure related to the biological activity of the enzyme ribonuclease which years later won them a Nobel prize.

where Nobel prizes have been awarded. This near identity of religious distribution in the two domains of science also holds for the Nobel prize-winners. Although there were no American physicist laureates of Catholic origin, the proportions of Protestant and Jewish laureates are practically the same in physics as in the biological sciences (*sans* medicine): seventeen of the twenty-nine physicist laureates and sixteen of the twenty-eight biologist laureates are of Protestant origin, while twelve of the physicists and eleven of the biologists are of Jewish origin. This works out in both domains of science to the Protestants' having slightly fewer laureates than their proportion among the professoriate in those fields might lead us to expect while Jewish laureates constitute about two and a half times their proportion of academics in those sciences.

Since the numbers are quite small, it is worth merely passing notice that in the case of the Jewish laureates in physiology or medicine the prizes have been awarded almost exclusively for work in the basic biological sciences. Only one of the twelve Jewish laureates—Selman Waksman, the discoverer of streptomycin—was honored for work bearing directly on medicine. This contrasts with at least six and perhaps as many as ten of the twenty-five Protestant laureates in physiology or medicine who received a prize for work on more directly medical problems: for example, Max Theiler for "his discoveries concerning yellow fever and how to combat it" and C. B. Huggins for "his discoveries concerning hormonal treatment of prostatic cancer" (Nobelstiftelsen, 1972, pp. 180, 190).

A somewhat similar pattern, again involving numbers too small to be significant, is found in physics. There half of the twelve Jewish recipients have received prizes for theoretical work and the other half for experimental work. In contrast, only three (18 percent) of the seventeen Protestant laureates were awarded prizes for contributions to theoretical physics, the rest being given for experimental physics.[20]

The figures for both physics and the biological sciences give the same general though necessarily weak impression that Jewish laureates tend more than Protestant laureates to work in the theoretical rather than clinical and experimental branches of their fields. This raises the question, requiring still unavailable information, whether these descriptive differences among the ultra-elite correspond to the distribution of

---

20. This is at least consistent with the observation that Albert Einstein personified the equation of great science and theoretical physics for a number of young Jews. The 7-year-old Joshua Lederberg (who later turned to genetics) announced his aspirations along these lines in a classroom essay written in 1932: "I would like to be a scientist of mathematics like Einstein. I would study science and discover a few theories" (private communication).

scientific interests among the rank and file of American scientists from differing religious origins.[21]

## Changing Religious Composition

The foregoing figures and ratios of religious representation are based on the composition of the American professoriate at the time of the Lipset and Ladd study in 1969. But, as these authors found, it is only in recent decades that Jewish academics reached current proportions: "The professorial generation which entered academe in the 1920s is today less than 4 percent Jewish; by the first post-World-War-II generation, however, the Jewish proportion had climbed to 9 percent, at which point it leveled off" (1971, pp. 92–93).

But the great majority of American-reared laureates—sixty out of seventy-one—entered academe long before the end of World War II—at a time, in other words, when the relative numbers of Jewish academics were much smaller than they have since become. When we take the historical change into account, we find that movement of Jews into the ultra-elite of science has proceeded at a higher rate than their movement into academe generally, as table 3–4 shows. For those born before 1920, the percentage of Jews among Nobel prize-winners is 3.3 times that among academics; for those born after 1920, 5.3 times.

Both the comparatively large proportion of Jews among the professoriate and the even larger proportion among Nobelists constitute end-products of the social selection and self-selection which funnels relatively more American Jews into higher education than Gentiles of comparable socioeconomic origins. This has reached the point, as we noted, that class origins now have little effect on college-going among Jews since four-fifths of those who are of college age (compared to 40 percent of Gentiles) are in college. When it comes to the professoriate, there are patterned differences in socioeconomic origins (table 3–5). Protestant and Catholic academics come in somewhat larger proportions than Jewish academics from both ends of the occupational distribution: 39.6 percent versus 33.8 percent from the strata of professionals, managers,

---

21. It is still too early to tell whether the religious origins of recipients of the Alfred Nobel Memorial Prize in Economic Science will follow the same patterns of underrepresentation of Catholics and overrepresentation of Jews that we observe in the physical and biological sciences since only eleven laureates in economics have been named since the prize was first given in 1969. Seven are American in the sense of having done most of their work here and at least four of the seven come from Jewish families. Thus Jews appear among Nobelists in economics 3.5 times as often as among professors of economics in general, a ratio that deserves to be given little weight since, obviously, it can be greatly changed by additions to the list of Nobelists over the next few years.

**Table 3–4.  Percent of Jews Among American-Reared Laureates (1901–72) and American Professoriate**

| | PERCENT OF JEWS | | |
| --- | --- | --- | --- |
| | Laureates | Professoriate [a] | Ratio of Laureates to Professoriate |
| Born before 1920 (age 50+) | 21.7 (60) | 6.5 (13,587) | 3.3 |
| Born 1920 & later (age to 49) | 54.5 (11) | 10.3 (42,396) | 5.3 |
| Ratio of Jews born post–1920 to pre–1920 | 2.5 | 1.6 | — |

a. Recomputed from Lipset and Ladd (1971, p. 92, table 1) on bases of the three major religious backgrounds.

and owners of big businesses and 25.8 percent versus 15.6 percent from blue-collar families. The greatest proportion of the Jewish academics come from families of small businessmen, 40.8 percent of them compared with 15.1 percent of the Christian academics.

What we saw before with respect to the religious composition in the various scientific disciplines we see once again with respect to socioeconomic origins: differences at the level of the professoriate at large become greatly magnified at the level of the ultra-elite. The contrast in the class origins of Christian and Jewish laureates is far greater (and, as a result of the small numbers involved, also far more unstable) than among the professoriate (columns 3 and 4 of table 3–5). Four of every five of the Christian laureates but only about a quarter of the Jewish laureates come from the upper reaches of the occupational distribution. More specifically, not only are smaller numbers of Jewish laureates than of Protestants and Catholics drawn from families headed by professionals—some 16 percent as against 67 percent—but just two of the Jews had fathers who, as physicians, were in science-related occupations compared with nineteen of the laureates from Protestant and Catholic families, whose fathers were physicians, engineers, or teachers of science. At the other extreme, more Jewish than Christian laureates come from blue-collar and lower white-collar families. Like Jewish academics generally, Jewish Nobel laureates in science come in disproportionate numbers from the petite bourgeoisie of small businessmen.

These comparisons help put the social origins of the American Nobelists in a new perspective. Although the ultra-elite in science, like elites in other spheres, came largely from the middle and upper occupational strata, this pattern is accounted for almost entirely by the Protestant and Catholic laureates. Among the laureates of Jewish origin, three-

Table 3–5. Socioeconomic Origins of American Professoriate and of American-Reared Laureates (1901–72), According to Religious Origins

| Socioeconomic Origins: Father's Occupation | PROFESSORIATE [a] | | LAUREATES | | TOTAL | |
|---|---|---|---|---|---|---|
| | Protestant & Catholic | Jews | Protestant & Catholic | Jews | Professoriate [a] | Laureates |
| Professionals | 21.7% | 19.6% | 67.3% | 15.8% | 21.4% | 53.5% |
| Managers and Proprietors of Large Businesses | 17.9 | 14.2 | 13.5 | 10.5 | 17.5 | 12.7 |
| Proprietors of Small Businesses | 15.1 | 40.8 | 7.7 | 36.8 | 18.0 | 15.4 |
| Farmers | 11.5 | 0.6 | 3.8 | 0 | 10.3 | 2.8 |
| Sales, Service, and Clerical Workers | 8.0 | 9.3 | 3.8 | 15.8 | 8.1 | 7.0 |
| Workers: Skilled and Unskilled | 25.8 | 15.6 | 3.8 | 21.1 | 24.7 | 8.5 |
| Total | (46,900) | (5,907) | (52) | (19) | (52,807) | (71) |

a. Recomputed from Lipset and Ladd (1971, p. 107, table 15) on bases of the three major religious backgrounds.

quarters came from the lower reaches of the stratification system, largely from the ranks of small business but also from clerical and blue-collar families.

Such descriptive statistics cannot of course be stretched into generalizations or forecasts. But it has been the historical case that although winning the Nobel prize represented some degree of upward mobility for practically all the laureates, this was especially so for the Jewish laureates who, having started at lower socioeconomic levels, had a longer distance to travel in the system of social stratification. However, compared with the general population and even with the professoriate at large, the American laureates in science came in greatly disproportionate numbers from the higher strata; as a result, many of them had advantaged access to the opportunity structure at the important, perhaps crucial, phase of their careers represented by formal education. We shall have to keep these socioeconomic patterns in mind as we examine the laureates' educational careers.

Turning briefly to geographic origins, we find that American laureates, like most other Americans born at the same time, came from rural areas, towns, and small cities. Just a third of them were big-city products. But within this general pattern, there is the subsidiary one of disproportionate numbers coming from New York City. Twenty-one percent (15 out of 71) were New Yorkers—that is, they were raised and educated there at a time when the city had no more than six percent of the nation's population (Rosenwaike, 1972, p. 188).

More detailed examination of the data show that the overrepresentation of New Yorkers among the laureates is almost exclusively an overrepresentation of Jewish New Yorkers. All but one of the New Yorkers were Jews, and they comprised 74 percent of all Jewish laureates. This apparent excess of Jewish New Yorkers is less surprising than it might seem at first. Jews have always been urban people, and, in the United States, they have continued to be urban; slightly more than 40 percent of American Jews lived in just one city—New York—at the time the laureates were growing up (*American Jewish Yearbook,* v. 42, 1941, p. 225; v. 50, 1949, p. 682). Therefore, on the basis of population distribution alone, about four out of ten Jewish laureates should have come from New York. But, as we have seen, 74 percent were New Yorkers, which is almost twice the expected proportion.

It may be, as one laureate speculated, that living in New York in the 1920s and 1930s fostered an interest in science—among bright and motivated boys at least. The reason for this was the availability of first-class education in science in many public schools and particularly in the city's elite high schools, which had predominantly Jewish student populations. If this is the case, a disproportionate number of scientists and

particularly of Jewish scientists should come from New York. Such
fine-grained data on the geographic and religious origins of scientists are
not readily available. Without comparative information of this kind, it
is not possible to say whether Jewish laureates turn up in about the same
proportion as Jewish scientists from New York generally or whether
there is an excess of New Yorkers among them.

## UNDERGRADUATE EDUCATION FOR THE ELITE

In tracing the educational careers of the laureates, it soon becomes
clear that recruitment into the elite involved their early concentration
within the social and educational structure. Largely drawn from middle-
class families to begin with, they were educated at comparatively few
colleges and universities. Much the same is true for the members of the
more extended scientific elite of the National Academy of Sciences, but
in the early phase of undergraduate education there are stratified differences even among future members of the elite. As we shall see, laureates-to-be had more homogeneous collegiate origins than the wider
elite of Academicians and both of these strata, in turn, were less dis-
persed than holders of doctorates in science and male baccalaureates in
general.[22]

A few comparative figures make the point. A mere ten colleges produced 55 percent of the 71 future laureates educated in the United
States, 33 percent of the academicians, but only 25 percent of the doctorates in science and 14 percent of the male baccalaureates in approximately the same period.[23]

Furthermore, of the seventy-one future laureates, forty-two went to
an Ivy League or other elite college (see Berelson, 1960, p. 280, for the
ranking procedure used).[24] To put this in another way, as is done in table

22. The comparisons are simplified but not much distorted by taking the collegiate
origins of those who earned their doctoral degrees between 1920 and 1939 and their
baccalaureates from the mid-1920s to the mid-1930s as the appropriate comparative
cohorts. Comparisons are limited to men because there were no women among the
American-reared laureates.
23. Data for doctorates were computed from Harmon and Soldz (1963, appendix 5,
pp. 120–36) and for baccalaureates primarily from Knapp and Goodrich (1952, appendix 4, pp. 1325–28). See footnotes to table 3–6 for other sources. Eisner (1973)
reports even more pronounced concentration of baccalaureate origins among Fellows
of the Royal Society of London.
24. There are of course marked differences among the colleges classified as elite. The
undergraduate colleges of distinguished state universities such as California, Illinois, or
Wisconsin were and are far less selective than the undergraduate colleges of the Ivy
League and other private universities such as Cal Tech or M.I.T. These differences
continue to be reflected in contemporary measures of student "quality" at various colleges, such as the one devised by Astin (1965, pp. 57–83).

3–6, five of the Ivy League colleges and three other elite colleges accounted for five times as large a proportion of future laureates as of male
baccalaureates generally, while another seven undergraduate colleges of
top-ranking universities graduated four times as large a proportion of
laureates. Fifteen elite schools accounted for 59 percent of the American
laureates but, at a roughly comparable time, produced only 12 percent of
all male graduates. Clearly, the clumping of future members of the
scientific ultra-elite in elite institutions begins early in the selective educational process.

In part, the concentration in elite schools reflects the relatively advantaged socioeconomic origins of most of the laureates. When the
majority of them were entering college, the social stratification of higher
education was even greater than it is now. Social class greatly affected
the chances of entering any college, let alone the elite ones. High school
graduates whose fathers were professionals, for example, were about
four times as likely to become college graduates as those coming from
the less well endowed strata of craftsmen, semiskilled and unskilled
workers, and farmers (Wolfle, 1954, pp. 158–63). As Dobzhansky has
recently reminded us, such stratified differences reflect a mix of environ-
mental and genetic "conditionings" (rather than full "determinations")
that involve both "aptitude aggregations" and class differentials in access
to higher education (Dobzhansky, 1973, pp. 10–49).

All the laureates went to college, of course, but they nonetheless
exhibit traces of differential access to educational opportunity related to
their socioeconomic origins. For they did not all get to the same kind of
college. Among the seventy-one laureates who did undergraduate work
in the United States,

68 percent of the forty-seven coming from professional, managerial,
        and big-business strata went to the fifteen elite schools,
        compared with

42 percent of the twenty-four laureates from lower socioeconomic
        strata.[25]

These and other figures of the same general kind suggest that, at
this stage of their careers, the social selection and self-selection of the
scientific ultra-elite involved both meritocratic and ascriptive criteria.
The allocative system of meritocracy is one in which the "best talents,"

---

25. The stratified differences in attendance at elite schools become even more marked
when we introduce finer distinctions of socioeconomic origins: 89 percent of the nine
laureates coming from the most affluent managerial and big-business families went to
elite schools compared with 63 percent from the thirty-eight professional families and
42 percent of the twenty-four laureates from small-business, clerical, sales, and blue-
collar families.

**Table 3–6. Baccalaureate Origins of American-Reared Laureates (1901–72) and Male Graduates (1924–34)**

| | Number of Laureates | Percent of Laureates [a] | Cumulative Percent | Percent of Male Graduates [b] | Cumulative Percent |
|---|---|---|---|---|---|
| *Ivy League Colleges* [c] | | | | | |
| Columbia | 7 | 9.9 | | .9 | |
| Harvard | 5 | 7.0 | | 1.1 | |
| Yale | 4 | 5.6 | | 1.1 | |
| Cornell | 1 | 1.4 | | 1.0 | |
| Dartmouth | 1 | 1.4 | | .6 | |
| | (18) | (25.3) | 25.3 | (4.7) | 4.7 |
| *Other Elite Colleges* | | | | | |
| Johns Hopkins | 1 | 1.4 | | .2 | |
| Oberlin | 1 | 1.4 | | .2 | |
| Swarthmore | 1 | 1.4 | | + | |
| | (3) | (4.2) | 29.5 | (.4) | 5.1 |
| *Undergraduate Colleges of Top Universities* [d] | | | | | |
| Berkeley | 4 | 5.6 | | 1.4 | |
| M.I.T. | 4 | 5.6 | | .8 | |
| Cal Tech | 3 | 4.2 | | .1 | |
| Chicago | 3 | 4.2 | | .6 | |
| Illinois | 3 | 4.2 | | 1.7 | |

| | | | | | |
|---|---|---|---|---|---|
| Wisconsin | 3 | 4.2 | | 1.0 | |
| Michigan | 1 | 1.4 | | 1.5 | |
| | (21) | (29.4) | 58.9 | (7.1) | 12.2 |
| | | | | | |
| *Other Colleges* | | | | | |
| C.C.N.Y. | 3 | 4.2 | | 1.2 | |
| Case Institute | 2 | 2.8 | | .2 | |
| Lafayette | 2 | 2.8 | | .3 | |
| 22 Others [e] | 22 | 31.0 | | 7.3 | |
| | (29) | (40.8) | | (9.0) | |
| | | | | | |
| *Total Graduates* | (71) | 100.0 | | (708,000) | 21.2 |
| | | | | 100.0 | |

a. These data cover the period stretching from 1879, when A. A. Michelson received his bachelor's degree from Annapolis, to 1953, when J. R. Schreiffer received his from M.I.T.

b. Data on the number of male graduates of each college are taken from Knapp and Goodrich (1952, appendix 4, pp. 325–28), supplemented by Robertson (1928, pp. 661–62), MacCracken (1932, p. 767), and U.S. Bureau of the Census (1960, p. 211).

c. No laureates were undergraduate alumni of Brown, Pennsylvania, or Princeton, the other members of the Ivy League.

d. Although there are obvious difficulties in ranking undergraduate colleges by the standing of their graduate faculties, that practice is adequate for purposes of rough classification. The Berelson ranking of graduate faculties (1960, p. 280) is based on Keniston's 1957 study that draws upon evaluation by department chairmen in twenty-five universities. It is methodologically less satisfactory than more recent surveys but it is the one most appropriate for these data; it is less anachronistic than recent surveys (Cartter, 1966; Roose and Andersen, 1970) and more systematic than earlier ones (Hughes, 1928; 1934). In spite of the procedural differences, these ratings have remained highly stable, especially within disciplines (Berelson, 1960, p. 98; Keniston, 1958; Cartter, 1966, *passim*).

e. These are California at Los Angeles, Earlham, Florida, Furman, Kenyon, Michigan College of Mining, Michigan State College, Missouri, Montana, Nebraska, New York University, Oregon, Oregon State, Purdue, Rutgers, South Dakota, U.S.N.A., Ursinus, Vanderbilt, Whitman, Washburn, and College of Wooster.

+ Less than .1 percent.

however they are gauged, are given preferred access to opportunities for developing their capabilities (Young, 1959). Meritocracy requires that these talents should be both facilitated in their development and rewarded according to their performance, whatever their origins in terms of race, sex, ethnicity, creed, socioeconomic origin, or any other status. But at this stage the social origins of future laureates did relate somewhat to the quality and standing of undergraduate institutions they attended. Still, the fact that as many as 42 percent of the future laureates from the lower socioeconomic strata graduated from the "best" colleges, more than three times the percentage among male baccalaureates generally, suggests that the allocative processes of meritocracy were also at work.[26]

There are other indications that a mixture of meritocratic and ascriptive processes of selection was operating. During much of the period covered by the data for the laureates, the quotas known as *numeri clausi* had placed severe restrictions upon the admission of Jews to American colleges, particularly the elite colleges (McWilliams, 1948, pp. 38–39). Nevertheless, as table 3–7 shows, comparatively more of the Jewish laureates, end products of a long chain of social selection and self-selection, than of the Gentile laureates got to the Ivy League colleges, although a significantly smaller proportion of them went to all kinds of elite undergraduate institutions taken together. The fact that almost half of the Jewish laureates studied in elite colleges takes on added point when we consider that they more often came from the lower socioeconomic strata.

Thus, whatever their socioeconomic beginnings, future members of the scientific elite were far more likely to go to elite colleges than the run of students in their age cohorts. Although the differential advantages associated with their origins were not completely erased by the time they left college, they were considerably reduced. This process of convergence in the course of mobility into the scientific elite continues in the next phase of their careers.

## GRADUATE EDUCATION FOR THE ELITE

With the growing institutionalization and professionalization of American science in the first quarter of the twentieth century, formal graduate study became a virtual prerequisite for admission into the ranks of scientists. Symbolizing the transition, the physicist A. A. Michel-

---

26. This pattern is consistent with the moderate correlation of 0.25 between socioeconomic origins and prestige of college attended reported by Hargens and Hagstrom for a sample of physical and biological scientists (1967, p. 29).

**Table 3–7.  Baccalaureate Origins of American-Reared Laureates (1901–72), According to Socioeconomic and Religious Origins**

| Baccalaureate Origins | JEWS | | | PROTESTANTS AND CATHOLICS | | |
|---|---|---|---|---|---|---|
| | Professional, Managerial, & Big Business | Small Business & Other | Total | Professional, Managerial, & Big Business | Small Business & Other | Total |
| Ivy League [a] | 80% | 21.4% | 36.8% | 26.2% | — | 21.2% |
| Other Elite Colleges [b] | — | 14.3 | 10.5 | 40.5 | 50.0% | 42.3 |
| Other Colleges | 20.0 | 64.3 | 52.6 | 33.3 | 50.0 | 36.5 |
| Total | 26.3 | 73.7 | (19) | 80.8 | 19.2 | (52) |

a. Harvard, Yale, Dartmouth, Cornell, Columbia (no laureates took undergraduate degrees at Brown, Pennsylvania, Princeton, the other Ivy League colleges).
b. Cal Tech, M.I.T., Berkeley, Chicago, Illinois, Michigan, Wisconsin, Oberlin, Swarthmore, and Johns Hopkins.

son, the first American to be awarded the Nobel prize, never bothered
to take a doctoral degree, although he did spend several years in what
he described as "postgraduate study" at Berlin, Heidelberg, and Paris.
All the other American-reared laureates took doctorates and all but one
of them, the chemist Irving Langmuir, took them in the United States,
as did five foreign-born laureates, giving (as of 1972) a total of seventy-
four with American doctorates (as of 1972 and eighty-five as of 1976).
However, not all of these were exclusively trained in the United States.
About twenty of the seventy-four followed the practice, quite common
among American scientists in the first part of the century, of going
abroad for postdoctoral training.

Graduate education is of course offered by fewer academic institu
tions than undergraduate education. At the time when most laureates
and members of the National Academy of Sciences took their doctorates—from 1920 through the 1940s—graduate training in the sciences
was even more heavily concentrated than it is now. Among the approximately one hundred universities then awarding doctorates, only five accounted for about 28 percent of doctorates in all fields of science, while
the ten largest accounted for about half (Harmon and Soldz, 1963,
pp. 10, 20–26). In the absence of more precise comparative data on
science doctorates, it is necessary to use those for 1920–49.

But greatly concentrated as graduate training then was, the concentration was still more pronounced among the future laureates and
members of the National Academy. All seventy-four laureates (chosen
between 1901 and 1972) with doctoral degrees from American universities were educated at only twenty-one universities.[27] In fact, 55 percent of them had studied at only five: Harvard (16.2%), Columbia
(14.9%), Berkeley, the Johns Hopkins, and Princeton (with 8.1%
apiece). Three of the same five—Harvard (17.9%), Columbia (9.6%),
and the Johns Hopkins (9.1%)—together with Chicago (8.2%) and
Yale (6.8%)—also trained more than half of the extended scientific
elite represented by the members of the National Academy.[28] But, as I
have noted, the five most prolific universities had accounted for only 28
percent of science doctorates generally, and these universities, just as we
found to be the case for baccalaureates, overlapped but did not coincide
with the schools producing the largest share of the scientific elite.

---

27. Moulin also found that future laureates did their graduate work in relatively few
great European universities—Cambridge, Munich, Göttingen, and Zurich foremost
among them (1955, p. 261). Other comparable data are provided by Gray (1949),
Fleming (1966), Cantacuzene (1969), and Silcock (1967).
28. Much the same concentration holds for the membership of the National Academy.
As of 1969, five universities—Harvard, California, Chicago, Columbia, and M.I.T.—
trained 48 percent of those who took doctoral degrees in the United States (computed
from Kash, White, Reuss, and Leo, 1972, p. 1077, table 2).

Duplicating the pattern found for undergraduate education, the doctoral work of the scientific elite was concentrated not only in a few places but particularly in the elite places. As can be seen in detail in table 3–8, the thirteen elite universities granted degrees to the following:

85 percent of the laureates

80 percent of the members of the National Academy

55 percent of the other scientists receiving their degrees at approximately the same time.[29]

By this relatively early point in their careers, the laureates and Academicians, as future members of the scientific elite, had converged into a pattern of concentrated affiliation with elite universities. Not only had almost equal proportions of them done their graduate work at the more distinguished universities but they began their professional careers by acquiring advanced degrees at the same comparatively early ages.

Although strictly comparable data are not available for the general population of scientists of the same ages, rough comparisons suggest that the future members of the elite get their doctorates appreciably earlier than the run of scientists, with the laureates being slightly more precocious than Academicians in the same fields of science. Laureates educated in American universities were a median 24.8 years old when they received their degrees as compared with a median of 26 years for members of the National Academy of Sciences and of 29.5 years for the run of doctorates in science in 1957 (see Harmon and Soldz, 1963, pp. 44, 51, for the data on doctorates). These sizable age differences between future members of the elite and the rank and file reflect the fact that the laureates started their graduate work earlier and also took less time to complete it. A quarter of them had begun their graduate studies

---

29. The Berelson ranking of graduate faculties is used here to distinguish elite universities from others offering graduate programs. It is, as I have observed, chronologically better suited to classifying data on laureates than the latter and more systematic appraisals sponsored by the American Council on Education (Cartter, 1966; Roose and Andersen, 1970). To Berelson's "Top Twelve" (California at Berkeley, California Institute of Technology, Chicago, Columbia, Cornell, Harvard, Illinois, Massachusetts Institute·of Technology, Michigan, Princeton, Wisconsin, and Yale), I have added the Johns Hopkins University, quite clearly one of the "elite" at the time its laureate alumni were taking their doctoral degrees and embarking upon their scientific careers. In Hughes's 1925 survey of graduate departments, fifteen of the seventeen Johns Hopkins departments that were ranked were among the "best" ten in their respective fields and five among the "best" five (1928, pp. 161–63), placing that university comfortably among the top ten in overall rankings. However, this early excellence was not maintained, so that by 1957 only six departments remained among the "best" ten and none among the best five. Similarly, Stanford University would now figure among the elite American universities, as both the Cartter (1966) and the Roose and Andersen (1970) evaluation studies testify, but this was not so for the 1920s to the 1940s, the period under examination here.

**Table 3–8.  Doctoral Origins of Laureates (1901–72), Members of National Academy of Sciences, and Other Scientists (1920–49)**

| Doctoral Origins | Laureates | Cumulative Percent | NAS Members [a] | Cumulative Percent | Science Doctorates [b] | Cumulative Percent |
|---|---|---|---|---|---|---|
| *Harvard | 16.2% | | 17.6% | | 3.6% | |
| *Columbia | 14.9 | | 9.2 | | 4.3 | |
| *Berkeley | 8.1 | | 6.6 | | 4.5 | |
| *Johns Hopkins | 8.1 | | 9.6 | | 3.8 | |
| *Princeton | 8.1 | 55.4 | 3.8 | 46.8 | 1.9 | 18.1 |
| *Chicago | 6.8 | | 8.6 | | 5.7 | |
| *Cal Tech | 5.4 | | 3.9 | | 1.9 | |
| *Illinois | 4.0 | | 1.9 | | 5.2 | |
| *M.I.T. | 4.0 | | 3.7 | | 3.6 | |
| *Yale | 4.0 | 79.6 | 6.5 | 71.4 | 3.5 | 41.7 |
| *Cornell | 2.7 | | 3.6 | | 6.3 | |
| Minnesota | 2.7 | | 1.9 | | 3.6 | |
| Rochester | 2.7 | | .3 | | .7 | |
| *Wisconsin | 2.7 | | 4.8 | | 6.4 | |
| George Washington | 1.3 | 92.1 | .2 | 82.2 | .3 | 59.3 |
| Indiana | 1.3 | | .9 | | .8 | |
| *Michigan | 1.3 | | 2.1 | | 4.1 | |
| Michigan State | 1.3 | | — | | .6 | |
| Pennsylvania | 1.3 | | 1.9 | | 1.6 | |
| Pittsburgh | 1.3 | | .3 | | 1.2 | |
| Washington U. (Mo.) | 1.3 | 100.0 | .4 | 87.8 | .6 | 68.1 |
| Other Universities | — | | 11.4 | | 31.9 | |
| | (74) | | (832) | | (33,960) | |

a. These are for all members of the National Academy of Sciences elected, except for Nobel laureates, from 1900 through 1967.
b. Harmon and Soldz (1963, pp. 10–11, 20–26, and appendix 2). These data do not include M.D.s granted by these universities. The five largest producers of science doctorates are Wisconsin, Cornell, Illinois, Berkeley, and Chicago.
* Elite universities.

by age 20, and well over half by age 21. At the extreme of precocity was the recipient of the chemistry prize in 1914, T. W. Richards: he was awarded a B.S. degree from Haverford College at the tender age of 17, an A.B. from Harvard a year later, and two years after that his Ph.D. Others who began early and finished even sooner than the average of laureates were the physicist Julian Schwinger (baccalaureate at 18, doctorate at 21), the chemist Robert B. Woodward (at 19 and 21, respectively), the molecular biologist James D. Watson and the physicist Murray Gell-Mann (both of whom got bachelors at 19 and doctorates at 22). In Woodward's case, this includes a year's obligatory absence from M.I.T. because the Institute decided at the end of his sophomore year in college that he should be "excluded for inattention to formal studies" (Nobelstiftelsen, 1965, p. 121).

Although the pattern of an early start was typical for the laureates, there were a few cases of delayed beginnings eventuating in focused and accelerated development in science only after a period of uncertainty. John F. Enders, who in 1954 shared the prize in medicine for discovering a simple method of growing polio virus in test tubes, was one of these late bloomers. He was graduated from Yale at age 23, having had his studies interrupted by World War I. It was not all that evident to him that he wanted to go into science. He tried the real estate business, found it boring, and went on to Harvard, where he spent four years studying English literature, German and Celtic. The prospect of spending his life as a teacher of English apparently became less and less inviting. At this time he came under the influence of the bacteriologist Hans Zinsser, an occupant of the forty-first chair who had come from Columbia to Harvard not long before. Zinsser urged Enders to go on to a doctorate in bacteriology and immunology. Enders did so in short order, but by then he was 33, well beyond the average age of the future laureates at the time of receiving the doctorate. That Enders was free to change his career plans so often and thus to become a late bloomer (at least among scientists) was probably facilitated by his family's relative prosperity (his father was a banker).

Another case of apparent late blooming and late recognition among the laureates reflects other social and economic constraints upon the scientific career. The biochemist and pharmacologist Julius Axelrod has been described as having immigrant parents "of modest means" (that is, they were poor). Axelrod nevertheless made his way through the tuition-free College of the City of New York, where he received the B.S. in 1933. Rejected from medical school, as we have seen, he worked at various menial jobs and managed to get a Master's degree at New York University, eight years after his baccalaureate. This helped him get a job as a technician at the Laboratory of Industrial Hygiene, where he had the good luck to encounter the pharmacologist Bernard B. Brodie, who would become a distinguished scientist. Brodie was then "developing his new concepts of drug metabolism which later revolutionized modern pharmacology. These influences rubbed off on Axelrod, who, by the early 1950s, had himself become an authority in drug metabolism" (Udenfriend, 1970, p. 422). These various vicissitudes delayed Axelrod's receiving the doctorate, from George Washington University, until he was 43. In retrospect, he felt that his work had been hampered by his not having the academic credentials and the access to facilities that sometimes went with them. "Although in some cases the Ph.D. has become a union card, it is important. I feel I wasted about 10 years: if I had possessed a Ph.D., I would have progressed much faster" (City University, *Graduate Newsletter,* 1971, p. 2).

Axelrod was a late bloomer only in the sociological sense of institutionalized recognition and career, not in the cognitive sense of scientific accomplishment. Long before his belated doctoral degree, he and Brodie worked on the metabolism of commonly available analgesics, specifically on why methemoglobin anemia developed when large amounts were taken, that is, why the drug caused people to turn blue. This research, reported in 1948, is still cited and is said to be the beginning of the first really systematic evaluation of drugs.

Enders and Axelrod were alone among the laureates to delay taking their doctorates until after the age of 30. As we might expect, a group as rigorously selected as the laureates exhibits little variability in such matters as the ages at which they finish their college work or their advanced studies. Their average age for the baccalaureate is 21.4, with a standard deviation of only 1.8 years; for the doctorate, the mean age is 25.8, with a standard deviation of 3.0.[30] Nor does this vary much by socioeconomic stratum: the thirty-eight laureates from professional and managerial families obtained their doctorates at an average age of 25.2; the twenty-four laureates from small-business, clerical, and blue-collar families, at age 26.4.

Thus, the deviant cases of Enders and Axelrod only emphasize by contrast the pattern prevailing for the scientific ultra-elite: whether owing to precocity or privilege or both, they get started earlier than the garden variety of scientists on their careers as scientific investigators, and they get their start in the more distinguished universities.

In these respects as in others, the selective process operating for this exceptional aggregate of scientists attenuates earlier differences of advantage among them as they head for their common destination in the

---

30. The difference of a year between the laureates' mean and median ages at the doctorate reflects of course the effect of extremes (here at the upper end) on calculations of means.

ultra-elite. From the following sequence of patterns, it appears that the differentials are soon narrowed, then erased.

First, class origins do count a bit when it comes to attending colleges of different standing. As will be remembered, the laureates from families headed by professionals or managers were somewhat more likely than the others to go to elite colleges (for example, 32 percent of the one compared with 19 percent of the other getting to Ivy League colleges, with a total of 68 percent versus 42 percent, respectively, going to the larger cluster of elite colleges.)

It is also true that, for the laureates as for students in general, going to one of the elite colleges paves the way for going to an elite graduate school. Thus,

100 percent of the seventeen laureates from Ivy League colleges taking advanced degrees in the United States took their doctorates in one of the thirteen top-ranked universities,[31] compared with

92 percent of the twenty-four who came from such other elite schools as Oberlin, Swarthmore, Chicago, M.I.T., and Cal Tech and

71 percent of the twenty-eight coming from colleges of less prestige such as Case, Furman, Lafayette, Ursinus, and College of Wooster.

Although the great majority of the laureates went on to the major universities for their graduate work, there remains a perceptible difference even within this destined ultra-elite that relates to the standing of the colleges they attended.

In light of the foregoing patterns, which show successive differentials related to social and collegiate origins, it may at first seem paradoxical that those coming from the upper-middle occupational strata were no more likely than those from the lower strata to get to the elite graduate schools (87 percent compared with 83 percent), as table 3–9 shows. This indicates that certain differentials of advantage associated with differences in class origins have somehow been not merely narrowed but practically erased by the time future laureates obtained the doctorate.

The seeming paradox is removed by examining the operation of the selective process in consecutive phases. Among the future laureates, the standing of the college they attended somewhat affects where they went

---

31. All five of the foreign-born laureates who did their graduate work in the United States after having completed their undergraduate work at home—Konrad Bloch, C. B. Huggins, Lars Onsager, C. N. Yang, and T. D. Lee—also did so in one of the top-ranked universities.

**Table 3–9. Socioeconomic and Baccalaureate Origins of American-Reared Laureates (1901–72) Holding Elite Doctorates** [a]

| Socioeconomic Origins [b] | PERCENT HOLDING ELITE DOCTORATES BACCALAUREATE ORIGINS | | |
|---|---|---|---|
| | Elite College [c] | Other U.S. College | Total |
| Upper Strata | 94% | 73% | 87% |
| | (31) | (15) | (46) |
| Lower Strata | 100 | 69 | 83 |
| | (10) | (13) | (23) |
| Total | 95 | 71 | 86 |
| | (41) | (28) | (69) |

a. See Chapter 3, footnote 29, for definition.
b. Upper strata are those having fathers who were professionals, managers, or owners of big businesses. Lower strata are those whose fathers were owners of small businesses, in sales or service occupations, farmers, and blue-collar workers.
c. See table 3–6 for definition.

to graduate school, with 95 percent of those from elite colleges finding their way to elite graduate schools as against 71 percent of those who went to other colleges. Where they went to college therefore counted more than their social origins in determining their chances for graduate work at a distinguished university. To put the same data in another way, when it came to attending a top-flight graduate school, the chances of overcoming the "handicap" of having gone to a college outside the elite circle were just as good for future laureates from the lower strata as for those from the upper strata (69 percent compared with 73 percent). And, since practically every one of the future laureates who went to an elite college also went to an elite graduate school (94 percent and 100 percent), the effects of social origins seem to have been eliminated. In short, by the time they had taken their doctorates, the future members of the ultra-elite who started on apparently less secure footing as far as socialization and education go had caught up with their apparently more fortunate peers.[32]

It is useful, however, to remember that these observations on future Nobel laureates tell us nothing about the impact of social origins on the

32. Hargens and Hagstrom, it will be remembered, report a correlation of 0.25 between family socioeconomic status and the prestige of undergraduate institution among a sample of physical and biological scientists. They also report a correlation of 0.13 between family status and prestige of doctoral institution (1967, p. 29). Roughly comparable figures for the laureates are 0.50 and 0.17, suggesting that their socioeconomic origins had an even stronger impact on where they went to college. When it came time for graduate study, that initial relation became attenuated. It appears that for both groups socioeconomic origins had less to do with recruitment to elite graduate schools than to elite colleges.

careers of other scientists. Crane's (1969) study of professors in all fields shows that social origins are directly related to the prestige of both the universities that trained them and those that later employed them.[33] Thus, there is reason to think that family affluence may be more strongly associated with access to first-class education and its associated benefits in the job market for the rank and file than for those judged to be highly promising scientists and scholars.

At this stage, then, future members of the scientific elite had moved toward homogeneity among themselves and differentiation from other scientists. The laureates-to-be and the future members of the National Academy of Sciences had converged upon essentially the same major universities for their graduate study, to a degree that differentiated them from other future scientists, who had gone to the same set of universities appreciably less often. These patterns of convergence and divergence, amounting to unplanned tracking in higher education, are consistent with our formulation of the early phases in the process of accumulative advantage. The composite of self-selection and social selection operates to enlarge opportunities for further effective work and role performance which, in turn, open up further opportunities. Although processes of accumulation of advantage do not operate for every future laureate at every phase, they do so in the aggregate, as chapter 4, on the scientific apprenticeships of future Nobelists, suggests.

---

33. Crane reports that the class origins of professors are related to academic success (defined as holding a position in a top-ranking university). She attributes this to the connection between class origin and the prestige of the university at which the doctorate was taken and the connection between the prestige of the doctorate-granting university and that of the later employing university (Crane, 1969, pp. 1–17. This interpretation of the data has been challenged by Gaston, Wolinsky and Bohleber (1976) who claim that class effects are smaller than Crane reports.

# Chapter 4

# MASTERS AND APPRENTICES IN SCIENCE

Nobel laureates in science constitute a functional, not a hereditary, elite, yet a sizable number of the 313 prize-winners named between 1901 and 1976 have in fact been related, a few through blood or marriage and a much larger number through ties that link masters and apprentices. Thus there is a good deal of inbreeding in the scientific ultra-elite, but it is primarily of the social rather than the biological variety.

The social ties between scientific masters and apprentices are, as we shall see, enduring and consequential, for it is in the course of apprenticeships that young scientists learn the scientific role. Socialization is plainly not confined to classrooms and lecture halls. But first we turn to the laureates' kinship ties as a special case of scientific inbreeding.

## KINSHIP TIES

By 1975 there were five sets of laureate parents and offspring on the Nobel roster. That year, Åage Bohr won the Nobel Prize in Physics; his father, Niels, had done so fifty-three years earlier, the year Åage was born. In addition to the Bohrs, there are three other father-son pairs. The first father and son to win Nobels were English physicists: J. J. Thomson, who received his prize in 1906 for work on the conduction of electricity through gases, and G. P. Thomson, who received his in 1937 for work on the diffraction of electrons by crystals. The next pair, also English physicists, are the only father-and-son team ever to have *shared* a Nobel prize: W. H. Bragg and W. L. Bragg, having been jointly recognized in 1915 for their basic studies of crystal structure by means of X-rays. The third pair emerged in 1970 with an award to the Swedish physiologist Ulf von Euler for his work on the chemistry of transmission of nerve impulses. His German-born father, Hans von Euler-Chelpin, received the prize in 1929 for his research on the chemistry of fermentative enzymes. Von Euler, it turns out, comes by his Nobel inheritance in two ways: not only as the son of a Nobelist father but also, by the principle of socioheredity, as the godson of Svante Arrhenius, laureate in chemistry for 1903.

Another set of laureate parents and offspring, unique in the annals of the Nobel award, consists of a quartet of laureates related by blood and marriage. In 1903 the Polish-born Marie (Sklodowska) Curie and her husband, Pierre Curie, shared (along with Henri Becquerel) the prize in physics for their discovery of radioactivity. Eight years later the indefatigable Madame Curie was again awarded the prize, this time alone (her husband having died in 1906) in Chemistry, for their earlier joint discovery of radium and polonium. A generation later, in 1935, their daughter Irène Joliot-Curie and her husband Frédéric Joliot shared the prize in chemistry for synthesizing new radioactive elements. Thus the four Curies—*mère, père, fille, et gendre*—accumulated a total of five Nobel prizes.

In 1973 Nobelist family ties were further augmented by the first pair of sibling prize-winners when Nikolaas Tinbergen won the prize in medicine, his brother Jan having done so four years earlier in economics.[1]

Again, as a functional rather than a hereditary elite, the Nobelists do not form an endogamous caste. There is, nevertheless, a fair amount of intermarriage between laureates and the kin of laureates,[2] influenced

---

1. Were the prizes a half century older, it would surely have been possible to identify still another genealogical formation among Nobelists. Justus von Liebig, who might be described as the father of both organic and agricultural chemistry—and thus an undisputed occupant of the forty-first chair—also happens to have been the great-grandfather of Max Delbrück, laureate in physiology and himself a father of molecular biology. Such family constellations are not of course unique to Nobel laureates. The galaxy of Bernoullis, seventeenth- and eighteenth-century Swiss mathematicians, astronomers, and naturalists, provide a conspicuous example of multigeneration scientific excellence, as do the nineteenth- and twentieth-century Darwins and Huxleys.
2. In his masterful examination of the social origins and family connections of the English intellectual aristocracy, Noel Annan (1955) shows the endogamous linkages

by propinquity of locale and scientific discipline rather than resulting from normative prescription or from the intention of perpetuating the "family business" by marrying the boss's daughter.

For the handful of female laureates, one can report the dramatic statistic that 60 percent have had laureate husbands, while fewer than 2 percent of all male laureates have had laureate wives. Less dramatically, it turns out that three of the five women who have won Nobel prizes were married to Nobelists: [3] Marie and Irène Curie and the biochemist Gerty Cori who, together with Carl F. Cori, received a prize in 1947 for studies of catalytic conversion of animal starch into sugar.

The physicist Maria Goeppert Mayer periodically collaborated with her husband, the physical chemist Joseph Mayer, though it was not for this work that she received a prize in 1963; rather, she won it for her research on the "shell model" of the atomic nucleus. The fifth and most recent female laureate, Dorothy Crowfoot Hodgkin, won her prize for painstaking "determinations by X-ray technique of the structures of important biochemical substances." Like the other female laureates, Hodgkin is the wife of a scientist (Thomas Hodgkin is an anthropologist) and was married when she got her award.

The elaborate networks of social interaction that develop among members of the scientific elite often lead to their families' getting to know one another. This contributes its share to intermarriage. There is no firm count of the marriages linking up members of this elite, but there have been a good many. Endogamy in this elite becomes especially marked when it extends beyond the small number of laureates to their peers, the occupants of the forty-first chair.

A few cases will illustrate the variety of family ties among these outstanding scientists. Sometimes laureates have married laureates' daughters. Thus, the German biochemist Feodor Lynen, who received his prize in 1964, is the son-in-law of the German chemist Heinrich Wieland, who received his in 1927; the American microbiologist Frederick C. Robbins, who shared the prize in 1954 with John Enders and

---

binding members of this more inclusive functional elite. Not surprisingly, as many as eight British laureates are tied into this network of Oxford and Cambridge dons by birth and marriage: Lord Adrian, Dorothy Crowfoot Hodgkin, Alan Hodgkin, A. F. Huxley, A. V. Hill, J. J. Thomson, G. P. Thomson, and R. M. Synge (who also happens to count the Irish playwright among his relatives.) And, as if they were following Annan's directions, G. P. Thomson's son, David, married W. L. Bragg's daughter, Patience, thus uniting the offspring of two generations of laureates.

3. The small number of women laureates partly reflects the small number of women at work, during the greater part of the century, in the three fields of science covered by prizes. The underrepresentation is most marked in physics, less so in chemistry, and least in those biological sciences designated in the Nobel terminology as "physiology or medicine" (Zuckerman and Cole, 1975).

Thomas Weller for growing polio virus in a variety of tissues *in vitro,* is the son-in-law of the American chemist John Northrop, who shared the prize in 1946 for preparing enzymes and virus proteins in pure form; and, in a gesture of Anglo-American amity, the British physiologist Alan Hodgkin, Nobelist in medicine, married the daughter of the American laureate Peyton Rous, who did not get his Nobel until 1966, three years after his son-in-law. The son-in-law of the laureate Maria Mayer is the astrophysicist Donat Wentzel, himself the son of the distinguished physicist Gregor Wentzel (whose work on the meson field theory overlapped the scientific interests of his co-parent-in-law).

The ties of kinship ramify through the siblings as well as the children of the ultra-elite. Take only one serial example: Paul A. M. Dirac, an English laureate in physics (1933), married the sister of Eugene Wigner, a Hungarian-born American laureate in physics (1963) who, in turn, married the sister of his Princeton colleague John A. Wheeler, himself a frequent collaborator of laureates—Niels Bohr among them— and a widely esteemed occupant of the forty-first chair.

It is not, however, the ties of kinship—through either blood or marriage—that most significantly involve the filiation of scientists in the ultra-elite. Instead, that filiation stems mainly from the social linkages between masters and apprentices (terms that have long been used by scientists to cover various role relationships, including senior-and-junior collaborators as well as teacher-and-student).[4]

## SOCIAL TIES: MASTERS AND APPRENTICES

To begin with, it is striking that more than half (forty-eight) of the ninety-two laureates who did their prize-winning research in the United States by 1972 had worked either as students, postdoctorates, or junior

---

4. Having been trained in a department with a Nobel laureate on its staff or even sitting in his classroom is not sufficient qualification for being called an apprentice of a laureate master. J. D. Watson is not counted as a student of Herman Muller nor is Ernest Lawrence counted as a student of Arthur Compton even though each might, in a loose sense, be considered as such. Were cases of this kind included in our enumeration, the number of laureates having laureate masters could have been greatly expanded.

The variation in the role relations of master and apprentice is indicated by the laureates' differentiated descriptions of their associations with several masters. Thus Emilio Segrè counts himself as Enrico Fermi's apprentice even though he spent time in Otto Stern's laboratory in Hamburg and Zeeman's laboratory in Amsterdam. Similarly, Hans Bethe declares himself a student of Arnold Sommerfeld even though he also worked in Rome with Fermi and at the Cavendish with Rutherford.

collaborators under older Nobel laureates.[5] The details of that genealogy [6] are reported in figures 4–1 and 4–2, which are arranged to have each pair of scientists appear in the decade in which they worked together.

What is more, these forty-eight future laureates worked under a total of seventy-one laureate masters. As can be pieced together from figures 4–1 and 4–2, eight of the eventual laureates—Alvarez, Bethe, Bardeen, Delbrück, Chamberlain, Rabi, Khorana, and Watson—had two laureate masters each, six more—Segrè, Davisson, Gasser, Kornberg, Pauling, and Wald—actually had three each, and Felix Bloch, a grand total of four. Thus, fifteen future laureates had the benefit of working with no fewer than thirty-eight laureate masters.

Having seen how many American laureates had served as apprentices to one or more laureates, we must now ask how many laureates helped to produce more than one American Nobelist. Which of them have been most prolific in reproducing their own kind through the processes of social selection and training? A gross measure can be derived from figures 4–1 and 4–2: ten laureate masters have helped to produce the hefty total of thirty American laureates. Enrico Fermi, producer of artificial radioactivity, stands alone with six American laureates to his credit. Ernest Lawrence and Niels Bohr each produced four; Nernst and Meyerhof, three; and five more seniors—Rabi, Schrödinger, Debye, Dale, and Enders—each helped to train two laureates. But their prolificity pales compared to the collective record of the Cavendish professors, J. J. Thomson and Ernest Rutherford, who between them trained seventeen future laureates of all nationalities. As figure 4–3 shows, their apprentices were themselves not exactly unproductive of still more laureates.

---

5. Although the pattern is marked in all three scientific fields covered by Nobel prizes, there are descriptive, if not analytically significant, differences in extent, as can be seen from these figures, which enumerate John Bardeen just once even though he has won two Nobel prizes.

|  | American Laureates with Laureate Masters |
|---|---|
| Physics | 61.3% (31) |
| Chemistry | 57.9   (19) |
| Physiology-Medicine | 42.9   (42) |

The figure for American laureates is not far off the mark for all Nobelists of every nationality: 41 percent of the 286 named between 1901 and 1972 have had at least one laureate master or senior collaborator. Even though the rate of laureates who have had laureate antecedents has remained at about 48 percent since 1925, it may decline as the science population and, more important, the number of occupants of the forty-first chair increase. More often than in the past, laureates in the future will have been trained by occupants of the forty-first chair and will in their turn train more scientists who qualify for Nobels but may never receive them.

6. For similar charts of filiation in organic chemistry and biology, see Pledge (1939, pp. 29, 106, 200); in psychology, Boring and Boring (1948), Wispé (1965), Ben-David and Collins (1966).

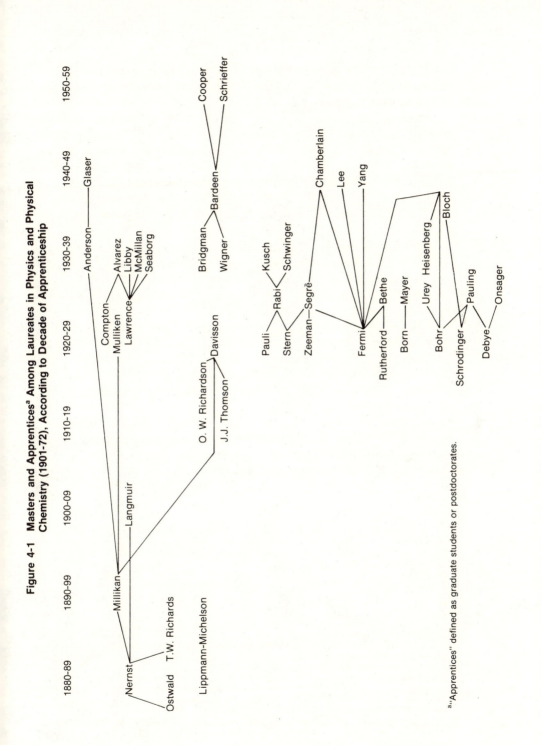

**Figure 4-1** **Masters and Apprentices[a] Among Laureates in Physics and Physical Chemistry (1901-72), According to Decade of Apprenticeship**

[a]"Apprentices" defined as graduate students or postdoctorates.

**Figure 4-2 Masters and Apprentices[a] Among Laureates in Biological Chemistry and in Biological Science (1901-72), According to Decade of Apprenticeship**

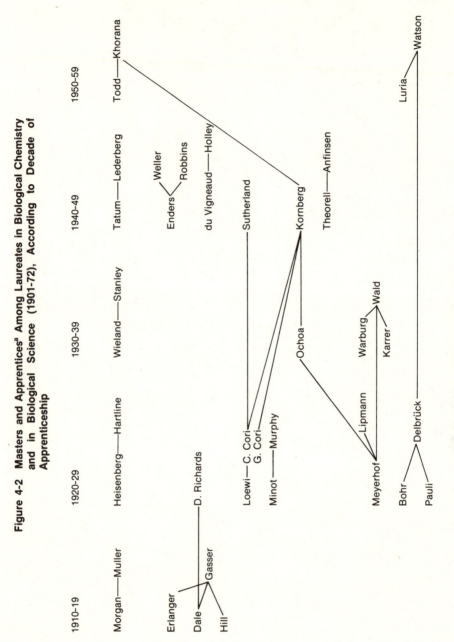

[a]"Apprentices" defined as graduate students or postdoctorates.

**Figure 4-3   Laureate Masters and Apprentices Associated with J. J. Thomson and E. Rutherford (1901-72)**

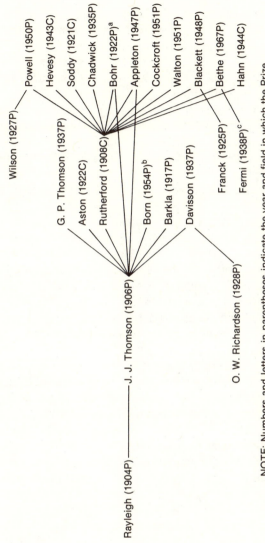

NOTE: Numbers and letters in parentheses indicate the year and field in which the Prize was awarded, P = physics and C = chemistry.

[a]Bohr was also a prolific master to future laureates. Seven worked with him at Copenhagen early in their careers: Felix Bloch, Max Delbrück, Werner Heisenberg, Lev Landau, Wolfgang Pauli, Linus Pauling, and Harold Urey.

[b]Pauli and Heisenberg, both students of Bohr, also studied with Max Born. Born served as master to two more laureates, Maria Goeppert Mayer and Otto Stern.

[c]Fermi served as master to Felix Bloch, Hans Bethe, and Emilio Segrè while still in Rome and to Owen Chamberlain, T. D. Lee, and C. N. Yang at Chicago.

The pattern of laureates' producing heirs to the prize becomes more marked, if we include the European laureates who were also trained by Nobelists, appearing in the genealogy of American laureates. For example, Otto Warburg, who turns up in figure 4–2 just once, as mentor of George Wald, also trained three other Nobelists, the Europeans Szent-Gyorgyi, Krebs, and Theorell.

The sociological inbreeding of the scientific elite becomes even more conspicuous when occupants of the forty-first chair are considered. Not surprisingly, we find that American laureates were trained by such members of the ultra-elite *sans* Nobel prizes as Arnold Sommerfeld and John Wheeler in physics, Michael Polanyi, Roger Adams, and G. N. Lewis in chemistry, and Selig Hecht and L. J. Henderson in the life sciences.[7]

The complementary pattern also holds: many occupants of the forty-first chair have been the students of junior collaborators of laureates. For example, T. W. Richards, the Harvard laureate chemist (and, incidentally, the father-in-law of the chemist and subsequent Harvard president James B. Conant), trained "a whole new generation of physical chemists," including L. J. Henderson, G. N. Lewis, and Roger Adams (all of whom were nominated for a Nobel more than once) (Ihde, 1969, p. 649). In their turn, as we have now learned to expect, these scientists went on to train new generations of the scientific elite, each having at least one laureate among his students. In this respect, as in others, occupants of the forty-first chair closely resemble their laureate peers except for the detail of not having been awarded the prize.

These patterns exhibit a fundamental difference between biological and social heredity. Until now, at least, biological parents cannot choose their children any more than children can choose their biological parents. But in the social domain generally, and specifically in the domain of science and learning, there is an option. To some extent, students of promise can choose masters with whom to work and masters can choose among the cohorts of students who present themselves for study. This process of bilateral assortative selection is conspicuously at work among the ultra-elite of science. Actual and prospective members of that elite select their scientist parents and therewith their scientist ancestors just as later they select their scientist progeny and therewith their scientist descendants.

---

7. Linus Pauling and Hans Bethe studied with Sommerfeld and Richard Feynman with John Wheeler. Michael Polanyi was mentor to Melvin Calvin and Eugene Wigner, Adams to Wendell Stanley, and G. N. Lewis to a quartet of laureates—Seaborg, Giauque, Urey, and Libby. Hecht trained Wald and Zinsser, John Enders.

In the assortative process involving American laureates at any phase, the longest sequence of Nobel mentors and students spans five "generations" of scientists. As can be seen by consulting figure 4–1, it begins with Wilhelm Ostwald, the German laureate in chemistry (1909), who taught the German physical chemist Walther Nernst (1920), who, in his turn, helped train the American physicist Robert Millikan (1923). When Millikan went to Cal Tech, Carl Anderson (1936) became his student. Anderson then went on to help train Donald Glaser, recipient of the prize in 1960 for inventing the bubble chamber to study subatomic particles. This five-step sequence of awards stretches over a half century.

Such a lineage perpetuating the scientific ultra-elite can easily be extended if the search is not arbitrarily confined to Nobel laureates. Then the lines of elite apprentices to elite masters who had themselves been elite apprentices, and so on indefinitely, often reach far back into the history of science, long before 1900, when Nobel's will inaugurated what now amounts to the International Academy of Sciences. As an example of the many long historical chains of elite masters and apprentices, consider the German-born English laureate Hans Krebs (1953), who traces his scientific lineage (Krebs, 1967, p. 1295) back through his master, the 1931 laureate Otto Warburg. Warburg had studied with Emil Fisher, recipient of a prize in 1902 at the age of 50, three years before it was awarded to *his* teacher, Adolf von Baeyer, at age 70. This lineage of four Nobel masters and apprentices has its own pre-Nobelian antecedents. Von Baeyer had been the apprentice of F. A. Kekulé (von Stradonitz), whose ideas of structural formulae revolutionized organic chemistry and who is perhaps best known for the often retold story about his having hit upon the ring structure of benzene in a dream (1865). Kekulé himself had been trained by the great organic chemist Justus von Liebig (1803–73), who had studied at the Sorbonne with the master J. L. Gay-Lussac (1778–1850), himself once apprenticed to Claude Louis Berthollet (1748–1822). Among his many institutional and cognitive accomplishments, Berthollet helped found the *Ecole polytechnique,* served as science advisor to Napoleon in Egypt, and, more significant for our purposes here, worked with Lavoisier to revise the standard system of chemical nomenclature.

From this summary, it appears that laureates are only continuing a longstanding historical pattern for replenishing the scientific ultra-elite. Seldom did American laureates study, at least in their graduate and postgraduate years, with scientists who were relatively unproductive in research. Instead, as we have seen, the composite roster of the laureates' teachers reads like a roll call of the world's elite in science.

Future members of the ultra-elite are known not only for the work they have done but also for the teachers they have had. As the economist laureate Paul Samuelson lightly but not frivolously ad-libbed in his own acceptance speech at Stockholm: " 'I can tell you how to get a Nobel prize. One condition is to have great teachers.' And I enumerated the many great economists whom I had been able to study under both at Chicago and Harvard" (1972a, p. 155).

The pre-Nobelian evidence of the self-perpetuation of the scientific elite can serve as a useful corrective to the unexamined assumption that it must be something about the process of selecting Nobel laureates that accounts for so many of them having been apprenticed to laureates. On the face of it, that assumption is plausible. For, under the rules, laureates are permanently eligible to nominate for a prize. It would therefore seem evident that they might be able to exert significant influence in putting their own outstanding apprentices into the competition. It is also known that, from time to time, laureates will mount campaigns and take to logrolling in order to advance the cause of scientists whose work they consider particularly meritorious—as, for example, in the case of Rutherford, pressing for the solo designation of his apprentice James Chadwick as laureate in physics for his discovery of the neutron. This sort of episode once again underscores the need to recognize that there is often no great difference with regard to the caliber of scientific achievement between the laureates and the also-rans who were nominated for the prize but did not get it. Both groups are members of the scientific elite.

The assumption that politicking may account for the inbreeding of laureates probably assumes too much. The laureates constitute a small minority of the nominators. Invitations to nominate for the prize are extended to scientists the world over, principally those working at the major universities and research institutes but at the minor places too. In recent years, the nominators canvassed in each of the three fields covered by the prize have run to a thousand, although in earlier years there have been only a hundred or so. Still, the procedures of nomination and election, specific to the Nobel prizes, may favor the scientific offspring of Nobelists in the limited sense that laureates are likely to be skillful advocates of their candidates in the first place and, having permanent rights to nominate, acquire experience and judgment about the kinds of documents that must be submitted to make their case. But this is probably not enough to account for the observed pattern of more than half of the American laureates having been apprenticed to other laureates, for the history of the scientific elite through the centuries exhibits the same kind of elite master-apprentice pedigrees (see Pledge, 1939, among others).

## The Process of Mutual Search

We have seen in chapter 3 that laureates, like other members of the scientific elite, studied in a comparatively small number of colleges and universities. This concentration resulted from the jointly operating processes of self-selection by the future scientists and selective recruitment by the academic institutions. Though sometimes subject to socio-economic constraints that limited their range of choice, the eventual laureates tended first to seek out the institutions of reputation for undergraduate study and then turned, in even greater proportions, to the outstanding departments in their field for graduate study. Postdoctoral study involved a third phase in this joint process of self-selection and selective recruitment, which contributed even more to the observed pattern of apprenticeship of the laureates-to-be to older laureates.

The filiation of laureates can thus be thought of as the result of mutual search by young scientists of promise and by their prospective masters. Both apprentices and masters were engaged in a motivated search to find and then to work with scientists of talent.

The most striking fact in the process of self-selection is that future members of the ultra-elite were clearly tuned in to the scientific network early in their careers. The laureates' reports on their own early experiences indicate how much this was the case. A laureate in physics contrasted the basis for his own choice of a potential master with that of his less knowledgeable classmates: "Many of the students were just silly about the way they choose professors. They just didn't know the professors of real quality. They were very innocent. I was far from innocent and made my own judgment of quality." Along the same lines, another physicist laureate reported: "I was attracted by his name. I knew enough about physics to appreciate him and to appreciate his style of work." (It happens that both were referring to Enrico Fermi who, it will be remembered, had no fewer than six graduate students and junior collaborators go on to become Nobel laureates. They did not come to him by chance.)

The laureates, in their comparative youth, sometimes went to great lengths to make sure that they would be working with those they considered the best in their field. The case of the physicist C. N. Yang is a bit extreme, but it illustrates the deeply motivated search for the particular master rather than only for the outstanding graduate department or university ambience.

At the age of 23, a student at the Southwest Associated University in Kunming, a provincial city in southwest China, Yang resolved to continue his graduate studies with Fermi and Eugene Wigner. Arriving in New York late in 1945, Yang went directly to Columbia University, in the belief that Fermi was still there. In fact, Fermi had during the war years been engaged in research on the atomic bomb at the University of Chicago and Los Alamos. That research being top secret, no one at Columbia could tell Yang of Fermi's doings or even of his where-abouts. Yang went on to Princeton University, only to discover that Wigner, a member of Princeton's department of physics, was also away (as it happens, also at work on the Manhattan project, first at Chicago and then at Oak Ridge). At Princeton, Yang heard the "rumor" that Fermi would head up a new institute at the University of Chicago. Con-tinuing his quest, Yang finally found himself sitting in Fermi's class in January 1946. He got to see a good deal of Fermi, who explained that he could not supervise Yang's thesis work since he himself was still occupied with "highly classified research." Fermi then directed Yang to Edward Teller, another physicist in the network of talented scientists at Chicago, who helped turn Yang's attention from experimental physics, at which he was plainly not his best, to theoretical physics, where he was plainly better than most (Bernstein, 1962, p. 49ff.).

This episode brings out a major aspect of the search for an appro-priate master. Clearly, Yang knew what he was about, having been socialized perhaps in the ways of science by his mathematician father. By the time Yang began his search, Fermi had been a laureate for seven years and so had acquired a heightened visibility that extended to bright young students of physics even in remote China. It is therefore not surprising that Yang knew of Fermi's distinctive contributions to the field and, further, that he elected to work with so conspicuous a star. What is perhaps more revealing is that he also wanted to work with Wigner who, though he had by then gone far with his ultimately prize-winning work on nuclear structure, was visible only among physicists and was not to receive his Nobel for another twenty years (ironically, a dozen years after Yang received *his* prize).

Thus the case of Yang illustrates a general pattern. The young scientists who would later receive Nobel prizes were tuned in early to the major channels of communication about new developments in their fields. They knew the most significant work that was being done, where it was being done, and by whom. That this was the case is strikingly reflected in table 4–1, which shows that in fully 69 percent of all cases of apprenticeship with laureates, the young laureates-to-be had chosen their masters *before* the masters' important work was conspicuously "validated" and made fully visible by the award of a Nobel prize. The fact that as many as thirty of the forty-eight Americans who had lau-reate masters did so as postdoctorates rather than as graduate students suggests that even they needed a little time to learn precisely who was working on what they wanted to study.

**Table 4–1. Laureate Master-Apprentice Pairs (1901–72): Status of Master at Time of Apprenticeship**

| Status of Masters | Physics [a] | Chemistry [a] | Physiology-Medicine [a] | Total | Percent |
|---|---|---|---|---|---|
| Number of future laureate master and apprentice pairs *before* master was laureate | 19 | 10 | 20 | 49[b] | 69 |
| Number of future laureate master and apprentice pairs *after* master was laureate | 12 | 3 | 7 | 22[c] | 31 |
| Number of pairs | 31 | 13 | 27 | 71 | 100 |
| Number of laureates having laureate masters | 19 | 11 | 18 | 48 | 52 |

a. Classified according to the field of apprentice (e.g., Delbrück, though trained by the physicist Bohr, is counted in physiology-medicine; Pauling, also trained by Bohr, is counted in chemistry).
b. Of these cases, six pairs shared a prize: Chamberlain and Segrè, Erlanger and Gasser, Minot and Murphy, Enders and Weller, Enders and Robbins, and Tatum and Lederberg. An additional master-apprentice pair, Kornberg and Ochoa, was awarded a prize jointly but for different and independent investigations.
c. Of the 22 pairs in which masters were already laureates, two—Bardeen and Cooper, and Bardeen and Schrieffer—shared a prize.

Thus table 4–1 suggests that the eventual laureates, in their youth, had a discriminating eye for the masters of their craft as well as for the major universities and departments doing work at the frontiers of the field (as noted in chapter 3). Early in their careers, the laureates-to-be searched out scientists who, as we now know, were destined to move into the Nobel elect. But at the time the majority had not yet done so. In this respect, these informed young scientists, at the time of their apprenticeships, were able to identify scientific talent of Nobel caliber. It is less clear whether their fellow students made the same choices. Especially at great universities, many students seek out scientific stars with whom they hope to work, and many are turned away. Thus self-selection is only one component in the process of linking masters and apprentices.

For various reasons, the choices of masters by would-be apprentices were not always reciprocated. We have already seen, for example, how Yang almost missed out on working with Fermi and actually did miss out on working with Wigner. In this case, the constraints that kept his preferred choices from being realized were external to the masters themselves. But in other cases, the potential masters themselves refused to accept as apprentices many of those who wanted to work with them, even those of obvious talent. For example, it was evident even to tyros that, long before he was awarded the prize, Percy Bridgman had "occupied the field of the physics of high pressure all by himself," as the laureate-to-be Edward Purcell put it. But Bridgman's solitary style of scientific work was such that neither Purcell, had he been so minded when he studied at Harvard, nor other young scientists of talent had much prospect of becoming his apprentice. As the *Dictionary of Scientific Biography* notes:

> The desire for full personal involvement in the experiment probably also accounted for Bridgman's reluctance to do joint research or to take on thesis students. He rarely had more than two at a time; the record shows fourteen doctoral theses on high-pressure topics, in addition to several on other subjects that he supervised. He was usually most pleased when least consulted [Kemble, Birch, and Holton, 1970, v. II, p. 459].

Such nonreciprocation of choices means that the selection of *future* laureate masters would have been even higher than the substantial 52 percent recorded in table 4–1 had the would-be apprentices had their way. The pattern of young talent identifying older talent is thus even more marked than the surface figures indicate.

Correlatively, some of the future laureate masters of future laureates contributed to the observed pattern by actively searching for youngsters of talent. If his lonely style of work helps to explain why Bridgman had "only" one future laureate as an apprentice—John Bardeen—then Fermi's style of work helps to explain why he had as many as six. For, as Segrè has observed, Fermi actively sought out younger co-workers of ability: "Fermi . . . decided that he needed some pupils—what we call graduate students. He decided also that they had to be seriously interested in physics and of reasonable ability so that his time would not be wasted. Given the right quality, he would see to it that they were taught" (Segrè, n.d., p. 2).

In other words, just as the younger laureates-to-be were able to assess the achievements of the older laureates-to-be, so the older ones were able to assess the potentialities of the younger ones. In some cases, the master served as scientific talent scouts or "truffle dogs" (Merton, 1960, pp. 308–309). The story told about I. I. Rabi's encounter with the young Julian Schwinger is a case in point. In 1936 the 18-year-old Schwinger, then an undergraduate at the City College of New York, chanced to accompany a friend to Columbia who wanted to transfer to the department of physics there. Rabi, still eight years away from his

Nobel prize, talked to the young men and soon turned his attention chiefly to Schwinger rather than to his friend, who had initiated the interview. Rabi found Schwinger knowledgeable in physics far beyond his years. More important, he noted Schwinger's capacity to think mathematically and his ingenuity in mathematical formulation. Rabi soon arranged for Schwinger to enter Columbia College and eventually to earn his doctorate at the University. By the time Schwinger was a graduate student, his gifts were evident to anyone who cared to look. As his fellow student Mitchell Wilson (1969) wrote in retrospect, Schwinger "was at once so obviously in a class by himself that no one bothered to envy him. One thing, each of us assured the others: eventually, he would earn a Nobel Prize." With his record of having identified many distinguished physicists, including the laureates Kusch and Schwinger, Rabi must be counted among the more successful scientific truffle dogs. Even so, the question remains open of how clear the signals of scientific promise actually are. The existence of ample numbers of "comers" who never made the grade and of smaller numbers of "late bloomers" suggests that there is considerable noise in the evaluation system and that judgments of scientific promise by the elite are far from perfect.

The joint processes of self-selection and social selection making for entry into the scientific ultra-elite did not always proceed smoothly. Error in early evaluation of scientific talent involved its own kind of vicissitudes. Thus James D. Watson, widely known as the co-discoverer of the structure of DNA, reports that in 1947 he had applied for admission to graduate school both at Cal Tech, whose "Biology Division was loaded with good geneticists," and at Harvard, where he had applied "without considering what he might find." He was refused admission to both. In retrospect, Watson found Harvard's rejection of him fortunate. For had he gone there, he "would have found no one excited by the gene and so might have been tempted to go back into natural history." Guided by his mentor at the University of Chicago, the human geneticist Herluf Strandskov, Watson applied to Indiana University, where he was to find the renowned geneticist H. J. Muller and two first-rate young geneticists, Salvador Luria and Tracey Sonneborn.

This episode has a particular interest for us in tracing the processes underlying the striking statistic that 68 percent of apprenticeships involved master scientists who had not yet gotten their prize. Having missed out on working with one laureate-to-be, Beadle at Cal Tech (who was not to receive his prize for another eleven years), it now appeared that Watson would be working with an actual laureate, Muller at Indiana (who had gotten a prize just the year before). Watson would then have slipped out of the category of those young laureates-to-be who recognized talent of a high order that had not yet been "validated" by a Nobel. But soon after he arrived at Indiana, Watson decided that although it first "seemed natural that I should work with Muller . . . I soon saw that Drosophila's better days were over and that many of the best younger geneticists, among them Sonneborn and Luria, worked with micro-organisms." We can continue to trace, in Watson's own words, the way in which his own judgment of comparative opportunities for training led to his ultimate choice of a master at Indiana:

> The choice among the various research groups was not obvious at first, since the graduate-student gossip reflected unqualified praise, if not worship, of Sonneborn. In contrast, many students were afraid of Luria who had the reputation of being arrogant toward people who were wrong. Almost from Luria's first lecture, however, I found myself much more interested in his phages than in the Paramecia of Sonneborn. Also, as the fall term wore on I saw no evidence of the rumored inconsiderateness toward dimwits. Thus with no real reservations (except for occasional fear that I was not bright enough to move in his circle) I asked Luria whether I could do research under his direction in the spring term. He promptly said yes and gave me the task of looking to see whether phages inactivated by X-rays gave any multiplicity reactivation [1966, pp. 239–40].

Watson's experiences thus delineate the kind of informed search that led many eventual laureates to be apprenticed to already arrived or prospective laureates. Had Watson been admitted to his first choice of Cal Tech, he would probably have studied with the laureate-to-be George Beadle and would thus have appeared as another case making up the dominant pattern recorded in table 4–1. Rejected at Cal Tech, Watson might have persisted in his trial choice of the laureate Muller at Indiana, thus contributing to the minor pattern found in the same table. But appraising the kinds of work that held most promise for fundamental advances in biology, Watson made the (reciprocated) choice of working under the yet-to-be-laureate, Luria. (This placed him definitely in the dominant pattern of laureate masters and apprentices.) Finally, through the sponsorship of Luria, Watson spent summers at the famous summer phage course instituted at Cold Spring Harbor by Max Delbrück (who later shared a Nobel prize with Luria). Thus Watson ends up as another laureate who had studied with laureates-to-be.

Watson's case is instructive in still another respect. Some laureates believe that luck played an important role in their having found congenial mentors and congenial places to work. And so it must seem in individual cases. Watson himself has not embraced the "luck" hypothesis, but his case perhaps may illustrate why that hypothesis is often unsound. Watson did not know about Sonneborn or Luria while he was still at Chicago and therefore could not have chosen Indiana to study with them. But the fact that a Sonneborn or a Luria happened to be there is not altogether accidental. As we have seen, excellent scientists tend to converge on a small number of places where other excellent scientists, such as H. J. Muller, are at work. And this was so even though Sonneborn and Luria may already have been convinced that they would be post-Mullerian geneticists rather than traditional *Drosophila* specialists. (It also turns out that Renato Dulbecco, laureate in medicine in 1975, was working in Luria's laboratory that year.) So the fact that Watson found a mentor at Indiana worthy of his mettle was probably not the result of luck but rather of the joint processes of self-selection and selective recruitment that we have repeatedly noted.

From such processes of self-selection and selective recruitment there results the observed pattern of more than half of the American laureates having been apprenticed to laureates, 69 percent of these apprenticeships before the masters had yet won the prize. These patterns of apprenticeship in turn link up with the careers of laureates-to-be.

## Types of Masters and Time of Award

The ninety-two American laureates received their prizes at the average age of 51. But this average obscures a wide variation. At the lower bound, the prize was awarded at the age of 31 to the physicists Carl Anderson, discoverer of the positron, and T. D. Lee who, with C. N. Yang, showed that parity is not conserved in weak interactions. At the upper bound, the prize was awarded to the virologist F. P. Rous at the venerable age of 87.

Examining more closely the age at which American laureates won their prizes, we find that those trained by laureate masters won their Nobel awards earlier than those whose masters never become Nobelists. And, as it turns out, this is not the spurious result of those with laureate masters having been trained at the great universities and thus having been part of the social network of elites from the beginning. Although it is the case that laureates who took degrees at the elite universities got their prizes earlier than the minority trained at less distinguished universities, having had a laureate master is consistently associated with winning an award earlier in every class of university, as table 4–2 shows.

It also turns out that not all laureate masters are of a piece in this regard. As table 4–3 indicates, apprentices whose masters already had Nobel prizes got their own awards decidedly earlier (by almost seven

**Table 4–2. Mean Age at Prize of Laureates (1901–72), According to Doctoral Origins and Master's Status**

| Status of Master | DOCTORAL ORIGINS | | | |
|---|---|---|---|---|
| | Elite University [a] | Other U.S. University | Foreign University | All Universities |
| Nobel Laureate | 47.4 | 43.3 | 53.9 | 48.6 |
| | (34) | (3) | (11) | (48) |
| Not Nobel Laureate | 53.7 | 54.7 | 54.7 | 53.8 |
| | (30) | (7) | (7) | (44) |
| All | 50.3 | 54.2 | 50.4 | 51.1 |
| | (64) | (10) | (18) | (92) |

a. See table 3–9 for definition.

years) than did apprentices whose masters never received Nobel awards. But the apprentices of masters then waiting in the wings did not get their prizes much earlier than those apprenticed to other scientists. Not quite four years separate these two groups, a difference that is not statistically significant.

It should be noted that the masters *sans* Nobel prizes were for the most part members of the scientific elite. If we adopt the broad criterion of those who were doing fundamental research, proposed by such observers as Warren Hagstrom (1965, p. 11), then practically all of them were in the elite. And, even by the most exacting criteria, the great majority of these masters were located in the upper reaches of the stratification system. At least eight—John Wheeler, Victor Weisskopf, and Robert Oppenheimer in physics, G. N. Lewis, C. R. Harington, and Michael Polanyi in chemistry, and Theodosius Dobzhansky and Jacques Loeb in the biological sciences—have been identified as occupants of the forty-first chair in the sense of being greatly esteemed by the community of science for having contributed as much to the advancement

**Table 4–3. Mean Age at Prize of Laureates (1901–72), According to Master's Status at Time of Apprenticeship**

| Status of Master | Mean Age at Receipt of Prize |
|---|---|
| (1) Already laureate | 46.9 |
| | (20) |
| (2) Not yet laureate | 49.9 |
| | (28) |
| (3) Never laureate | 53.8 |
| | (44) |

$F = 3.7355$, Sig. .05, 89 $df$;
$M_1 - M_3 = 7.01$, Sig. .05;
$M_1 - M_2 = 3.01$, Not sig.;
$M_2 - M_3 = 4.01$, Not sig.

of science as some of the laureates, and perhaps more. The elite character of these nonlaureate masters becomes even more evident when we add to these eight (who were Academicians) the twenty-seven others who were also members of the National Academy of Sciences or of one of the European academies. Thus, 63 percent of the fifty-six identifiable [8] nonlaureate masters were Academicians. By way of comparison, fewer than 2 percent of American scientists are members of the National Academy. Altogether then, in addition to the forty-eight American laureates who had laureate masters, twenty-five were apprenticed to Academicians, making a total of 79 percent whose masters were members of the scientific elite. In view of this, no great differences in age at receipt of a prize (or other attributes of careers) should be expected.

The age difference between those who had masters with Nobel prizes and those who did not, revealed in table 4–3, is nonetheless significant. It can be given at least two distinct though not mutually exclusive interpretations. Apart from possible differences in the distribution of talent of apprentices, resulting from self-selection and selective recruitment, laureate masters are in a position to facilitate the careers of their apprentices, some only by enhancing their visibility, others by outright logrolling in their behalf. As we have noted, Lord Rutherford lobbied openly and zestfully for his Cavendish apprentices, eleven of whom eventually became laureates. This interpretation suggests that apprentices whose masters already had their prizes were launched with greater thrust than those whose masters still occupied the forty-first chair. It may also be that, on the average, laureate masters provided better training than other masters.

On the other hand, on the assumption that scientists who are in the throes of research of Nobel prize-winning caliber or about to embark on it are in their scientific prime, it should be the case that apprentices working with them at this time do so more closely, receive better training and so win their prizes earlier than those trained by scientists who, having already won their prizes, are more likely to have passed their prime.

The evidence turns out to be consistent with the first rather than with the second interpretation. The scientists who had been apprenticed to established laureates got their awards, on the average, three years before those trained by laureates-to-be and almost seven years before those whose masters never became laureates.

---

8. In eight cases, it was not possible to identify mentors of Nobel laureates. There were also nine cases in which masters' names appeared in the public record but could not be traced further for birth dates or other biographical data. These tabulations do not include apprenticeships of laureates with nonlaureate masters if they also worked under laureate masters.

But these differences, though substantial, can easily become misleading. For the question is not whether the one-time apprentices of laureates win the prize at an earlier age. Rather, it is whether, once having done the work ultimately honored by the prize, they spend less time waiting in the wings. Table 4–4 provides three related findings bearing directly on this question. First, it shows that scientists who had laureate masters did not wait as long as the others for their awards: an average of eleven years for the former and thirteen years for the latter. Thus the difference of more than five years between the two groups in age at the time of an award is reduced by almost half, to a difference of just two years between the time of doing the ultimately prize-winning research and the award itself. Second, we now see that the greater youthfulness at the time of their prizes of laureates apprenticed to laureate masters [9] is the joint result of their having done their prize-winning research earlier and of having been given the award somewhat more expeditiously. It is still not clear why those with laureate masters tended to do the prize-winning research earlier. And, third, those whose masters already had prizes did have a shorter wait in the wings than the rest, a fact consistent with the interpretation of sponsorship mentioned earlier.[10]

## Master-Apprentice: Sharing the Prize

Throughout, we have noted that whether or not laureate masters have yet won their prizes makes for only negligible differences in the age at which their apprentices are honored. But there is one subset of Nobelists for whom this is distinctly not the case. This is comprised by the six laureates—Chamberlain, Gasser, Murphy, Robbins, Weller, and Lederberg [11]—who shared their awards with their one-time mentors and two more, Cooper and Schrieffer, who shared in John Bardeen's second Nobel. They received their prizes very early indeed: at an average age of 41.3 years compared with 50.1 years for all the others with laureate

---

9. When we examine the role of masters' status at the time their apprentices do prize-winning research (instead of the time of the apprenticeship) and the age at receipt of the prize, the pattern of findings remains the same.

10. Although the status of master is related to the age of apprentices at receipt of the prize and to the time that elapses between the publication of prize-winning research and the award, there is no relationship between the age of masters and the time their apprentices win Nobel prizes.

11. Joshua Lederberg, who shared the prize with his teacher, E. L. Tatum, and with George Beadle is included in this subset but is nonetheless a unique case. The work for which Lederberg won his Nobel, the demonstration of sexual recombination in bacteria, began when he was 21, just before he went to work with Tatum, but his later research with other co-workers in bacterial genetics was also cited for his prize. Tatum won his prize for earlier work, done in collaboration with Beadle, on genetic transmission of hereditary characters in *Neurospora*.

**Table 4-4. Mean Age of Laureates (1901–72) at Prize-Winning Work and at Prize, According to Master's Status**

| Status of Master at Time of Apprenticeship | Age at Prize-Winning Work | | Age at Prize | | Interval between Time of Work and Time of Prize | | Number of Apprentices |
|---|---|---|---|---|---|---|---|
| Master already laureate | 37.2 | } 37.6 | 46.9 | } 48.6 | 9.7 | } 11.0 | 20 |
| Master not yet laureate | 37.9 | | 49.9 | | 11.9 | | 28 |
| Master never laureate | 40.8 | | 53.8 | | 13.0 | | 44 |

masters. Understandably, the members of this subset were also younger than the others at the time they did their prize-winning work, and they had a shorter waiting time as well. They were, on the average, only 30.8 at the time of the subsequently honored research compared with an average age of 39.0 for the others who studied with laureates still to receive an award. And, on the average, they waited for 10.5 years after that work compared with 11.1 years for the rest.

All this reflects a process of socialization, at least in the upper strata of science, where, perhaps more often and sooner than in other disciplines and other strata within science, the roles of teacher and student become transformed into the role of collaborator (see Zuckerman and Merton, 1972, pp. 330, 338–41, for discussion of this kind of role transformation). In these cases, the status transition from novice to research colleague occurred in the course of the work that eventually brought the prize both to them and to their one-time masters. Thus the evidence points to an institutional structure providing for mobility into the scientific ultra-elite in which the recognition of precocious achievement appears to take precedence over sponsorship by influential masters.

## Relative Ages of Masters and Apprentices

That the relative ages of people involved in various kinds of social relationships differ in both normative prescription and social fact is obvious but often forgotten. In Western society, for example, for reasons not immediately evident, wives are expected to be and generally are younger than their husbands. Friends are expected to be and generally are of about the same age and teachers, as agents of socialization, are expected to be and generally are older than their students. (For the observation and some of its implications, see Riley, Johnson, and Foner, 1972, pp. 410–11, 536–48.)

Among the future laureates, apprentices were on the average about 26 years of age and their masters about 43. But the average discrepancy in age of some seventeen years obscures a great deal of variation, as can be seen from figure 4–4. Among the 121 apprenticeships on which we have information, that difference ranged from two cases in which the apprentice was actually older than the master (to be sure, by only one or two years) to the other extreme in which the master was forty-four years older than the apprentice. Turning from the extremes to the entire distribution of ages in figure 4–4, we note that in just under a third of the cases, the age discrepancy between master and apprentice was

**Figure 4-4    Age Differences Between Masters and Laureate Apprentices (1901-72)**

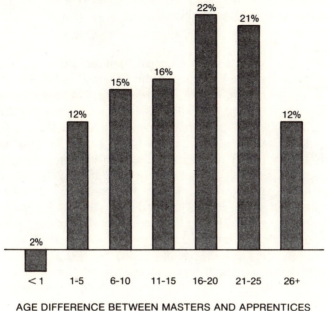

AGE DIFFERENCE BETWEEN MASTERS AND APPRENTICES
(IN YEARS)

less than ten years; in a little more than a third, between eleven and twenty years; and, in the remaining third, more than twenty-one years.

In the absence of comparable data for the population of scientists generally or for the various major branches of learning, we can only speculate on whether this pattern among the American ultra-elite of science differs from the others. We would expect the age difference between master and apprentice to be somewhat greater in the humanities, where accumulated erudition counts more heavily in achieving a major place in the field. For this elite in science, however, it appears that the chronological, as distinct from the professional, age of the master played no great part in the decision of apprentices to seek out this or that scientist to study and work with. They tended to go to the known centers of major advance in the fields of their interest, and what mattered for most was whether the scientists with whom they worked were central to those advances, not whether they were old or young.

Thus, extreme youth of the master was plainly no barrier in the case of the theoretical physicist Wolfgang Pauli, who was all of 27, two years younger than Rabi when Rabi elected to work with him. The prodigy Pauli was by then a major figure in the field, having three years before proposed the fundamental Exclusion Principle (which won him a Nobel prize twenty-one years later). Nor was ripe old age a barrier in the case of the physiological chemist J. J. Abel, who was 71 at the time that the future laureate Vincent du Vigneaud went to Johns Hopkins to study with him.

The story of du Vigneaud's varied apprenticeships over a span of seven years perhaps best testifies to the apparent irrelevance of the age of the master relative to that of the apprentice. Du Vigneaud reports no fewer than eight scientists who trained him, from the 30-year-old C(arl) S(hipp) Marvel, seven years older than du Vigneaud, and the 32-year-old C(harles) R(obert) Harington to, as we have noted, the 71-year-old Abel, forty-four years older than his apprentice.

There is, however, a distinct pattern of relative age that distinguishes the established laureate masters from the masters who had yet to get their prizes and these in turn from the masters who (thus far) have not received the prize. The age of apprentices did not vary much among those who sought out one or another type of master; all of them were 26 or 27 at the time. But the age of the masters of differing status tended to vary far more. As might be expected, the established laureates were the oldest (an average of 48) and the laureates-to-be the youngest (40) with the other masters in between (43). Since the ages of apprentices were much the same in all cases, this results in an average discrepancy of twenty-one years between the laureate masters and apprentices, thirteen years for future laureate masters and apprentices, and seventeen years for nonlaureate masters and apprentices.[12]

The occasionally small differences in age of laureate-to-be masters and apprentices have contributed to a pattern in which a handful of apprentices have leapfrogged into the ultra-elite over the backs of their masters. These five apprentices, as is shown in table 4–5, were, except for Watson, all pretty much of an age with their masters and anticipated them by one to seven years in obtaining a prize. This meant that on the average they received their prize at the comparatively early age of 46, some five years younger than the rest of the American laureates and eleven and a half years younger than their respective masters. Dramatic events in scientific discovery of the kind that are caught up in these figures no doubt contribute their share to the widespread belief that "science is a young man's game." It should be noted, however, that the relative ages of masters and laureate apprentices as a whole show no determinate relation to the age at which the prize is received.

12. For the cases in which one-time apprentices shared prizes with their masters, the average age of the apprentices during the period they were being trained was 29.7 and of the masters, 44.5.

**Table 4–5. Laureate Apprentices Whose Awards Preceded Their Master's**

| Cases | No. Years by Which Apprentice's Award Preceded Master's Award | Relative Age of Master to Apprentice | Age of Apprentice at Time of Award | Age of Master at Time of Award |
|---|---|---|---|---|
| Michelson—Apprentice | | | | |
| Lippman—Master | 1 | 7 | 55 | 63 |
| Bardeen—Apprentice | | | | |
| Wigner—Master | 7 | 6 | 48 | 61 |
| Rabi—Apprentice | | | | |
| Pauli—Master | 1 | −2 | 46 | 45 |
| T. W. Richards—Apprentice | | | | |
| Nernst—Master | 6 | 4 | 46 | 56 |
| Watson—Apprentice | | | | |
| Luria— | 7 | 16 | 34 | 57 |
| Delbrück— Masters | | 22 | | 63 |
| Mean age | | | 46 | 57.5 |

## SOCIALIZATION FOR THE SCIENTIFIC ELITE

The importance that these members of the scientific elite attach to their apprenticeships is perhaps better indicated by their behavior at the time than by what they said about them in our interviews much later. As we have seen, they often exerted great effort to find their way to masters at the forefront of their fields, typically masters who were already in the aristocracy of science or would become members of it before long. We now turn to the evidence provided in our detailed interviews to find out what actually went on in the interaction between masters and apprentices, what this meant for the quality of scientific work the apprentices went on to do, and how it contributed to their mobility into the elite.

### Education, Training, and Socialization

The testimony in the interviews is in some ways surprisingly uniform. This similarity need not be assumed to reflect the piety of disciples for masters for, in the main, the laureates have a strong sense of their own abilities, a strong ego (at times expressed in undisguised egotism), and a critical tendency that does not spare even their admired teachers and colleagues. Moreover, the similarity of report extends beyond generalities to specifics with regard to what they individually wanted from their teachers, what they found in actual experience, and what they now use as implicit criteria for appraising that experience as they report it.

One point on which the laureates are largely agreed is that the least important aspect of their apprenticeship was the acquiring of substantive knowledge from their master. Some even reported that in the limited sense of information and knowledge of the scientific literature, apprentices, focused on one or another problem, sometimes "knew more" than their masters. A laureate in chemistry speaks for many of them:

> It's the contact: seeing how they operate, how they think, how they go about things. [Not the specific knowledge?] Not at all. It's learning a style of thinking, I guess. Certainly not the specific knowledge; at least not in the case of Lawrence. There were always people around who knew more than he did. It wasn't that. It was a method of work that really got things done.

A physicist sums up the difference between the learning of specifics and the learning of a style of qualitatively distinctive thought as a difference between the techniques and the traditions of major scientific work:

> I knew the techniques of research. I knew a lot of physics. I had the words, the libretto, but not quite the music. In other words, I had not been in contact with men who were deeply imbedded in the tradition of physics: men of high quality. This was my first real contact with first-rate creative minds at the high point of their power.

Thus the laureates testify that for them the principal benefit of apprenticeship was a wider orientation that included standards of work and modes of thought. They report, in effect, that the apprenticeship was a time of what social scientists call socialization. Socialization includes more than is ordinarily understood by education or by training: it involves acquiring the norms and standards, the values and attitudes, as well as the knowledge, skills, and behavior patterns associated with particular statuses and roles.[13] It is, in short, the process through which people are inducted into a culture or subculture.

Like other scientists, then, the future laureates were continuing their socialization in the culture of science. But, unlike most other scientists, they were also being prepared to take a distinctive place in the forefront of science. In terms of social stratification, they were being socialized for a position in the aristocracy of science. Seldom explicit, this was often tacitly understood. Although it was far from evident, of course, during the time of their apprenticeship that they would become Nobel laureates, it was largely understood that, with few exceptions, they would become very good rather than run-of-the-mill scientists. For, by the time they got to their masters, they had already given evidence of uncommon scientific ability.

Throughout this book, the reader should remember that we are dealing with a highly selected part of the scientific elite. At this point in our study we consider only those apprentices who did in fact enter the aristocracy of science. We can say nothing in detail about the careers and contributions of the other apprentices *of the same masters* who, for one reason or another, did not later move into the upper strata of the scientific community. But there were probably few young scientists apprenticed to these same masters who, by the standards of judgment commonly employed in the community of scientists, did not develop into reasonably good or excellent working scientists. For, as we have

---

13. See Merton (in Merton, Reader and Kendall, 1957) and Brim (1968) for further discussion of the adult socialization process and Goslin (1969) for a comprehensive review of the literature on socialization.

seen in detail, the master scientists used demanding standards in deciding to accept a postdoctoral apprentice or a junior collaborator. These were generally young scientists whose previous work or whose sponsorship was already enough to distinguish them from most of their age peers. The "errors" in this phase of the selective process would therefore be expected to be primarily of the kind in which young scientists of real talent somehow did not come to the attention of elite masters and only secondarily errors of the kind in which the youngsters who had been accepted as apprentices proved to be mediocre scientists.

It is not possible to compare the patterns of advanced socialization of these future members of the scientific elite at the hands of their elite masters with the patterns of socialization of other young scientists during their graduate and postdoctoral years, for there has been little investigation of socialization in science.[14] But for the special aggregate of future laureates it can be reported that the socialization during advanced apprenticeship helped shape their styles of scientific work, their conception of the role of the working scientist, and their self-image in that role. It did so by transmitting standards of performance, inculcating scientific taste, and reinforcing a sense of confidence in their work.

## Standards of Performance

Even before they found their way to the masters with whom they studied and worked most closely, these young scientists had acquired fairly demanding standards for judging scientific work.[15] Still, with few exceptions, the future laureates report that their standards for assessing performance, their own as well as others', became considerably more exacting in the course of their advanced socialization. The biochemist Hans Krebs reflects:

> If I ask myself how it came about that one day I found myself in Stockholm, I have not the slightest doubt that I owe this good fortune to the circumstance that I had an outstanding

14. Studies that touch upon changes in norms and values in the course of scientific training are Becker and Carper (1956a; 1956b) and Underhill (1966).

15. Self-selection by apprentices can of course involve purposeful avoidance as well as purposeful association. Thus styles of scientific and scholarly work by exceedingly demanding masters of their craft can put off prospective apprentices. It has been said of George Sarton, founding father of the modern history of science, that he conceived of the field as an encyclopedic discipline, requiring a command of many and sometimes esoteric languages, wide-ranging scientific knowledge, and thoroughgoing historical skills. That "immense commitment served to discourage potential apprentices. The master did not offer the easily learnt and transferable technique on which disciplines and more especially 'schools' have usually been built. Instead he set forth an attitude and a vision beyond all but the bravest hearts" (Thackray and Merton, 1972, p. 481).

teacher at the critical stage in my scientific career. . . . Otto
Warburg set an example in the methods and quality of first-
rate research. Without him I am sure I would never have
reached those standards which are prerequisites for being considered by the Nobel committees [1967, p. 1244].

From the interviews with laureates, it appears that the acquiring of
these elevated standards came about in three mutually reinforcing ways:
the masters' own performance provided a model to be emulated; the
masters evoked excellence from the apprentices working with them,
and they were severe critics of scientific work.

To begin with, the masters generally served as role models, teaching
less by precept than by example. By themselves adhering to demanding
standards of work, they sustained the moral authority to pass severe
judgments on work that failed to meet comparable standards. As one
physicist remembered his teacher: "You tried to live up to him. It was
wonderful to watch him at work. Sometimes I eventually did things the
way he did."

The authority of masters gained or reinforced through their own
exemplary behavior enabled them to serve as "evokers of excellence":
bringing out the best in others and, by their own report, eliciting better
performance than ordinarily occurs (Merton, 1960, p. 341ff.).

The evoking of excellence by these elite masters had its own dynamics. As role models for the younger scientists, the masters sometimes
led them to levels of accomplishment they could not ordinarily imagine
for themselves. A physicist laureate and apprentice to Enrico Fermi
described the experience in this way:

If you worked with him, it was as if you were a mediocre tennis
player and you were playing with a champion. You would do
shots that you had never dreamt of. He got quite a few suggestions of important things from his collaborators but there
people were just—I don't know, the solution developed in ten
minutes with him, you see.

In part, the elite masters evoked superior performance by conveying
through their own behavior a sense of how much could be achieved in
scientific inquiry and what it was like to do scientific work of importance.
In part, they did so by inducing a feeling of obligation, a sense of
reciprocity requiring the apprentice to justify through the quality of his
work the master's decision to invest time and effort in training him.
The need to reciprocate in this way was reinforced by periodic signals
from masters, not least through their comments on other students, that
they had little interest in continuing to work with apprentices who were
satisfied with routine performance.

A laureate in medicine described his reinforced motivation to live up to the exceedingly high standards of performance set by H. H. Dale and L. J. Henderson:

> Both were men of great intellectual and personal stature. Each one was so tremendous an individual that you could not help but be impressed. You worked much harder because you felt it was the only thing that was fair to them.

The scientific achievements of many elite masters were seen as evidence of their charismatic character. Apart from the sanctions they could impose, these charismatic figures could make great demands of their apprentices in terms of quality of work expected. As severe task-masters and tough critics, the masters were intolerant of what they took to be shoddy work, laziness, or plain stupidity and of course they transmitted a sense of appropriate standards to their often suffering and sometimes grateful apprentices. A member of Max Delbrück's "Phage Group" describes a characteristic episode:

> After my seminar [Delbrück] took me by the arm to his office to tell me in confidence that it was the worst seminar he had ever heard. It was not till years later, after I had met dozens of people, each of whom had been told that *his* had been the worst seminar Delbrück had ever heard, that it dawned on me that Delbrück told this to (almost) everyone [Streisinger, 1966, p. 336].

Delbrück not only had his apprentices toe the mark while he was on the spot but he served, even in his absence and through the years, as a reference figure whose standards were adopted by the sometime apprentices during their later careers. Gunther Stent, a former junior colleague, depicted this role of reference figure:

> Delbrück managed to become a kind of Gandhi of biology who, without possessing any temporal power at all, was an ever-present and sometimes irksome spiritual force. "What will Max think of it?" became the central question of the molecular biological psyche [Stent, 1969, p. 481].

To summarize, the elite apprentices of elite scientists internalized exacting standards of work through several related processes. They emulated the masters whose own work exemplified those standards; they were led to see things they did not know they knew and to have ideas of a kind they had not had before through the evocative behavior of the masters; and they experienced these elevated standards in practice by having their own work severely evaluated. In the process, they acquired scientific taste, another mark of the scientists being prepared for their roles at the frontiers of scientific development.

## Scientific Taste and Styles of Work

Like other departments of culture, science has its own esthetic. Among the elite scientists, the prime criteria of scientific taste are a sense for the "important problem" and an appreciation of stylish solutions. For them, deep problems and elegant solutions distinguish excellent science from the merely competent or commonplace.

Looking back on their apprenticeships, the laureates typically emphasize that they were able to acquire a better sense of the significant problem. This occurred through the same processes of socialization we have been examining. For example, here is a laureate in physics referring to a role model:

> A great scientist is someone who is working on the right thing and the important thing. Millikan was an outstanding example of this, a man who could pick out the important things. He opened up what later turned out to be large and very important fields in physics. He had a knack for finding what was important to look into.

It is one thing to observe this "knack," another to acquire it oneself. Again, it appears that this is not a matter of didactic instruction but, rather, one of intuition. For this set of creative scientists, learning to identify good problems occasionally involved a growing awareness of the importance of timing and feasibility in selecting significant questions for investigation. Michael Polanyi, occupant of the forty-first chair and master to two American laureates (Wigner and Calvin), remembers his own apprenticeship with the laureate in chemistry Fritz Haber and, in particular, Haber's admonishing him:

> "Reaction velocity is a world problem." . . . He thought I had taken on a problem that was not yet ripe, and in any case, too large for me. . . . Discovery requires in fact something beyond craftsmanship, namely the gift of recognizing a problem ripe for solution by your own powers, large enough to engage your powers to the full and worth the expenditure of the effort [1969, p. 98; first published in 1962].

In the scientific genealogies of the laureates, we noted the numerous sequences of elite masters and apprentices extending over several generations. We now see that aspects of scientific taste are transmitted along chains of masters and apprentices, aided by the apprentices' strong identification with their teachers (sometimes involving what is, for them, a thorough hero worship). Thus, the German laureate in medicine (1931) Otto Warburg reminisces about his master:

> The most important event in the life of a young scientist is personal contact with the great scientists of his time. Such an event happened in my life when Emil Fischer, the second scientist to be awarded the Nobel prize in chemistry, 1902, accepted me in 1903 as a co-worker in protein chemistry, which at that time was at the height of its development. During the following three years, I met Emil Fischer almost daily.... I learned that a scientist must have the courage to attack the great unsolved problems of his time, and that solutions usually have to be forced by carrying out innumerable experiments without much critical hesitation [1965, p. 531].

What the laureate-to-be Warburg reports having learned from his master, the laureate Fischer, is in turn largely reiterated by Warburg's own apprentice, the British laureate in medicine (1953) Hans Krebs:

> If I try to summarize what I learned in particular from Warburg I would say he was to me an example of asking the right question, of forging new tools for tackling the chosen problems, of being ruthless in self-criticism and of taking pains in verifying facts, of expressing results and ideas clearly and concisely and of altogether focusing his life on true values [1967, pp. 1245–46].

Identifying a good problem takes these scientists some distance toward making an important contribution but, as they testify, not the whole way. They must also have a sense of how good solutions look. An excess of concern with precision can easily be mistaken for experimental elegance. As a laureate student of the chemist G. N. Lewis reported: "He led me to look wherever possible for important things rather than to work on endless detail or to do work just to improve accuracy."

Many of the laureates identified "simplicity" of solutions as a mark of scientific taste. Thus, C. N. Yang has publicly described Fermi's emphasis on getting to the essentials:

> He always started from the beginning and treated simple examples and avoided ... "formalisms." (He used to joke that complicated formalism was for the "high priests.") The very simplicity of his reasoning conveyed the impression of effort-

lessness. But this impression was false. The simplicity was the result of careful preparation and deliberate weighing of different alternatives of presentation. The emphasis was always on the essential and practical part of the problem. . . . We learned that *that* was physics. We learned that physics should not be a specialist's subject. Physics is built from the ground up [1961, p. 5].

In composite, the interviews with the American laureates indicate the ways in which association with elite masters form scientific taste. A sense for "the right kind of question" and for the character of its solution develops during the interaction between masters and apprentices and among the apprentices themselves as they pass judgment on the quality of scientific work, new and old, their own and that of others. It develops also as they speculate about the direction their field "should take," identify gaps in basic knowledge, and argue about which problems are "ripe" for solution at the time and which are not. These matters are of evident and prime interest to scientists who intend to help shape the fields in which they work; they might be of less interest to scientists who see themselves playing a more modest role. The substantive aspect of the process of socialization, involving a concern with such basic issues and problems, is congruent with the self-images of the future laureates as scientists located actually or potentially on the advancing frontiers of their special fields.

Thus, the elite masters shape their apprentices and prepare them for elite status by inculcating and reinforcing in them not only cognitive substance and skills but the values, norms, self-images, and expectations that they take to be appropriate for this stratum in science.

## Socialization and Self-Confidence

The laureates who have lived through the process of stringent socialization with elite masters typically emerge as scientists even more confident of their own abilities than before. Without much self-consciousness, they declare that they almost always know a good problem when they come upon it, that they can cope with failure and on occasion transform it into success, and that they are not deeply disturbed by periods of scientific infertility, which they tend to see as fallow times before a new harvest.

Their self-confidence, the laureates typically report, was increased through study with master scientists. In part, it is a result of early success, as these young scientists discover that they are really quite good at their job. As apprentices to scientists for whom good research is commonplace and exceptional research not unusual, these future laureates

have often had the opportunity to take part in important work. One of
them put it this way: "I learned that it was just as difficult to do an
unimportant experiment, sometimes more difficult, than an important
one." And the experience of success, validated by the judgments of
demanding masters as well as of peers, heightens confidence in their
own capabilities as it reinforces their commitment to science.[16]

Their confidence is also reinforced in the course of a demanding
apprenticeship, by comparing themselves, as young scientists, to "the
best" in the field and finding that they measure up reasonably well. Along
these lines, a laureate in physics remarked on his association with two
older Nobelists: "I'm quite sure that I would have been greatly handi-
capped if I had not developed that kind of confidence which one gets by
being able to talk to and measure oneself against the leaders in the
field."

Such reference group behavior, in which people assess themselves
by comparison with selected others, can of course be devastating rather
than reassuring (see Hyman and Singer, 1968; Merton, 1968c).[17] But
in the special case of the future laureates, who were beneficiaries rather
than victims of the process, this leads to their accepting the idea that
they need not be *primus inter pares,* considering that their equals are
themselves first class. They make the comparisons without damage to
their self-esteem. Thus, a physicist laureate who had watched Enrico
Fermi at work reported:

> Knowing what Fermi could do did not make me humble. You
> just realize that some people are smarter than you are, that's
> all. You can't run as fast as some people or do mathematics as
> fast as Fermi. You can't do everything.

This matter-of-fact acceptance of difference is almost echoed in a
published observation by Paul Samuelson, laureate in economics:

> When one heard the late John von Neumann lecture spontane-
> ously at breakneck speed, one's transcendental wonder was re-
> duced to mere admiration upon realizing that his mind was
> grinding out the conclusions at rates only twice as fast as what
> could be done by his average listener. Is it so remarkable that a
> few leading scholars will again and again lead the pack in the

---

16. Having experiments turn out badly and papers rejected by journals or, once published, unnoticed cannot be reinforcing experiences. As the Coles (1973) have found
for the field of physics, scientists who receive no recognition in any form tend to be
discouraged from continuing their work, and this is presumably also true for artists,
writers, and other creative workers.
17. For an incisive examination of reference individuals and reference idols, see Hyman
(1975).

conquering of new territory? If you think of a marathon race in which, for whatever reason, one clique gets ahead, then you will realize that they need subsequently run no faster than the pack in order to cross each milestone first [1972a, p. 157].

For young scientists as talented as these destined laureates, great advantages accrue from being apprenticed to elite masters. Once internalized, standards of performance and scientific taste, unlike material facilities, do not generally depreciate with use. Combined with the special access to resources that often comes from being sponsored by eminent scientists, these basic orientations and the skills and knowledge that go with them contribute greatly to the process by which advantage accumulates for young scientists moving into the elite.

Before turning to the concrete advantages that accrue to apprentices through sponsorship by elite masters, we note that the socialization of young scientists often involves more than the master-apprentice pair. It is not the master who produces outstanding young scientists but, rather, the master in conjunction with the ambiance he creates and the other people he attracts to his laboratory or department. In chapter 6, more will be said about "evocative environments" and their effects on scientists' work. Here we focus on the fact that elite masters usually recruit cohorts of apprentices of unusually high quality and that the young scientists advance one another's development as well as, each of them, learning from the master. Thus, in examining the genealogies of elite scientists, it is evident that especially prolific masters benefited by having a multiplicity of able apprentices at the same time and that these apprentices benefited from close association with one another. The Cavendish Laboratory, first under J. J. Thomson and then under Rutherford, and the Copenhagen Institute for Theoretical Physics under Bohr provide two compelling cases of the joint impact of great scientific masters and cohorts of exceptionally talented apprentices. The roster for 1932 at Bohr's institute, for example, included three who were later to become Nobelists (Landau, Delbrück, and Bloch) and many future occupants of the forty-first chair (among them Edward Teller, Chandrasekhar, Weisskopf, and Bhabha). From retrospective reports, it seems quite clear that in such milieus the young scientists would drive themselves and one another hard (their competitiveness is almost legendary). Thinking back on that year, Weisskopf says:

By the way, I should say that these people, the group, "us"— we were not all yes-men, absolutely not. On the contrary. Out of sport, rather, we liked to contradict the great man. Also in order to start a good discussion apart from the fact that we really wanted to understand the things. We were not yes-men at all, and Bohr had a very hard time at that time with us. I was still very young, but the great "fighters" against Bohr were people like Landau and like Bloch. I'm not sure whether Bloch was involved in this, but he could have been, and Teller. There was not at all a "yes" spirit, and Bohr had to fight extremely hard [Kuhn and Weisskopf, n.d., p. 17].

## ELITE SPONSORSHIP

The system of apprenticeship to elite masters in science involves a mixture of the two types of social mobility identified by Ralph Turner (1960): "contest mobility," based on the quality of role performance, and "sponsored mobility," in which the elite recruit their successors.

As authorities and influentials in their field, elite masters can mobilize resources on behalf of their apprentices, by arranging for fellowships and jobs and facilitating access to research grants and publication in major journals. Even when they do not actively intervene, the fact that they have accepted certain young scientists to work with them provides credentials recognized by other members of the scientific community. To have been a student of Fermi or Warburg or Bohr, as the case may be, speaks volumes even for those who do not later rise into the topmost stratum of scientists.

But, more often than not, the masters do intervene. The role of master includes the role of sponsor, particularly for those apprentices who are judged to be among the best of the lot and promising to contribute notably to science. Masters differ, of course, in the extent to which they perform the role of sponsor, sometimes engaging in it without great awareness that they are doing so. Sponsorship often appears as a latent rather than manifest function when the established authorities appraise the work and performance of young scientists, both in correspondence and in conversation. Their judgments spread to other scientists at the centers of influence, whose recommendations are in turn sought by still others within the national and international community of science.

Thus, throughout much of his forty-year-long correspondence with Einstein, Max Born in effect introduced various young scientists to Einstein as he described and appraised the work they were doing with him. He referred, for example, to "my private assistant" Dr. Brody as "a very clever man."

But sponsorship is not enough to ensure passage into the ultra-elite; early promise must develop into substantial achievement, as was the case for Born's other students and assistants, such as the laureates

Pauli, Heisenberg, Stern, and Goeppert Mayer (but not for Brody). We can see both the expanding network of sponsorship and its impact on scientific careers in this passage of a letter from Einstein to Born:

> First of all, the matter of our young colleague Dehlinger, whom you have written about to Berlin. We are now getting a lot of money for astronomical research which I have entirely under my own control. Would he like to work in astrophysics? I could appoint him for the time being at a salary of approximately 6,000 M a year, possibly more if the present adverse conditions require it. He would then work with Freundlich. Photometric investigations on star spectra. If, however, he prefers a post in technology, I have other connections who could try to find something for him. It is difficult nowadays to make a living by scientific work alone. Let me have further details as soon as possible [Born, 1971, pp. 20–21].

But the "gifted young physicist" Dehlinger, though sponsored by the highest authority, nevertheless disappeared from view, so that Born reported, much later, "I know nothing of his subsequent fate" (1971, p. 24).

Other students sponsored by Born had a different history. Thus Born's efforts to help Wolfgang Pauli along seem almost superfluous. By age 21, Pauli had already written his basic monograph on the theory of relativity, about which Born wrote in retrospect:

> Sommerfeld [then 51] was originally supposed to write it. He got Pauli to help him with it, but Pauli made such a good job of it that Sommerfeld handed the whole thing over to him. It is truly remarkable that a young student of 21 was capable of writing so fundamental an article, which in profundity and thoroughness surpassed all other presentations of the theory written during the next thirty years—even, in my opinion, the famous work by Sir Arthur Eddington [1971, p. 65].

Born and Sommerfeld were, it appears, generous in their sponsorship of young Pauli. From the perspective of these elite masters, sponsorship is an obligation that comes with established position. As they see it, they are engaged principally in contributing to the advancement of science by providing opportunities for early-identified talent. This is illustrated most vividly in the case of laureate-to-be George Beadle. As a postdoctoral fellow at Cal Tech, Beadle, a geneticist, spent considerable time working with Boris Ephrussi, an embryologist who was a visitor at Cal Tech. The two agreed that their joint interest in the embryology of *Drosophila* would be best served by Beadle's going to work in Ephrussi's laboratory in Paris. This seemed a sensible plan to

Nobelist T. H. Morgan, chief of Cal Tech's laboratory of biological science, and, as Beadle reports, "Morgan arranged to continue my Cal Tech salary, then $1500 annually." This was more or less routine but, Beadle continues, "Only years later did I find that this stipend was almost surely provided by Morgan personally" (1974, p. 6). Apart from his own research, Morgan did indeed contribute to the advancement of science by sponsoring young Beadle. It was in Paris that Beadle began to develop the "one-gene-one-enzyme" hypothesis that would provide the groundwork for the new field of biochemical genetics and also for Beadle's Nobel award, years later.

In the meritocratic system of science, elite masters can do nothing for scientific talent that does not come their way or is not at least called to their attention by others in the social system of science whose judgment they respect. Some of the masters have expressed concern about the problem of talent that might never be developed and so is lost from view altogether, urging, as a step toward solution, the most open access to educational opportunity.

A few elite masters find their position in the power structure of science rewarding in its own right. They take pleasure in the exercise of influence that their achievements in science have made available to them. A chemist laureate, for example, is thoroughly self-aware, even enthusiastic, about his position within the international organization of science:

> I've accumulated over the years a lot of what I call wealth. Many, many connections. I keep them all extremely active. All for one purpose: to do my research and teaching better. Science in this day is a world-wide proposition. I have three worldwide empires which I can call on: research, teaching and industry. So if it's necessary to send my students to Bangkok or somewhere else, I can do it easily. If it's necessary to get someone to teach one of them who isn't here but in Europe, I can do it. All of this is done for this reason, taking care of the students. There aren't many people "rich" enough to do that: rich in the sense of connections, power, financing for research, influence.

This laureate's plain-spoken satisfaction with having accumulated influence to exercise is not typical. For the most part, sponsorship by the elite masters is subdued. But recognition of its importance, particularly under conditions of stress, is widespread among them. Just as the laureate chemist we have encountered above "takes care of his students," so laureates were often taken care of when they were young. In retrospect, J. D. Watson speaks from the perspective of an unproven young scientist when he remarks: "I think it is extraordinarily important that you have a scientific patron because there'll be times when you are bound to strike it bad and you'll need somebody to convince people that you are not irresponsible" (Bradbury, 1970, p. 59). And, as *The Double Helix* (1968) testifies, Watson was taken care of, more or less consistently, even when he seemed most irresponsible.

Eminent sponsors are not only better equipped by their power and influence to look after their apprentices; they can also increase the visibility of those apprentices. Young scientists are often known, if not finally judged, by the distinction of their masters. And with the growth of big science that brings with it more anonymity, visibility may become increasingly important in the early stages of developing a professional reputation. The visibility conferred by having a well known master means, among other things, that the young scientist who has not yet acquired a scientific identity will have a better chance of having his work noticed, read, and used than other scientists doing work of the same quality.[18]

Greater visibility in general and to key influentials in particular should serve to bring the protégés of laureates more resources for research, better jobs, and perhaps more recognition than they would have gotten on their own. All this is no substitute for talent, but it does serve to augment the advantages that accrue to recognized talent. If nothing else, apprentices of elite scientists are less likely to be lost in the crowd of newcomers. It should be noted, however, that this visibility also means that their weaknesses and foibles, as well as their strengths and accomplishments, are exposed to scrutiny by demanding scientists.

## REENACTMENT OF THE MASTER'S ROLE

The scientific genealogy of the laureates indicates, among other things, that the elite apprentices in due course themselves became elite masters. This sequence allows us to examine the other side of the reciprocal relationship between master and apprentice which, until this point, has been examined mainly from the perspective of the apprentice.

As they take on the role of the master, elite scientists tend to reproduce in their own attitudes and behavior some of the same patterns they witnessed when they were apprentices. This is only one of several reenactments of role-defined patterns that occur at various stages in the laureates' careers, as we shall see in later chapters (see also Zuckerman and Merton, 1972, p. 342). It does not mean, of course, that the laureates, in their capacity as masters, repeat every detail of their own

---

18. See Cole and Cole (1973, pp. 99–110) for an analysis of the determinants of visibility and S. Cole (1970) for the effects of visibility upon the reception of a scientist's work.

masters' idiosyncratic styles of role performance. Some of them in fact report that they turned away from role behaviors that, as apprentices, they found aversive. Those trained in the great chemical laboratories of Germany in particular were determined not to import the autocratic style of the *geheimrat* professor into American laboratories. But, in general, laureates tended to define the role of master much as their own teachers had done. Thus a physicist laureate, reflecting on the fact that his master had made no great effort to teach him substantive details, went on to say:

> It's the attitude which is the great benefit. With all due modesty, I think my students will probably say the same thing. I don't believe that they learn so much in detail. It's not that I've taught them how to use certain equipment and so on. I'm not very good at that myself. But perhaps the approach, the way I look at a problem, what it is that comes out [of the work] and how— maybe this is what they benefit from.

Laureates adopt other aspects of their masters' style in their own relationship with apprentices. In an impressively sustained figure of speech, a geneticist described himself as acting much like his teacher in "letting them sink or swim, though not giving them enough rope to drown themselves" and taking care, when assessing the scientific poten- tial of students, to observe their "buoyancy" when left on their own.

A common theme in the laureates' perceptions of the teaching role is their frustration with what they regard as the frequent necessity for misallocating their limited energies. They see that role, just as they say their elite masters generally did, as primarily that of training talented students who later will significantly advance scientific knowledge. In this sense, they adopt an elitist orientation to training. They see little merit in merely multiplying the numbers of scientists.

As successors to their elite masters, the laureates in effect define their own role as teachers as that of preparing *their* successors. But they often remark on finding themselves confronted by the ironic necessity to spend more time and effort with the many less promising students than with the few truly gifted ones. Their elitist role conception, which leads them to focus on the few, therefore conflicts with the teacher's role that is actually assigned to them most of the time. One geneticist said:

> It's a social duty to put graduate students through a normal course of training. But it's a bit doubtful that the average gradu- ate student produces anything that is, by itself, worth the efforts of the professor it's done with. It takes more effort on the pro- fessor's part to see that it gets done right than if he did it with- out the student.

Then, expressing his ambivalence toward the role, he continued, "Of course, the payoff sometimes comes later when the student goes on to work by himself."

But the elite tradition of scientific knowledge, passed on through the genealogy of elite masters and apprentices, continues to be reinforced by the emergence of a few young scientists who measure up to that tradition. A laureate reports the great satisfaction that comes from teaching able and autonomous students:

> Some of them take off and are so wonderfully self-running that they simply drop in every week to give you a new surprise and to find out what is going on. Others, you have to hold their hands and keep close tabs and make suggestions or they get into trouble. I've had a few very wonderful ones. You just sit back and admire what goes on and just get out of their way.

Just as apprentices take pride in having had distinguished masters, masters take pride in apprentices who become distinguished scientists. Among other things, this vindicates their judgment. But the motivation to concentrate on the most promising young scientists is reinforced by more than personal satisfaction with the result. The reward system of science confers esteem upon masters whose students develop into first-rate scientists. Thus direct personal reward and secondary social reward coalesce to reinforce the elite tradition and the strong interest in having outstanding students.

Within the social system of science, the number of such outstanding students is adopted as a measure of the masters' intellectual (as distinct from merely organizational or political) influence. The laureates take that measure as self-evident. Yang, for example, uses it as a yardstick in referring to Fermi:

> Enrico Fermi . . . as a teacher at Chicago had directly or indirectly influenced so many physicists of my generation that it suffices to let the record speak for itself. The following is a list of names of some of the physicists who received their graduate education in Chicago in the years 1946–49 . . . H. M. Agnew, H. V. Argo, O. Chamberlain, G. F. Chew, G. W. Farwell, R. L. Garwin, M. L. Goldberger, D. Lazarus, T. D. Lee, A. Morrish, J. R. Reitz, M. N. Rosenbluth, W. Selove, J. Steinberger, R. L. Sternheimer, S. Warshaw, A. Wattenberg, L. Wolfenstein, H. A. Wilcox, C. N. Yang [1965, p. 678].[19]

---

19. By 1972, about twenty years afterward, seven of Fermi's students of the Chicago period had been elected to the National Academy of Sciences: Chew, Garwin, Goldberger, Steinberger, Chamberlain, Lee, and Yang. The last three are also laureates.

The yardstick is explicitly used in another instance when the physicist Luis Alvarez unself-consciously notes that "one indicator of Ernest
Lawrence's influence is the fact that I am the eighth member of his
laboratory staff to receive the highest award that can come to a scientist—the Nobel prize" (1969, p. 1090).[20] Thus, although a very few
laureates forgo the costs and benefits of training their successors (the
British physicist Dirac is nearly unique in reputedly never having had a
student), most do not. This is so in part because the reward system provides incentives for what would otherwise be an unattractive role requirement and of course some enjoy the process in spite of its costs.


## AMBIVALENCE AND CONFLICT

As we might expect, the relationship between master and apprentice
did not always run smoothly. Both for the apprentice about to enter
upon this important phase of his career and for the master wanting to
get on with his work, the relationship was typically fraught with strong
affect.

For the young scientists, intense feelings were often generated even
before they entered on their apprenticeship, for the selective recruitment
of apprentices involved the risk of their being rejected by the masters
they preferred to work with. Having emerged more or less successfully
from this competition—as we have seen, some laureates were not accepted by masters of first choice—they then faced the problem of establishing a satisfying pattern of training and work. The expectations of
master and apprentice were not always congruent, with resulting
ambivalence and conflict.

Disparate expectations about the closeness of the relationship provided one source of conflict. Some apprentices felt that they suffered
from neglect: "Sometimes it was weeks that I didn't see any of these

---

20. The eight laureates include four who were apprentices or junior collaborators of
Lawrence—Alvarez himself, Willard Libby, E. M. McMillan, and Glenn Seaborg—and
four who were otherwise associated with the Radiation Laboratory—Donald Glaser,
Emilio Segrè, Owen Chamberlain, and Melvin Calvin. Pride in being a member of the
Berkeley group and of the Radiation Laboratory, in particular, is symbolized by the
passing on of a white waistcoat from E. M. McMillan to Emilio Segrè to Donald Glaser;
each wore it at his Nobel ceremony. As Glaser put it, it is regarded as "a very valuable piece of equipment" (Nobelstiftelsen, 1960, p. 53). In his reference, Alvarez was
plainly using the common practice, examined in chapter 2, of using the number of
laureates connected with an organization to gauge its scientific distinction. He also
adopted the practice, not uncommon, as we have seen, of maximizing the tally by including all those laureates associated with the organization at any time and in any
capacity rather than confining the list to those who were "products" of Lawrence's
laboratory.

people." In other cases, the satisfaction that came with working closely with a leading figure in the field was mixed with concern over being dominated and deprived of intellectual and personal autonomy. Thus Victor Weisskopf, an occupant of the forty-first chair in physics, pondered:

> It is a strange feeling when you work directly under Bohr as his collaborator for a short time. You lose your personality; you have no responsibility for anything; you know you have to be at the institute at ten o'clock, your mind is blank until you come and then it is his mind that fills your head and you discuss with him and help him express things. Actually you are only there as an echo, still a deep echo that goes through your mind.... Therefore I always thought it was a wonderful but also a terrible thing to be caught by Bohr. Of course, if Bohr says "will you work with me?" you say yes with enthusiasm. But I considered myself lucky that this didn't happen too often. I do not think I would be what I am if it had happened a little more [Kuhn and Weisskopf, n.d., p. 15].

More often, with masters less overwhelming than Niels Bohr, apprentices wanted close attention from them, not regarding their autonomy as threatened by temporary direction or even domination. A biochemist reported having been quite calculating about this:

> When I came to his laboratory, he said. "What would you like to do? Do you have a problem that you'd like to work on?" I said: "There are things I had in mind that I could do but I would rather work with something you are interested in." This is advice I'd give to anyone. You get far more out of the experience of training—no matter what his intentions are at the time you start—if you work on something he is interested in.

The feeling of apprentices of too little or too much attention from the masters has its counterpart feeling among masters that apprentices either expect too much of them or fail to respond adequately to the great deal that is being provided for them. We saw that laureates in their capacity as master report that some students make excessive demands upon their energies, keeping them from their work. In some extreme cases, as with Bohr, an established in-group of disciples would constitute themselves gatekeepers to protect the great man from overwhelming demands. As reported by one observer (who was himself not excluded in this fashion):

> It was very difficult to get into Copenhagen. I have seen cruel things happen if you come and cannot get through the "guard."

> Bohr was surrounded by five or six, maybe more, of his disciples,
> who were a very arrogant crowd. If you were not accepted by
> them, you would have a very difficult time with him.

Collaboration between master and apprentice on research provides
another source of strain upon the relationship as well as satisfaction
with it. Apprentices find that, even when they are granted co-authorship,
their joint work is often credited to their eminent masters. As one laureate summarized it: "The man who's best known gets more credit, an
inordinate amount of credit." Although junior authors may benefit from
joint publication with well-known scientists—it is one way of increasing
the chances that their papers will be read—credit for the important
ideas, if not the burdensome work, often goes to the senior author. A
laureate in medicine observed his own response to jointly authored
papers:

> You usually notice the name that you're familiar with. Even
> if it's last, it will be the one that sticks. In some cases, all the
> names are unfamiliar to you, and they're virtually anonymous.
> But what you note is the acknowledgement at the end of the
> paper to the senior person for his "advice and encouragement."
> So you will say: "This came out of Greene's lab, or so-and-so's
> lab." You remember that rather than the long list of names.

The "Matthew Effect," in which greater recognition accrues to the
collaborating scientists of greater reputation (see Merton, 1968f, based
on the interviews with laureates), thus involves a major cost for many
apprentices whose contributions get lost from view. But it is less costly
for the relatively few apprentices who ultimately enter the ultra-elite as
Nobel laureates or as undisputed occupants of the forty-first chair. For
the scientists who go on to do first-class work of their own (or with
apprentices whose own contributions are in turn being obscured) are
often credited retroactively for their early collaborative work. A laureate
in medicine cited a variety of such cases in which once unknown youngsters "who have been identified with such joint work and . . . then go on
to do good work later on, get the proper amount of recognition." From
the privileged perspective of one who actually had the experience, a
laureate in physics even suggests that the unknown collaborator can
eventually profit in the process of retroactive crediting: "The junior
person is sometimes lost sight of, but only temporarily *if* he continues to
do important work. In many cases, he actually gains in acceptance of
his current work, and in general acceptance, by having once had such
an association."

Still, disagreement over the extent of contribution by masters and
apprentices is sometimes the source of outright conflict between them.

This involves not only the operation of the Matthew Effect, which tends to give disproportionate credit to the distinguished master, but also a sense on the part of the apprentice that his contributions to collaborative research were not even noticed or adequately valued by his master.

In perhaps the most extreme case of this kind of conflict, we find the eventual laureate Herman Muller, more than two decades after his apprenticeship, writing with bitterness that T. H. Morgan, his laureate master, only reluctantly accepted the "modern" view of inheritance embodied in Mendelism and did so largely in response to the growing corpus of experimental data and interpretations contributed by his younger colleagues, among them Muller himself. Muller observes:

> The great bulk of the facts of real significance subsequent to 1911, and practically all after 1913, were found by younger workers quite independently of guidance from him in experiments which they had planned on the basis of their own more advanced viewpoints. Their results and interpretations were, however, later accepted by Morgan [1934, quoted in Carlson, 1966, p. 91].

But it was not only Morgan's scientific conservatism (or what Muller called his "reactionary" attitude) that created difficulties in the Drosophila Room, the small crowded laboratory appropriately named after the fly that was the prime research material used. More important from Muller's point of view, Morgan never properly recognized the contributions of his apprentices.

> [Our] results and interpretations were, however, later accepted by Morgan and presented chiefly by him to the scientific and lay public, so that these developments have sometimes been referred to, especially in circles farthest removed from contact with the original work, as "Morganism" [1934, quoted in Carlson, 1966, p. 91].

Muller was nothing if not consistent in his version of the relative contributions of Morgan and his apprentices. In our interview, more than a half century after his apprenticeship, Muller could still summon up his anger in recalling his years with the master. Echoing the two themes in his earlier account—Morgan's resistance to the ideas later credited to him and his insensitivity to matters of priority and credit—Muller observed:

> Morgan didn't play fair with his young students. He did not have the point of view which is represented here. [He held a copy of *The Mechanism of Mendelian Heredity*.] He later put up the appearance of having pretty much originated it and proved it.

But he did not accept this point of view—which [E. B.] Wilson had and which we also got from Locke. It was the toughest sort of fight all the way through. He finally did change his mind when the results made it unmistakably clear but he really held up our progress before he accepted it. It was irritating especially when controversies emerged later with people like Castle who were much more positively negative, much more openly set against the clear-cut ideas of Mendelian chromosome inheritance. And then Morgan got the credit for being the big hero who set them straight, you see. In many parts of the world . . . they speak of the Mendel-Morgan point of view.

Emphasizing once again the theme of Morgan's insensitivity to the recognition of scientific priority and the climate of competitiveness and fear of plagiarism that this created in the Drosophila Room, Muller continued:

It was Morgan's idea that the only things you had to credit were experimental findings, not ideas. What was carefully recorded were new experiments which proved an idea but not who it was who suggested the idea or how it might be proved, what experiments could be done. I know Sturtevant [one of the four Morgan apprentices] [21] has said that one should acknowledge anything that's published or any data that anybody gets but if it's just a matter of ideas, that's free. I couldn't go along with that. Ideas were one of my main stocks in trade at the time since I had to put in so many hours being an assistant in undergraduate classes and didn't have much time to get the actual data. What happened in this case, it seems to me, is that the major man—if I may speak freely—stole from the younger men and then some of the younger men stole from each other and were able to get away with it.

For Muller, the paradigmatic episode in the misallocation of credit was the authorship of *The Mechanism of Mendelian Inheritance* and Morgan's subsequent behavior:

We had the project of the book just before Sturtevant went into the army and most of it was worked out in his absence. [He was second author, wasn't he?] Morgan insisted on that. He said, "Between you and Bridges, it doesn't matter," so we decided to toss a coin for that and my name happened to be first in the coin toss. [The sequence of authors was thus T. H. Morgan, A. H. Sturtevant, H. J. Muller, and C. B. Bridges.] Sturtevant has said a number of times that it seems to him that the book was written by me. I wrote quite a lot of parts of it—parts Morgan

---

21. A. H. Sturtevant was also one of Morgan's junior collaborators and went on to a distinguished career in genetics.

wouldn't have written. Yet as you perhaps know, within a few years of that book, Morgan got out a book entirely under his own name, *The Physical Basis of Inheritance,* which is practically the same thing. It was this book which was so widely translated into foreign languages. Lots of people have cited it, only citing Morgan's name. So that's why I got out of that lab the first opportunity I could.

Life in the Drosophila Room was somewhat less Hobbesian for Sturtevant, a reminder that different perspectives on social reality are afforded to those occupying different social statuses.

There was a give-and-take atmosphere in the fly room. As each new result or new idea came along, it was discussed freely by the group. The published accounts do not always indicate the sources of ideas. It was often not only impossible to say, but it was felt to be unimportant, who first had an idea. . . . I think we came out somewhere near even in this give and take, and it certainly accelerated the work [1965, pp. 49–50].

For Sturtevant, Morgan's attitude was "compounded of enthusiasm combined with a strong critical sense, generosity, openmindedness, and a remarkable sense of humor" (Sturtevant, 1959).

These conflicting reminiscences suggest that the accounts by other laureates of idyllic apprenticeships with their masters could be supplemented by reports of conflict from others who studied under the same masters. It is not only that time erodes memories of bitterness and conflict between masters and apprentices and that one-time apprentices are reluctant still to speak of them, but also that laureate apprentices probably had particularly benign apprenticeships owing to early recognition of their scientific promise and to their having contributed their share to collaborative research.

More will be said about patterns of scientific collaboration in later chapters. Here we have noted only the bearing of these patterns upon the relationship of elite masters and apprentices. As with significant human relations generally, this one has its own complement of tension and release, of losses and gains, of ambivalence comprised of conflict and agreement. In this respect, as throughout this chapter, our principal focus has remained the social processes that make for the formation of a scientific ultra-elite—constituted here by the Nobel laureates and their functional equivalent, the occupants of the forty-first chair—and how that ultra-elite is perpetuated.

# Chapter 5

# MOVING INTO THE SCIENTIFIC ELITE

As is true in every sphere of human activity, popular myth notwithstanding, few scientists have been thrust from obscurity into the ultra-elite. The youngest laureate by far was the English physicist W. L. Bragg, who was 25 when he shared the prize with his famous father, W. H. Bragg. But even by that early age, he had acquired a public identity in physics through several years of collaborating with the elder Bragg on their pioneering work in X-ray crystallography. Similarly, the most youthful Americans who received the prize—Joshua Lederberg, T. D. Lee, C. N. Yang, and James Watson, for example—had already been widely recognized for their work. The great majority who received the prize at a later age—such as George Beadle, R. B. Woodward, Linus Pauling, Richard Feynman, and Eugene Wigner—had of course been acknowledged members of the scientific elite for years.

We shall therefore find few surprises in tracing the upward mobility of the laureates from the time of their early research and first jobs through their prize-winning work to the eve of the Nobel prize itself.

Their ascent into the ultra-elite follows an almost commonplace script. The future laureates begin their careers by working hard and long at their research. Consequently, they produce a good deal of it. This is generally judged as being of superior quality by specialists in their field, with the result that they acquire a growing reputation in the wider community of scientists who depend on such judgments since they cannot assess the specialized work for themselves. Growing recognition tends to bring better facilities for research, better students and colleagues who are tuned in to the social networks through which scientific reputations are transmitted, and still more rewards. These in turn help the scientists to do more and better work. This chapter, in an effort to understand how a subset of the scientific elite comes to receive a Nobel prize, examines the ways in which these interacting advantages accumulate.

## EARLY ROLE PERFORMANCE

Practically from the beginning of their careers as working scientists, the behavior of those future laureates who did their prize research in the United States, deviates from the mean of their age peers. Most were energetic producers of scientific publications from the start. While still in their twenties, they published an average of 13.1 papers, strikingly more than the entire lifetime average of 3.5 papers that has been attributed to the general population of scientists (Price, 1963, p. 45). When we compare them not with the run of scientists, many of whom publish next to nothing, but with a matched sample of the relatively productive scientists of the same age drawn from *American Men of Science*,[1] we find that in these early years the future laureates published at about twice the rate of these fairly productive age peers. Later on, as we shall see, the gap widens.

In part, the future laureates publish more than other scientists when they are young by starting to publish earlier. They are not quite 25 years old, on the average, at the time of their first scientific papers while the

---

1. A subsample of forty-one laureates (those who had been interviewed) and forty-one other scientists listed in *American Men of Science* were matched in terms of (1) age, within a five-year span; (2) field of specialization; (3) organizational affiliation at the time of the award (university, government, independent nonprofit and industrial laboratories); and (4) first letter of last names, for reasons that will become clear in chapter 6. The matched sample is itself selective in terms of research productivity, since inclusion in *AMS* requires a doctoral degree or its equivalent and some research activity. As it turns out, the sample includes two members of the National Academy of Sciences. This suggests that the sample may be biased toward inclusion of productive scientists but, since this bias works against the hypotheses developed in the chapter, it only adds weight to the confirming data.

matched sample are past 28. But, though the average age at first publication differs by a significant three years, the range does not. The youngest author among the "American" laureates was Carl Cori at age 19 and the oldest was André Cournand at 35—not much different from the comparable ages of 21 to 35 in the matched sample. But, as we shall see later in this chapter, contrary to the belief that the major discoveries are disproportionately the product of quite young scientists (Lehman, 1953), in only a few cases did these early papers report the research that was to bring them the prize. Still, the fact remains that the future laureates were then producing far more work and, as assessed by their fellow scientists then and later on, work that was far better than that being done at the same time by the matched sample of productive age peers.

## EARLY PUBLICATION:
## INDIVIDUAL AND COLLABORATIVE

In large part, the early productivity of future laureates in the form of publications can be attributed to their working under scientists who were themselves productive. As we have seen in chapter 4, their masters were role models and taskmasters and this contributed, in its way, to their establishing patterns of work conducive to doing significant research. Enrico Fermi's apprentices among the laureates, for example, invariably refer to his enormous energy, intense concentration on research, and considerable scientific output. (See Holton, 1974, for an account of Fermi's Rome group.) Along somewhat different lines, the collaborators of the energetic inventor of the cyclotron and leader of scientific teams, Ernest Lawrence, focus on his style of "getting things done," a style that not a few of them emulated. And although some future laureates reported having learned the art of "cutting corners," they were also typically warned about the penalties of rushing into print with half-baked ideas.

The laureates derived other, more immediately visible benefits from their early association with scientists of the first rank. One of these, as has been noted, was in the form of collaborative publications with their masters. Much of the difference in early published output between laureates-to-be and the matched sample of scientists resulted from such collaboration, with the laureates having their names on 7.9 jointly authored papers while still in their twenties as compared with 2.9 for the matched sample.

Eminent masters exercised *noblesse oblige* [2] not only by lengthening the bibliographies of their young associates (granting them joint authorship) but often by heightening the visibility of the junior's contributions to the research in arranging to have their names appear first in the list of authors.[3] Sometimes, the laureates report, their mentors even removed their names from research papers so as to give the up-and-coming youngsters a better chance for recognition (although this is not without its own costs in diminished visibility to the scientific audience). In the absence of data on the early senior collaborators of scientists in the matched sample, we assume that, on the average, they less often enlarged the record of publication of their young associates in this way. Since these senior scientists were presumably less productive than the masters of laureates described in chapter 4 and their standing in the field was presumably less elevated and secure, they probably were not so willing to share authorship with their young co-workers, much less to give them highly visible listing when authorship was shared and least of all to give up authorship altogether. As a result, future laureates probably received greater proportionate recognition for their work in these early years than did many of their age peers.

Apart from the substantial difference in the number of joint papers, there was no great difference between the average number of single-authored papers published by future laureates in their twenties (5.2) and those published by the matched sample of productive scientists (4.3). Such purely quantitative differences were surely not large enough to be noticeable to specialists, let alone to the larger community of scientists. In a way, it is not the fact that there was a difference in the number of early papers published by the two groups that calls for explanation but, rather, why the difference was so small. The answer may lie in the more exacting socialization and social control experienced by the youthful laureates-to-be. As we have observed in chapter 4, they had been exposed to their masters' standards of what was worth publishing and what was not—demanding standards that shorten the list of publications from what it might otherwise have been but strengthen the impact of what is put into print. Thus, the biologist Seymour Benzer, an occupant of the forty-first chair, reports Max Delbrück's insistence that he stop publishing so much, a friendly suggestion finally communicated to Benzer's wife in this note:

---

2. See Malinowski (1961) and Homans (1974, pp. 214–17, *passim*) on *noblesse oblige* as a social pattern.

3. The rapid growth of co-authored papers in science has accentuated the use of various devices for allocating intellectual property rights among collaborators. Procedures adopted for this purpose by laureates in comparison with other scientists are discussed in chapter 6 and in Zuckerman (1968).

> Dear Dotty, please tell Seymour to stop writing so many papers. If I gave them the attention his papers *used* to deserve [presumably when Benzer limited his output], they would take all my time. If he *must* continue, tell him to do what Ernst Mayr asked his mother to do in her long daily letters, namely, underline what is important [Benzer, 1966, p. 157].

With this in mind, we might conjecture that the quantitative difference in output between the future laureates and their fairly productive counterparts would have been even greater had the laureates not curbed their production in order to improve the qualitative value of their research. There is evidence to support this conjecture. The significance or quality of scientific work can be gauged by the extent to which scientists make use of it, and a rough measure of use is provided by the number of citations it received.

By this measure, the earliest published work of laureates was far more significant than that of even the most productive scientists in the matched sample, to say nothing of the general run of scientists. It had enough scientific interest to remain in active use for quite some time. For example, papers published while they were still in their twenties by laureates who received the prize in the years 1965–69 were still receiving an annual average of eight citations some thirty years later, in the scientific literature of 1965. This is sixteen times the average of .5 citations of papers of the same vintage published by the matched sample. Their relatively infrequent current use suggests that these early papers by nonlaureates were less consequential or had outlived whatever usefulness they once had.

The count of citations to the early papers by laureates, it must be emphasized, deals only with those made *before* the prize was awarded. The long-continuing citations therefore cannot represent retroactive interest in the early work of newly elevated Nobelists. Thus, the *Science Citation Index* shows that papers published by the physicist Hans Bethe while he was in his twenties (1926–35) were still cited twenty-eight times in 1965, two years before he was crowned a laureate (although many years after he had become a notable). By contrast, even the most prolific scientist in the matched sample had a total of only four citations in the same year to his comparably early papers.

It is not only the early research by laureates that tends to have great staying power; this is true of research throughout their careers. In this respect, their scientific papers differ from most others, which enter into limbo soon after publication, practically unmentioned in the work of other scientists. And generally those that are mentioned do not survive long. The "half life" of papers published in the fields recognized by Nobel awards—that is, the age of papers receiving half of all references

being made in the field—runs from about 4.6 years for the physics literature to 7.2 and 8.1 years for physiological and chemical publications
(Price, 1963, pp. 79–81; Burton and Keebler, 1960). These figures
contrast sharply with the half life of publications by laureates named
between 1965 and 1969. The papers of the physicists had a half life of
7.6 years, the life scientists 12.5 years, and the chemists 17.7 years.
Thus the published work of the laureates has greater durability as well
as greater intensity of use. To take a reading for one year, the "most
cited paper" published by each of the recent recipients of the prize
(1965 to 1969) received a lofty average of 42.3 citations in 1965.
Further, these most heavily cited papers had been in print for an average
of 12.2 years when the reading was taken.

In short, although the youthful research of most laureates did not
include the prime contributions for which they received their prize, as
we shall see when we examine that phase in their careers, the papers
reporting their early research had unusual staying power in a scientific
literature characterized by rapid obsolescence.

## FIRST JOBS

Their copious productivity made the future laureates attractive
recruits to the staffs of universities and research institutes. But, just as
young scientists differed in attractiveness to these organizations, so the
organizations differed in attractiveness to the young scientists. And, just
as we observed an early accumulation of advantage for the relatively
few individual scientists on their way to entering the ultra-elite, so we
shall now observe an accumulation of advantages for the relatively few
elite organizations which identified, attracted, and retained these
talented scientists. The processes leading to the concentration of prospective laureates in a few places, observed with regard to their formal education and apprenticeships (see chapters 3 and 4), continue to operate
for their first jobs. Organizations and individuals are involved in a
process of mutual attraction, social selection by organizations and self-
selection by individuals continued to reinforce one another in this phase
of the developing careers of the laureates.

Results of these selective mechanisms are shown in table 5–1, which
relates the location of their first jobs to the academic origins of laureates
and the matched sample of scientists. The table contains a number of
interesting patterns. To begin with, it suggests that the elite organizations—the top-ranked universities and such prestigious research organizations as the Institute for Advanced Study and the Rockefeller Insti-

Table 5-1. Doctoral Origins and First Jobs of Laureates (1901–72) and Matched Sample of Scientists

PRESTIGE OF INSTITUTION OF FIRST JOB

| | LAUREATES | | | | MATCHED SAMPLE OF SCIENTISTS | | | |
|---|---|---|---|---|---|---|---|---|
| Doctoral Origins | Elite Institution [a] | Other U.S. University | Non-academic | Total | Elite Institution | Other U.S. University | Non-academic | Total |
| Elite University | 69% | 22% | 9% | (64) | 35% | 65% | — | (17) |
| Other U.S. University | 50 | 20 | 30 | (10) | 16 | 63 | 21 | (19) |
| Foreign University | 61 | 28 | 11 | (18) | — | 100 | — | (5) |
| All Universities | 65 | 23 | 12 | (92) | 22 | 66 | 10 | (41) |

a. Elite institutions are the thirteen elite universities, the Rockefeller Institute (later, University) and Foundation, the Institute for Advanced Study at Princeton, and the Mayo Clinic.

tute—were able to distinguish between young scientists of differing talent and to attract the most promising ones to their staffs. The category of elite institutions, numbering just sixteen altogether, appointed about three times the proportion of the ninety-two future laureates (those who did prize-winning research in the United States) to their first jobs than the category made up of all the hundreds of other American universities: 65 percent as against 23 percent, with the remaining 12 percent being appointed by other private and public research organizations.

The pattern is just the reverse for the matched sample of fairly productive scientists who, we know from hindsight, did not end up in the ultra-elite. Here, the elite institutions appointed only one third the proportion appointed by the other American universities: 22 percent compared with 66 percent.[4]

Table 5–1 also provides clues to the selective mechanisms that may have brought about this concentration of future laureates in the small number of elite institutions. For various reasons, these institutions were in a favorable position to identify potential elite scientists early. As we saw in chapter 3, throughout most of the period under review the elite universities accounted for a disproportionately large share of doctorates in the sciences, all the more so in the case of the future laureates. Thus, as we see again in table 5–1, sixty-four, or 70 percent, of the ninety-two future laureates who did their prize-winning research in the United States took their doctorates in the top-ranking universities in contrast to ten, or only 11 percent, in all the other degree-granting American universities; the remaining eighteen, or 20 percent, took their degrees abroad. This means that the few elite universities had a distinct advantage in getting to observe and assess these talented young scientists during an important formative phase of their careers.

In fact, the educational concentration that provided special opportunity for close-up assessment of these young scientists was even greater than these figures suggest. Since we are trying to account for the distribution of their first jobs in the United States, it seems reasonable to confine ourselves for the moment to the seventy-four future laureates who received their doctorates there. Eighty-six percent of these took their degrees at elite universities as well as almost half the scientists with American doctorates in the matched sample, reflecting the considerable concentration of doctoral production in these universities. Once again, this means that a small number of elite universities were in an excep-

---

4. To put this in another way, of the seventy-four future laureates whose first appointments were in a university, 69 percent were affiliated with elite universities, but these universities had only 9 percent of all faculty members on their staffs in 1940—the approximate midpoint in the period in which laureates took their first jobs (Marsh, 1940; U.S. Bureau of the Census, 1960, p. 210).

tionally advantageous position for observing young scientists and assessing their potentialities.

They used that advantage. Confining our attention to the elite universities, we find that they appointed almost twice as large a proportion of the future laureates to their first jobs as of the matched sample of fairly productive scientists among their graduates: 61 percent compared with 35 percent. These figures cover a span of years and are therefore not closely comparable to data on the retention of graduates in any one year. Nevertheless, that the elite institutions managed to discriminate between the potentialities of their graduates is suggested by the contrast between the retention of 61 percent of the future laureates and the 28 percent of all graduates reported as having been retained by the elite universities in 1957 (Berelson, 1960, p. 113; Hargens, 1969, p. 29).

The elite universities also managed to recruit scientific talent beyond the boundaries of their own graduates. This compensated somewhat for their imperfect, though substantial, capability of spotting and keeping their own future laureates, an imperfection roughly measured by their having allowed 39 percent of them to get away, at least temporarily. (As we shall see, some of these returned to the fold later.) Having missed out on some of their own, they reached out to recruit future laureates among graduates of the other universities. There were only ten of these, all told, and the elite universities managed to appoint three of them to their first jobs. Altogether, then, a substantial proportion of the ninety-two future American laureates, educated in the United States and abroad, began their careers in elite institutions, and this is particularly the case for those who studied in universities of high rank. As we noted earlier, by this time in the lives of these scientists, academic origins come to be more important than social origins in determining organizational affiliations.

Yet the effects of social class continue to be detectable. Surprisingly, the impact of future laureates' class origins, which was attenuated at the stage of their graduate training, reemerged when they entered the job market. Confining our attention again to the seventy-four who were educated in the United States, we note that two thirds (66 percent) were first appointed to the faculties of elite universities and research institutes—more than seven times the proportion expected given the size of the science faculties of these institutions—but that future laureates from upper-strata families were *somewhat* more likely to have gained prestigeful appointments than those from the lower strata—71 percent as against 57 percent. Closer examination of the data shows that this difference is confined to the small number of future laureates who took doctorates at universities outside the top rank. Future laureates from lower-strata families who found their way to the great universities for doctoral training were *not* disadvantaged in the job market by their class origins. Having been trained at a center of scientific learning was what counted in getting their first jobs. It was only among the graduates of the less prestigeful universities, presumably less effectively linked to the communication and job-allocation network, that coming from the lower strata seems to have made some difference. And the difference is that they were steered to or chose or had to settle for jobs outside the academy.

A substantial part of the concentration of future laureates in the elite universities at this stage is accounted for by "inbreeding," that is, by appointments of academics to the staff of the same university that trained them. In this respect also, the elite universities have the edge. Since future laureates are scarce, the numbers are small, but it is clear that the rates of both *inbreeding* (the proportion of a university faculty who are its graduates) and *retention* (the proportion of its graduates kept on by a university) favor elite universities. Of the fifty-three laureates whose first jobs were in one of the elite American universities, 40 percent (twenty-one) were inbred in the strict sense. This proportion is much higher than the 5 percent (one out of twenty-one) whose first jobs were in other American universities. By way of crude comparison, Hargens and Farr (1973) report for a small sample of young scientists in their first jobs in 1966 that 25 percent of those affiliated with "distinguished" universities—much the same group as our "elite" thirteen—were inbred as were 6 percent of the others holding academic positions elsewhere. The Hargens and Farr data remind us that young scientists affiliated with universities of the highest prestige tend more than others to have graduated from these very same universities.[5] At the same time, these comparative data also indicate that overall the tendency for elite universities to appoint future laureate alumni to faculty posts seems even greater than their tendency to appoint their own graduates.

Turning from rates of inbreeding to rates of retention of graduates, we find that twenty-one, or a third, of the sixty-four laureates who graduated from the elite universities stay on where they took their degrees as against only one of the ten laureates trained elsewhere in the United States. By these two measures, then, the elite universities have greater success in holding on to scientists who later win Nobel prizes than do other American universities.

As with their home-grown products, the top-ranking American universities were also advantaged in recruiting future laureates who took

5. Other studies also find that the top-ranking universities are more given to inbreeding than others (Berelson, 1960, pp. 113–16; Hargens, 1969, p. 29; Crane, 1970, p. 958; Hagstrom, 1970, p. 12). As Hargens and Farr (1973, p. 1390) point out, however, there is no indication that scientists inbred at the most distinguished universities are any more productive than are graduates of other universities on the same staffs.

their doctorates abroad. That advantage resulted from their occupying the same kind of privileged position in the international, as in the domestic, social structure of science. Elsewhere, as in the United States, identified scientific talent is concentrated in elite centers of research and training. Since networks of personal relations and informal communication are socially stratified, members of the mobile international elite in various fields of science often have close personal ties. Information about promising young scientists passes along these networks, linking elite American scientists with their foreign counterparts. The American scientists visiting these centers or meeting with foreign peers in small international conferences often encounter the bright young men and women who are their friends' protégés. This differential access to young scientists abroad has worked to the advantage of the elite universities, particularly during the Great Migration of scientists from fascist-dominated countries in the 1930s (compare Weiner, 1969, pp. 152–234).

The structural advantage is reflected in the distribution of the first jobs of the eighteen laureates who were trained abroad, although they did their prize-winning work in the United States and thus qualify as "American laureates." All but two were foreign-born-and-bred, comprising that part of the contingent of illustrious immigrants who have expanded the impressive number of "American laureates" in science (Fermi, 1968; Fleming and Bailyn, 1969). The two exceptions are the physicist Albert A. Michelson who, although German-born, qualifies as the first American recipient of a Nobel prize (1907); he did his undergraduate work at the U.S. Naval Academy and, after graduate work in Germany and France, all his later scientific work in the United States, and the chemist Irving Langmuir, who was trained in Germany but born in Brooklyn.

As the detailed list in table 5–2 shows, eleven of the eighteen foreign-trained scientists destined to become American laureates had their first American jobs in one or another of the top-ranking universities. Another five came to other universities—Stanford, Vanderbilt, and Washington (St. Louis)—which, while not at the very top at the time, were in the upper reaches of the stratification of American universities. (For a succinct analysis of early studies of the quality of universities, see Berelson, 1960, pp. 124–128.)

Thus, it was not exactly by chance that, while still in their twenties, such outstanding European scientists as Eugene Wigner and Hans Bethe found their first American posts in top-ranking universities (Princeton and Cornell). They were known by colleagues at these elite universities who were interested in working with them. Eugene Wigner illustrates how the informal network of acquaintanceship brought him to Princeton in 1930. "I was invited on the advice of Paul Ehrenfest [very

**Table 5–2.   First Jobs of Foreign-Trained Laureates (1901–72)**

FOREIGN-TRAINED AMERICAN LAUREATES

| Elite Institutions | Name of Laureate | Age at Appointment | Date of Appointment |
|---|---|---|---|
| Columbia University | André Cournand | 35 | 1930 |
|  | Salvadore Luria | 28 | 1940 |
| Cornell University | Hans Bethe | 29 | 1935 |
|  | Fritz Lipmann | 40 | 1939 |
| Harvard University | Max Theiler | 23 | 1922 |
|  | Georg von Békèsy [a] | 50 | 1949 |
| Johns Hopkins University | Maria Goeppert Mayer | 24 | 1930 |
| University of California, Berkeley | Emilio Segrè | 33 | 1938 |
| University of Chicago | Alexis Carrel | 32 | 1905 |
| Princeton University | Eugene Wigner | 27 | 1930 |
| University of Wisconsin | Har G. Khorana | 38 | 1960 |
| *Other American Universities* |  |  |  |
| Case Institute of Technology | A. A. Michelson | 31 | 1883 |
| Stanford University | Felix Bloch | 29 | 1934 |
| Stevens Institute of Technology | Irving Langmuir | 33 | 1906 |
| Washington University (St. Louis) | Severo Ochoa | 36 | 1941 |
| Vanderbilt University | Max Delbrück | 34 | 1940 |
| *Nonacademic Institutions* |  |  |  |
| State Institute for the Study of Malignant Diseases, Buffalo, N.Y. | Carl Cori | 26 | 1922 |
|  | Gerty Cori | 26 | 1922 |

a. Von Békèsy came as a mature researcher to Harvard from Budapest by way of the Caroline Institute in Stockholm.

much an occupant of the forty-first chair]. . . . He was in Leiden but traveled around a great deal and . . . he advised Princeton to invite two [of us]—namely, John von Neumann and me. He knew that we were close friends. He knew that if somebody comes to a new place he feels at times lonesome, and it is a very good thing if he has some old friend around to talk to" (Walsh, 1973, p. 529).

Since positions at the ranking institutions were then generally filled through informal procedures of spotting and sounding out rather than by formal solicitation of applications, it is not surprising that the accumulation of advantage holds across national boundaries as well as within them. The differentials in observability resulting from the stratification of personal relations among elite scientists and the concentration of recognized talent in elite institutions operates elsewhere as in the United States. This is shown by the experience of the "British" laureate in medicine Boris Chain. Having fled Berlin for the usual reasons in 1933 at 27, he arrived in London with no money but with his work known to J. B. S. Haldane. Haldane arranged for a temporary post with Sir Charles Sherrington at University College. "A few months later, he got me a more permanent post under Hopkins at Cambridge and from there I went to work with Florey at Oxford. In fact, the whole of my career in England is really due to Haldane" (Clark, 1968, p. 106). The episode shows how the elite network operated. It resulted in the young and talented Chain's finding himself in elite institutions working with members of the scientific ultra-elite: F. G. Hopkins had received a Nobel prize four years before, Sherrington one year before, and a dozen years after, Howard W. Florey was to share the prize with Chain for helping to isolate penicillin. (Haldane himself remained an occupant of the forty-first chair throughout his extraordinary life.)

Social selection by the elite organizations and self-selection by the future laureates thus converged in the allocative process, with each seeking out the other and for compatible reasons. Organizations want to maintain or upgrade the quality of their staffs; young scientists want to work in superior research environments. For, as the laureates report in retrospect, it was not so much the prestige of the elite institutions, welcome as that was, that attracted them as the quality of their scientific departments and the access to superior facilities for research. In fact, they typically did not have to choose between prestige and superior work environment. The two tended to go hand in hand. As reports on the distribution of research funds among universities uniformly show, they are heavily concentrated in the prestigious institutions (Gustafson, 1975): in the years 1960–63, for example, just ten academic institutions—all but one of these (Minnesota) among our top-ranking universities—out of the more than 2,000 universities and colleges received 38 percent of the federal funds allocated for research and development in such institutions (Hirsch, 1968, pp. 105–106; National Science Foundation, 1976, pp. 13–14). Abundant resources do not ensure that significant science will be done, but they obviously do no harm and help to attract able young scientists.

At any rate, there was a growing concentration of young future laureates in the top universities over the years, even though the staffs of the same universities have comprised a declining proportion of the total aggregate of American faculty members: 51 percent of the future laureates acquired their first jobs in the top universities before 1930 and 65 percent afterward, but these same universities had 12 percent of all faculty members on their staffs in the earlier period and just 7 percent later on (Robertson, 1928; Brumbaugh, 1948; U.S. Bureau of the Census, 1960, p. 210).[6]

## UPWARD MOBILITY IN ACADEME

Although there was some movement of future laureates into and out of academe, by far the greater number remained in universities for all or most of their careers, the rest being largely affiliated with such academic research institutions as the Rockefeller Institute (later University) and the Institute for Advanced Study. In fact, only six of the ninety-two count their principal affiliation as non-academic.

In this respect, the laureates are much like the larger scientific elite represented by members of the National Academy of Sciences: 93 percent of the laureates and 82 percent of the Academicians were affiliated with universities at the time they won their respective honors. Since both groups are so heavily weighted with academics, we limit our examination of their upward mobility to the academic hierarchy.

Practically all scientists and scholars who remain in academic life can count on eventually attaining the rank of full professor. They need only pay a modicum of attention to their teaching and departmental chores, publish a modicum of research, and live long enough, and the principle of seniority in a tenure system will generally ensure their elevation. Thus, attaining this rank need not mean distinction of performance or special recognition; it requires only persistence for the faculty member to remain on the academic escalator (Wilson, 1941; Caplow and McGee, 1958). For promotion to be used as a valid indicator of academic recognition of role performance, it must therefore be related to the age at which it occurs. With some variations resulting from differing conditions of the academic marketplace, early promotion to the rank of full professor—that is, promotion long before the average age—generally indicates special institutional recognition; late promotion, long after the average, is little more than the reward for endurance. This holds, of course, only in the aggregate. In any individual case, either early or late promotion may only reflect local errors in judgment or departmental politicking. But, as Caplovitz (1960) and others have shown, age-specific *rates* of academic promotion do differentiate between

---

6. Enlarging the comparison to include laureates whose first jobs were in elite research organizations as well as in the top twelve universities, we again find that elite institutions—now in the extended sense—recruited a smaller share of future laureates entering the job market before 1930 than afterward; they hired 60 percent of the earlier entrants as against 74 percent of the later ones.

grades of role performance, and this is especially so when the prestige of the institutional affiliation is taken into account.

Comparisons of such age-specific rates in table 5–3 indicate in a variety of ways that the detection-and-monitoring system of science has operated to identify future members of the scientific elite. That the aggregate figures on speed of promotion do not sharply mark off future laureates and members of the Academy from the fairly productive scientists listed in *American Men of Science* is due to their differential distributions among elite institutions and other universities, and to the apparent fact that the elite institutions raise scientists to full professorships more slowly than do less distinguished ones. We shall return to this complex interrelation between mobility in rank and mobility in institutional affiliation later. Here we note that a larger share of the future laureates than of Academicians or other scientists were promoted to full professor before they became 35 and that this is so in the elite institutions as well as the others.

The eight most precocious laureates experienced this form of institutional recognition while still in their twenties: the physicists Arthur Compton, Ernest Lawrence, Felix Bloch, Eugene Wigner, Julian Schwinger, and Murray Gell-Mann, and the geneticists T. H. Morgan and Joshua Lederberg. Each got off to an early start by earning the doctorate long before the median ages of 28 and 30 at which the aggregate of physical and biological scientists in the United States obtained theirs (Harmon and Soldz, 1963, p. 44); the median age does not differ much among the various age cohorts of scientists.

As we should expect by now, the occupants of the forty-first chair exhibit the same pattern of early achievement, early recognition, and early institutional reward: the mathematicians John von Neumann and Kurt Gödel, for example, earned their doctorates at 23 and 24 while the physicists Edward Teller and John van Vleck had theirs at 22 and 23, all being promoted to top academic rank while still in their twenties.

Since science, like other social institutions, operates in a historical and social context, it is subject to the disruptive or stimulative impact of major events external to the institution itself. More specifically, although the reward system of science tends to provide for the rapid promotion of talented scientists, historical events can delay that process, by changing the character of their research and other aspects of their scientific role. World War II had this side effect for nearly a full cohort of scientists. We observe, by way of example, the pattern of delayed mobility for one laureate, Richard Feynman, and his mentor and occupant of the forty-first chair, John Wheeler. Both physicists completed their doctorates in short order, Wheeler at 22 and Feynman at 24. Soon after, both interrupted their programs of research to take part in what

Table 5-3. Age at Full Professorship: Laureates (1901–72), Members of National Academy of Sciences, and Sample from *American Men of Science*, According to Prestige of Institution Making Appointment

| Age at Promotion | NOBEL LAUREATES | | | MEMBERS OF NATIONAL ACADEMY OF SCIENCES [a] | | | SCIENTISTS IN AMERICAN MEN OF SCIENCE [b] | | |
|---|---|---|---|---|---|---|---|---|---|
| | Elite [c] Institutions | Other Universities | Total | Elite Institutions | Other Universities | Total | Elite Institutions | Other Universities | Total |
| 34 or younger | 33% | 47% | 36% | 18% | 39% | 25% | 22% | 33% | 32% |
| 35–39 | 35 | 32 | 34 | 37 | 32 | 35 | 34 | 25 | 26 |
| 40 or older | 32 | 21 | 29 | 45 | 29 | 39 | 44 | 42 | 42 |
| Number | (66) | (19) | (85) | (629) | (326) | (955) | (41) | (324) | (365) |
| All age groups | 78% | 22% | (85) [d] | 66% | 34% | (955) | 11% | 89% | (365) |
| x̄ age | 37.9 | 35.4 | 37.3 | 39.3 | 37.0 | 38.5 | 42.7 | 41.8 | 41.9 |

a. Members elected 1900–67, excluding Nobel laureates.
b. A random sample was drawn from the fifth and eleventh editions of *American Men of Science*. Those holding engineering degrees were excluded.
c. Elite institutions here are the thirteen elite universities, the Carnegie Institution, the Institute for Advanced Study, the Mayo Clinic, and the Rockefeller Institute (later University) and Foundation. The equivalent of the full professorship was used when a system other than professorial ranking was in force in an institution.
d. Although von Békésy worked at Harvard until his retirement, he held the post of Senior Research Fellow and was not a professor. He is therefore excluded from this tabulation.

came to be known as the Manhattan Project. Feynman was assigned to
Los Alamos to work on the A-bomb itself and Wheeler to a succession
of installations to help in developing the means to produce the fissionable
material needed for the bomb. By the time they were demobilized,
Feynman was 27 and Wheeler 34. Neither had had an opportunity to do
much of his own work and to make it known in this usually productive
phase of the scientific life. Wheeler returned to Princeton and became a
full professor at 36. Feynman, having begun his postwar career as an
associate professor at Cornell, accepted a full professorship at Cal Tech
five years after demobilization, at 32. For both, as for others, the war
retarded what would prove to be a distinguished career.

Apart from such disruptions, however, we see in table 5–3 that the
differences in promotion rates noted earlier between the laureates and
members of the Academy were nearly evened by the time these elite
scientists reached the age of 40. Nevertheless, the academic reward
system makes fine-grained distinctions in another way between the ultra-
elite and the wider scientific elite. As the next-to-last line of table 5–3
shows, the laureates were more likely to have been first appointed to
full professorships in the thirteen top-ranking universities and elite research institutes: 78 percent of them compared with 66 percent of the
Academicians.

Comparison of the mobility experience of the academic laureates and
their counterparts in the matched sample of scientists points to another
aspect of the reward system in operation. As we noted earlier in this
chapter, the future laureates were far more likely to obtain their first
jobs in the elite institutions. It now appears that the proportion of laureates in these elite universities and research institutes increased from
65 percent at the time of their first jobs to 78 percent when first appointed full professors while the proportion of the rank-and-file actually
declined from 22 to 11 percent. In the nature of the case, the numbers
are small since the laureates are few, but the data suggest a continuing
process of sifting and sorting in which a greater proportion of the scientists destined to enter the ultra-elite moved into the top universities while
the matched sample, much closer to the rank and file, tended to move
and be moved away from the elite institutions.

In organizational terms, this finding means that the selective process
in this phase of their careers continues to increase the concentration of
the potential ultra-elite scientists in the elite organizations. Having overlooked, or at least having failed to recruit, one third of the future laureates at the outset of their careers, the elite institutions make up for the
oversight by appointing a fair number of them to full professorships.
Typically, this is at the expense of the other universities. Thus, even before such good though delayed organizational judgment has been con-

firmed by the award of the Nobel prize to these additional recruits, four of every five of the destined laureates are concentrated in the top-ranking institutions. These observations leave open the question of whether being associated with a major university helps scientists to get Nobel prizes. Without good data on the university affiliations of occupants of the forty-first chair, we cannot say. But it is evident that Nobel prizes are not confined to scientists on the staffs of elite universities nor do they come to them any sooner than to others.

In terms of the careers of individual scientists, the pattern appearing in table 5–3 has another, correlative meaning. For some of the future laureates who were not initially appointed to the top universities, it means a pattern of upward mobility in terms of both organizational affiliation and academic rank. For some of the rank-and-file scientists who *were* initially appointed to these universities, it means a mixed pattern of upward mobility in terms of academic rank and downward mobility in terms of organizational affiliation (see Hargens, 1969, for similar findings). In other words, the reward system provides for a trade-off of full professorships in lesser universities for lower rank in major ones. The data suggest that this kind of tradeoff occurs more often among the run of scientists than among the prospective laureates.

The bottom line on table 5–3 makes the contextual implication of this tradeoff clear: the rank of universities affects the rate of promotion within them. As we noted earlier, promotions to full professor occur more slowly in the top-ranked universities and research institutes, and this is somewhat less the case for future laureates than for Academicians. Since, as we have seen, the scientific elite are concentrated in such universities, this tendency reduces to 4.6 years aggregate differences between the ultra-elite and the others in the average age at which members of each group are promoted to the rank of full professor. But the academic reward system within each class of institution does discriminate between the potentialities of scientists in such a way as to promote future laureates earlier than future members of the larger elite of Academicians and both groups earlier than the rank and file. The differences between the groups are small, and while they are statistically significant they are not likely to provide the basis for perceived differences between the elite and the rest that would gratify the former and make the latter feel deprived. Thus, the system appears to "work" in the sense of recognizing differences in past performance and potentials, but not with great precision. The same findings appear in table 5–3 in another way; only 29 percent of the future laureates in contrast to 42 percent of the less eminent scientists had to wait until at least the age of 40 to become full professors.

These figures can also be taken to suggest "inefficiency" in that reward system. Instead of saying that, by contrast with other scientists, "only 29 percent" of the laureates had to wait until they were 40 for full professorships, we can say that "as many as 29 percent" had to wait this long for such recognition. This directs our attention to organizational slippage, lags and errors in the assessing and rewarding of scientific potential and achievement. One way to gauge the extent of such inefficiencies is to relate the time of promotion to the time at which the future laureates did the work that was ultimately judged important enough to win the prize. Having examined the variability in the ages of promotion to full professorships, we must now examine the variability in the ages at which the future laureates did their prize-winning research. As we shall see, the timeliness of rewards is no small matter for their recipients as well as for the system as a whole.

Chapter 6

# THE PRIZE-WINNING RESEARCH

The same historical hindsight that has allowed us to refer confidently to "future laureates" also allows us to refer to their Nobel "prize-winning research." Before the actual award, there can be hope or even strong expectation but no certainty. But after the fact, there can be no ambiguity about what constitutes the prize-winning research since this is designated in the official citations at Stockholm.

For mobility into the scientific ultra-elite, unlike academic mobility, seniority without prime research achievement is not enough. In this connection, we inspect the laureates' age when they did the prize-winning research; the place where it was done; what the research was on; whether it was done in collaboration or solo; the devices used to highlight or register who has done it; simultaneous independent discoveries in prize-winning research and their significance; the impact of the research before an award; and, finally, the extent and kinds of recognition accorded the laureates prior to their receiving the prize.

## TIME OF PRIZE-WINNING RESEARCH

Widely accepted mythology holds that scientists do their best work very early in their careers and little of consequence after that. P. A. M. Dirac, one of the more powerful minds in theoretical physics and himself a laureate, has versified this belief (quoted by Jungk, 1958, p. 27):

> Age is, of course, a fever chill
> that every physicist must fear.
> He's better dead than living still
> when once he's past his thirtieth year.

In the light of this statement, it is only fitting that Dirac formulated the mathematical theory describing the positron when he was 26, became a Fellow of the Royal Society at 28, and received a Nobel prize just past his announced watershed age, at 31. The similarly precocious molecular biologist James Dewey Watson, who was 25 when he did his prize-winning research with Francis Crick, raises the watershed by five years, declaring that "almost every important new discovery comes from someone under thirty-five" (*New York Times,* 22 March 1973, p. 43).

A special version of the myth maintains that it is particularly the truly transforming ideas in science, the fundamental reconceptualizations, that are the work of youthful minds. T. S. Kuhn, whose essay on scientific revolutions brought about its own small revolution among philosophers and historians of science, suggests that the creators of fundamental new paradigms are almost always young or very new to the field (1962, pp. 89–90). And, of course, we can easily assemble a roster of cases to illustrate that observation. Newton wrote of himself at 24, when he had begun his work on universal gravitation, the calculus, and the theory of colors: "I was in the prime of my age for invention, and minded Mathematics and Philosophy more than at any time since." Darwin was 22 when he made the historic voyage on the *Beagle* and 29 when he formulated the essentials of the theory of natural selection. Einstein was 26 when he developed three major scientific ideas, among them the special theory of relativity. As a last collective example, eight of the ten physicists generally credited with doing most to overturn classical mechanics and to establish quantum mechanics—all of whom eventually received a Nobel prize—were less than 30 when they produced this work, just as Dirac, who was in that group, tells us to expect (Gamow, 1966).

These arresting illustrations are not enough to show that it is especially the young scientists who revolutionize scientific thought. As Kuhn himself says, this is a generalization that "badly needs systematic investigation."

Moreover, influential studies of the relation of age to scientific achievement, such as Lehman's (1953), are faulty in two respects. First, they do not take into account the age structure of the scientific population. As is known from the exponential growth in the numbers of scientists (Price, 1963) the young make up a hefty percentage at any given time, and so they will produce a large aggregate of contributions. What is required are data not on the proportions of scientific contributors in the various age strata but on the proportion of each age stratum making contributions. This is what the Lehman studies fail to do.

Second, Lehman does not take into account, as Wayne Dennis (1956, 1958, 1966) has demonstrated, the biasing effects of differing life spans on the distribution of achievements at various ages. The fact that short-lived scientists are cut off from making any contributions in later years factitiously enlarges the proportion assigned to younger scientists in data that do not take longevity into account. The essential point is expressed in Newton's lament over the premature death at age 34 of his protégé, the mathematician Roger Cotes: "If he had lived, we might have known something." Or the similar observation by John Maynard Keynes about the brilliant young logician and mathematician Frank Ramsey, robbed of his future at the age of 27.[1]

The ideology of the role of youth in science thus includes three related ideas: first, that contributions to science, and especially the path-breaking ones, come disproportionately from the ranks of young scientists; second, that having done their best work in their youth, scientists burn out soon afterward; and, third, that the many scientists who have done little of significance in their early years can count on doing even less later on. According to this imagery, the years after 30 or so are a gloomy, unproductive period for research scientists. (On age-specific rates of scientific productivity, see S. Cole, 1972, and appendix D.)

For the Nobel laureates, this is at best a partial and somewhat misleading truth. Typically, they did their prize-winning work at a fairly early age, but not so early as the general belief would lead us to expect. As table 6–1 shows, the aggregate of laureates did the work that earned them the prize at an average age of almost 39. The physicists were the youngest of the lot, averaging 36 years; the chemists, 39; and the laureates in medicine and physiology, 41. The pattern is almost identical for the subset of American laureates.

This does not mean that the norm is a higher rate of significant discovery by youthful scientists. We may need to be reminded that reports of median ages at time of discovery tell us that half of the

---

1. The preceding paragraphs are from Zuckerman and Merton (1972).

**Table 6–1.  Mean Age [a] of Laureates (1901–72) at Time of Prize-Winning Research, According to Year and Field of Prize**

| Year of Award [b] | Physics [c] | Chemistry [c] | Physiology or Medicine | All Fields |
|---|---|---|---|---|
| | | | **FIELD OF PRIZE** | |
| 1901–25 | 36.7 | 35.6 | 39.1 | 37.1 |
| | (30) | (22) | (23) | (75) |
| 1926–50 | 33.9 | 40.3 | 43.8 | 40.0 |
| | (24) | (29) | (36) | (89) |
| 1951–72 | 37.0 | 39.7 | 39.4 | 38.7 |
| | (42) | (32) | (48) | (122) |
| All Years (all laureates) | 36.1 | 38.8 | 40.8 | 38.7 |
| | (96) | (83) | (107) | (286) |
| All Years (American laureates) | 36.8 | 39.4 | 41.1 | 39.2 |
| | (32) | (19) | (42) | (93) [b] |

a. Age at prize-winning research refers to age in the year of publication of that research or, if a number of publications were involved, age at the midpoint of the period covered.
b. Data for the period 1901–50 are drawn from Manniche and Falk (1957, p. 302ff.).
c. Marie Curie and John Bardeen are counted twice in these computations since each did prize-winning research on two separate occasions.

discoveries were made *after* the median as well as before. One of Lehman's findings emphasizing the youth of productive scientists could be reported, for example, as showing that "fully half" of the discoveries important enough to be listed in Magie's *Source Book of Physics* were made by scientists *over* the age of 38, or that "fully half" of the discoveries listed in genetics were by scientists over 40.

What may be both the source and the product of the ideological accent on youth in science are the occasional dramatically precocious scientists such as Joshua Lederberg and Brian Josephson, Rudolf Mössbauer, Svante Arrhenius and J. R. Schrieffer, all of whom reported their prize-winning research in their doctoral dissertations.[2] The Joshua Lederberg who discovered sexual recombination in bacteria at 21 attracted far more attention than, say, John Enders, who was an undramatic 52 when he and his co-workers, T. H. Weller and F. C.

---

2. Five other laureates-to-be presented prize-winning research in their doctoral dissertations: the physicists Nicolai Basov, Marie Curie, Louis de Broglie, and Leo Esaki, and the chemist Victor Grignard. All were more than 30 years old at the time, and Curie and de Broglie were candidates for the advanced and demanding Doctorate of Physical Science at the University of Paris, not for the ordinary Ph.D.

Robbins (both in their early thirties), developed their method of growing polio virus *in vitro*. Such selective perceptions result in large measure from the practice of limiting comparative evaluations to scientific age peers, especially among the young. Relative to other 21-year-olds, Lederberg's achievements *were* noteworthy in the literal sense of the word. Not too many years later, his contributions were substantial enough for his teacher and collaborator, Francis Ryan, to suggest (not entirely in jest) that a *Festschrift* ought to be put together in celebration of Lederberg's twenty-fifth birthday. The few extreme cases of this kind are diverting and come to be regarded as the typical ones.

Although Dirac's verse in praise of youth in physics was hardly meant to be taken literally, it is nevertheless in point that only six of the thirty-two American laureates in physics had even begun to publish, let alone completed, their prize-winning research before Dirac's watershed age of 30. Nor does it turn out that these young physicists were all theoreticians, as popular myth would have it.

Although Schwinger, Gell-Mann, Cooper, and Schrieffer *are* theoretical physicists, and their prize-winning investigations are theoretical contributions, neither Carl Anderson, the discoverer of the positron, nor Donald Glaser, the inventor of the bubble chamber, can be considered members of that genre. It also turns out that the difference in age between theoretical and experimental physicists at the time of their prize-winning research is not great, the former having been 35.3 on the average and the latter 37.6, which is a difference "in the right direction" but not statistically significant.

The precocious laureate physicists, particularly the theoreticians, it would seem, have attracted more notice than, for example, the laureate Emilio Segrè, who was 50 when he and his co-workers, Owen Chamberlain (then 35) and Clyde Wiegand (then 40), discovered the antiproton. It is true that Segrè had made other important contributions to physics before his antiproton work—even more important, some observers suggest, than his prize-winning research. But if he began early, he evidently did not finish early. His is not the only case suggesting that first-class scientific ability need not waste away soon after the magic age of 30. Max Planck himself, who began the quantum revolution that, as we have seen, was largely advanced by young physicists, was 42 when he made the fundamental (and, incidentally, prize-winning) discovery of the quantum of energy that started it all.

Table 6–1 sums it up by showing that throughout the first seventy-year span of the Nobel awards, the average age at which the prize-winning work was done remains in the late thirties. But, as summary measures, averages can conceal great variations in underlying distributions. Is it possible that combining the ages of laureates whose prize-winning research covered a span of years with the ages of laureates who made just one contribution results in blending two distributions with quite different means and thus underestimating the contributions of young scientists? The answer is yes—but not to the extent of modifying our earlier observations on age at prize-winning research.

Confining our attention to recent awards, between 1951 and 1972 (for reasons that will become evident), we find that laureates whose prize-winning research extended over a period of years were, on the average, 39.9 at the midpoint of that research compared with 37.6 for the other laureates, neither group being nearly so young as the ideological accent on youth would have us believe. Putting the same data in the form of distributions rather than averages, we find, as table 6–2 shows, that 12 percent of the laureates did their prize-winning research before they reached Dirac's watershed age of 30 and a total of 34 percent before Watson's 35.

The question still remains whether Nobel laureates are disproportionately young at the time of their prize-winning research or whether the figures merely reflect the overall age distribution of scientists. For the purpose of answering this question let us limit our observations to recent laureates. The essential comparison is between the age distribution of Nobelists and that of the general population of scientists (which has changed considerably over the seventy-year span covered by the Nobel prizes). Lacking information on the age distribution of the worldwide population of scientists, we can make inferences from recent data about the population of American scientists for comparison with American laureates, since the age distributions of American and all other laureates at the time of doing the prize research are practically the same. As the last two columns of table 6–2 show, a representative fraction of the recent American (and of other) laureates did their Nobel research before they were 35: 35 percent of them were in this age category compared to 37 percent of all American scientists. If anything, it is not the young who turn up disproportionately often among those making prize-winning contributions but the middle-aged: 23 percent of the laureates were 40 to 44 years old when they did their prize-winning research but only 14 percent of the run of scientists fall into this age cohort.

Thus, the question of the most creative age in science is not so simple as the ideology of youth would have it. Although the young turn up fairly often among contributors of Nobel prize-winning research, so do those in the middle years. Among Nobel laureates at least, science is not exclusively a young person's game; evidently it is a game the middle-aged play as well.

**Table 6–2. Age Distribution of Laureates (1951–72) at Time of Prize-Winning Research and of American Scientists (1970)**

| Age at Prize-Winning Research | TYPE OF RESEARCH CITED FOR THE PRIZE [a] | | | | Age Distribution of American Scientists [c] |
|---|---|---|---|---|---|
| | Single Discovery | Line of Research | All Laureates [b] | American Laureates | |
| 20–29 | 16% } 41% | 6% } 25% | 12% } 34% | 12% } 35% | 19% } 37% |
| 30–34 | 25 | 19 | 22 | 23 | 18 |
| 35–39 | 22 } 43 | 18 } 46 | 20 } 44 | 15 } 38 | 15 } 29 |
| 40–44 | 21 | 28 | 24 | 23 | 14 |
| 45–49 | 7 } 14 | 20 } 29 | 13 } 21 | 15 } 24 | 12 } 21 |
| 50–54 | 7 | 9 | 8 | 9 | 9 |
| 55+ | — | 2 | 1 | 2 | 12 |
| N = | (67) | (55) | (122) | (65) | (312,644) |
| Mean age | 37.6 | 39.9 | 38.7 | 39.8 | |

a. Age at prize-winning research is taken as of the year of publication of the research cited or, when a number of publications are involved, age at midpoint of the period covered. The former is identified as a "single discovery" and the latter as a "line of research."
b. These data cover laureates of every nationality.
c. Drawn from National Register of Scientific and Technical Personnel, 1970, in U.S. Bureau of the Census (1971, p. 515).

## PLACE OF PRIZE-WINNING RESEARCH

Just as at every other phase in their careers, the laureates were concentrated in a few institutions when doing their prize-winning research.
Table 6–3 shows that somewhat more than half of the ninety-two laureates who did that research in the United States did so at just six institutions: Harvard, Columbia, Berkeley, the Rockefeller, Chicago, and
Washington University (in St. Louis). The thirteen institutions that
were the sites of more than one prize-winner's work account for more
than four fifths of the prizes, and just thirty institutions account for them
all. Considering the several thousand universities, colleges, and research
organizations in the United States, table 6–3 suggests marked concentration in the small number of elite institutions. This is decidedly the
case for academic laureates: 76 percent were on the staffs of elite universities (including the Rockefeller) at the time they did their prize
research while these same universities had just 7 percent of all faculty
members in American colleges and universities in 1948—the approximate midpoint of the period during which the laureates were doing their
research (Brumbaugh, 1948; U.S. Bureau of the Census, 1960, p. 210).
And, while the numbers are small, concentration of future laureates is
also evident among the nonacademic research laboratories that were
sites of prize-winning research. Although just nine out of ninety-two
laureates worked outside of universities, four of them did so at the
Bell Laboratories and three at the National Institutes of Health, making
each incomparably productive of laureates among industrial and governmental research installations.

Table 6–3 also shows that the more elite institutions were not equally
productive of laureates in the three fields covered by the prizes. Thus
Harvard takes first place in the life sciences (followed closely by Washington University, whose award winners are almost wholly confined to
this field); Berkeley in chemistry; and Columbia in physics.

As we have seen to be the case for sites of doctoral training, so the
sites of prize-winning research have two patterns of distribution. One
involves clusters of prizes at a particular center for related investigations. Thus, all but one of Berkeley's eight physicists and chemists did
much of their research in Ernest Lawrence's Radiation Laboratory.
Similarly, three of Columbia's laureates in physics—Willis Lamb, Polykarp Kusch, and Charles Townes—were attracted to Columbia's Radiation Laboratory early in their careers by I. I. Rabi, whose own research
on magnetic properties of atomic nuclei had made him a Nobelist.

In a contrary pattern, the awards made to the eight life scientists
who had done their work at Harvard were spread over a period of
many years, a variety of specialties, and a number of laboratories. The

**Table 6–3. Sites of Prize-Winning Research: American Laureates (1901–72)**

| Site of Research | Physics | Chemistry | Physiology or Medicine | Total | Cumulative Percent |
|---|---|---|---|---|---|
| *Harvard | 3 | 2 | 8 | 13 | 14.3 |
| *Columbia | 5 | 1 | 3 | 9 | 24.1 |
| *Berkeley | 4 | 4 | — | 8 | 32.8 |
| †Rockefeller Institute & Foundation | — | 4 | 4 | 8 | 41.5 |
| *Chicago | 2 | 2 | 3 | 7 | 49.1 |
| Washington University (St. Louis) | 1 | — | 5 | 6 | 55.6 |
| *Cal Tech | 3 | 1 | 1 | 5 | 61.0 |
| Bell Laboratories | 4 | — | — | 4 | 65.3 |
| Stanford | 2 | — | 2 | 4 | 69.8 |
| *Cornell | 1 | 2 | — | 3 | 72.9 |
| *Illinois | 3 | — | — | 3 | 76.2 |
| National Institutes of Health | — | 1 | 2 | 3 | 79.5 |
| †Mayo Clinic | — | — | 2 | 2 | 81.7 |
| Brown | — | 1 | — | 1 | 82.8 |
| †Carnegie Institution | 1 | — | — | 1 | 83.9 |
| Case Institute of Technology | — | — | 1 | 1 | 85.0 |
| General Electric | — | 1 | — | 1 | 86.1 |
| Indiana | — | — | 1 | 1 | 87.2 |
| †Institute for Advanced Study | 1 | — | — | 1 | 88.2 |
| *Michigan | 1 | — | — | 1 | 89.3 |
| New York University | — | — | 1 | 1 | 90.4 |
| *Princeton | 1 | — | — | 1 | 91.5 |
| Rochester | — | — | 1 | 1 | 92.6 |
| Rutgers | — | — | 1 | 1 | 93.7 |
| St. Louis University | — | — | 1 | 1 | 94.8 |
| Texas | — | — | 1 | 1 | 95.9 |
| ‡U.S. Plant, Nutrition, & Soil Lab. | — | — | 1 | 1 | 96.2 |
| Western Reserve | — | — | 1 | 1 | 97.3 |
| *Wisconsin | — | — | 1 | 1 | 98.4 |
| *Yale | — | — | 1 | 1 | 100.0 |
| | (32) | (19) | (41) | (92) § | |

* Elite universities.
† Elite research institutes.
‡ Robert Holley also held a part-time appointment at Cornell in the Department of Biochemistry.
§ James D. Watson, it will be remembered, did his prize-winning research at Cambridge and not in his native United States. John Bardeen's appointments at the time of both prize-winning investigations are included.

pattern here is not of a particular scientist being at the center of more or less related research but rather one of sustained excellence across the biological sciences.

The number of laureates who did prize-winning research in a particular laboratory or university does not, of course, indicate how many were there at the same time or whether others were on hand who would do their Nobel research elsewhere. Thus, the official tally of future prize-winners in the biological sciences who worked at Washington University (St. Louis) is five, but they are not the same five who were there in 1947, the year that Carl and Gerty Cori were to win their prize and that Arthur Kornberg, Earl Sutherland, and Christian de Duve, all biochemists, were then associated with its School of Medicine.

Affiliation with major institutions may contribute to the visibility of scientists and their work, as we learn from the Coles (1968, p. 40), although we do not know that it thereby affects the distribution of Nobel awards. We can, however, report that institutional affiliation is *not* related to the speed with which the prize is awarded. The sixty-five investigators whose prize-winning research was conducted at the elite institutions not only did not get their prizes earlier than the rest, but got them slightly later; the difference between the two groups, however, is statistically insignificant. Although affiliation with visibility-enhancing institutions may contribute to differential access to resources and the consequences that flow from it, once having done scientific research of Nobel caliber, future laureates wherever they are receive their prize at about the same time.

## Evocative Environments

Do social contexts of scientific research affect its scientific quality or significance even though they do not affect the timing of the prize? Next to nothing is known about the ways in which particular universities or laboratories promote intellectual achievement over and above what would have been the case were the same individuals working elsewhere. Scientists have frequently noted, almost always in reflecting on the "golden" days of one or another laboratory or institute—the Cavendish or the Pasteur being good examples—that people did especially good work in these places, better than they would otherwise have done and sometimes better than they ever did later on. Accounts of these "evocative environments" often take note of the presence of several generations of distinguished and promising scientists rather than just one generation. They also note the presence of intensive interaction and competition. It may be that evocative environments enhance opportunities for doing excellent science in ways that are formally akin to the mutually reinforcing effects of environments with high crime rates where vulnerable individuals become criminals. Similarly, recent studies of how student-body composition and achievement affect achievement scores of minority students (Coleman et al., 1966, p. 302ff.) suggest that the influence of evocative environments on scientists is a question of generic sociological interest rather than specific to the sociology of science.

Finally, the data on the sites of prize-winning research show that twenty-three laureates-to-be (one fourth of them) did not work in an elite university or research institute. Rather, as Paul Samuelson observed in commenting on the award of the Nobel to his fellow economists Hicks and Arrow, "One need not be at the outstanding university of the moment to make one's scientific mark. Hicks, at LSE and Manchester, helped to elevate those places to distinction in economics. Stanford gave Arrow his chance before he was famous. He rewarded it by creating the Stanford school of economic theorists" (1972b, p. 489). Evidently, individual scientists affect their environments as well as being creatures of them. Some draw upon the intellectual and social reserves of the places where they work; others manage to enlarge them and in the process to bring good institutions into the very first rank. Exchanges between individuals and organizations apparently involve intellectual capital as well as social prestige. In the absence of systematic data on the location, attributes, and effects of "evocative environments" in science, these observations must remain preliminary and unsatisfying.


## CHANGING FOCI OF
## PRIZE-WINNING RESEARCH

Although there have been few discernible changes in the ages of American scientists at the time they did prize-winning research and in the places where they did it, the substantive foci of research honored by the awards, as in science more generally, have changed enormously since the turn of the century.

These shifts in research attention are not germane to the processes of career mobility examined thus far. They do, however, allow us to gauge, in a rough way, whether considerations other than scientific merit have affected selections for the prizes—in particular, whether efforts to distribute the awards among scientific specialities can be detected.

Recent studies of scientific growth indicate that the sciences and their component specialties develop unevenly and at different rates. In part, this results from self-augmenting tendencies in scientific growth—

more is apt to be learned about subjects about which more is known (see Lederberg et al., forthcoming, for further discussion). As a consequence, if the prizes are given for significant scientific advances, in the short run they should more often go to investigations bunched in one or another subfield than to a wide spectrum of specialties. To get a sense of the extent to which these developmental features of science have been reflected in investigations selected for awards, prize-winning investigations were classified according to specialty and date of award according to categories used in the official Nobel history (see appendix C). They reveal two patterns in selection: great diversity of specialties covered over the more-than-seventy-year span of the prizes and fairly marked concentration of awards in two or three specialties in every period—after the earliest awards were distributed to the varied contributions by the nineteenth-century scientific giants. Thus the great advances in the recent period in nuclear physics and quantum electrodynamics, molecular biology and biochemistry are registered along with diminishing interest in such areas as vitamins and radioactivity. The distribution of awards reflects, at least in a crude way, the kind of clumping to be expected from the growth patterns of the sciences. If the Nobel committees make stringent efforts to spread the awards among specialties so as to achieve equality among them, these efforts are not captured in this rough survey. More systematic and thorough study of the changing character of prize-winning research would of course require much finer substantive and temporal classification than the one used here.

The temporal distribution of prize-winning research also indicates why many scientists complain that certain specialties have received excessive attention from Stockholm. Given the near domination of the chemistry prizes in recent years by biochemists (exacerbated by the fact that the prize in medicine and physiology has also frequently gone to biochemists) and the even greater share of awards in the biological sciences that have gone to molecular biologists, it is not surprising that scientists at work in other chemical and biological specialties feel neglected.

Complaints like these set social processes in motion to redistribute the awards among specialties. Nominators are likely to call attention to research in areas not recently recognized by awards, and the committees for their part are likely to consider these nominations more carefully than they would have before the concentration of awards became evident. Over time, the pressures mount and, given the fact that there are always more meritorious candidates than there are awards, candidates in "unprized" areas come to be preferred. One consequence of this process is that scientists who work in a specialty early in its development stand a better chance than those who enter it later on and do work of comparable quality. Among these lines, a laureate in chemistry remarked about some of his own work:

> There's no denying it. At a certain stage in the development of a field, Nobel awards start being given out for work of the same caliber as what I did on vitamin X. . . . A certain field gets recognized, you see. Then after a while they stop giving awards in that field, and work that is just as good doesn't get prizes.

As we have seen, the fact that there is an excess of deserving candidates means that criteria other than merit are brought to bear in selecting or, in this case, rejecting candidates for the prize.

The same is true when it comes to choices that must be made between contributions to fundamental science and to technology. In spite of Nobel's explicit inclusion of "inventions" and "improvements" within the purview of the prizes, the committees have given preference to fundamental science from the beginning. Altogether, just thirty-nine scientists (14 percent) have won prizes for inventions and improvements, more or less broadly defined. Among these, twenty-five, or almost two thirds, have been for technological advances primarily useful in research. Thus, Moissan was honored for his electric furnace, which enabled him to analyze a variety of chemical compounds; Grignard for the development of Grignard reagents, used in organic synthesis; Wilson for the Cloud Chamber; Martin and Synge for partition chromatography; Lawrence for the cyclotron; and Zernicke for the phase contrast microscope.

The preference for fundamental science is also expressed in the emphasis given to the scientific implications of discoveries that might just as well have been honored for their technological importance. Thus the discoverers of the transistor, which has revolutionized the production of electronic equipment, were cited in the physics award "for their investigations on semiconductors and the discovery of the transistor effect" (Nobelstiftelsen, 1972, p. 417). Similarly, Charles Townes, Nicolai Basov, and Alexandr Prochorov, whose efforts culminated in the maser and later in the laser, received their award "for their fundamental work in the field of quantum electronics which has led to the construction of oscillators and amplifiers based on the maser-laser principle" (*New York Times,* 30 October 1964, p.23).

Reasons for this preference are deeply rooted in the normative structure of science. Scientists are, as Robert Merton has observed, constrained by institutional norms to give prime attention to contributions that extend certified knowledge and to appraise contributions on that basis and not on their technological merits (1973, p. 267ff.). Since committee members are trained scientists, as are, typically, members of the Academy and the Royal Caroline Institute, which approve the award selections, it is not surprising that inventive activity has been given comparatively short shrift. As Denis Gabor, laureate in physics for his invention of holography, remarked on the occasion of his award, "I feel that I am very, very lucky. . . . What I did was not pure science. I consider it an invention, [and inventions] rarely get Nobel prizes" (*New York Times,* 3 November 1971, p. 28).

### Social Organization of Prize-Winning Research: Collaborative and Individual Work

Quite contrary to the twin stereotypes that scientists, especially the better ones, are loners and that important scientific contributions are the products of individual imagination (see Jewkes et al., 1959, and White, 1957, by way of example), the majority of investigations honored by Nobel awards have involved collaboration. Altogether, as many as 185 or almost two thirds of the 286 laureates named between 1901 and 1972 were cited for research they did in conjunction with others. (That is, the majority of papers reporting the prize-winning research were co-authored, or the laureates unambiguously gave credit to collaborators in published work such as their Nobel addresses.) During the first 25 years of the awards, however, just 41 percent of the laureates were honored for collaborative work, as table 6–4 shows. During the second quarter century the proportion jumped to 65 percent, and it now stands at 79 percent of all prize-winners. This marked trend toward collaborative research in investigation cited for prizes is part and parcel of the secular shift to joint research in all the sciences, a shift which began before the turn of the century, as I have noted elsewhere (Zuckerman, 1965). But, as table 5–4 also indicates, the laureates were trend setters and were more assiduously engaged in this pattern of work than were other authors of journal articles in the same sciences published at the same time. This is especially so for Americans who, consistent with their reputation among continental scientists, find collaboration a congenial way to get things done.

We also note an increasing proportion of joint awards, those going to scientists who have worked together. Almost a third of the laureates (32 percent) overall have shared their prizes with co-workers—an aggregate statistic which conceals an increase from 19 percent during the first twenty-five years of the prizes to 24 percent during the second twenty-five years, and 41 percent during the third. These data suggest at the very least that the Nobel committees are discovering increasingly often that the "prime movers" behind research are collaborating scien-

**Table 6-4. Percent of Laureates (1901–72) Cited for Collaborative Research and Multiauthored Papers Published in Comparable Years**

| Date of Prize | PERCENTAGE COLLABORATIVE | | | |
| --- | --- | --- | --- | --- |
| | Prize-Winning Investigations by | | | Papers in Journals [a] |
| | American Laureates | Other Laureates | All Laureates | |
| 1901–25 | 50% (4) | 41% (71) | 41% (75) | 25% (2,598) |
| 1926–50 | 88 (24) | 57 (65) | 65 (89) | 51 (6,786) |
| 1951–72 | 85 (65) | 72 (57) | 79 (122) | 71 (14,913) |
| All | 84 (93) [b] | 55 (193) [b] | 65 (286) | 60 (24,297) |

a. Since prize-winning research was published an average of thirteen years before the award, comparative data from the journals are drawn from papers published a decade earlier than when the prizes were given. E.g., journal articles enumerated in the row 1901–25 were published between 1891 and 1915 so as to provide for comparability. The number of authors of articles appearing in the following journals in two out of every ten years was counted and classified: *Journal of Biochemistry*, *Biological Bulletin*, *Genetics*, *Human Biology*, *Journal of Morphology*, *Journal of the American Chemical Society*, *Analytical Chemistry*, *Physical Review*, *Journal of Chemical Physics*. Since the last was not published until 1933, *Abstracts of the American Physical Society* was substituted for the years 1900–29.

b. Both of John Bardeen's prize-winning investigations are counted here as are those of Marie Curie.

tists. Also implicit, as we will see in the following chapter, is that a sizable number of contributors to prize-winning research do not win Nobel prizes. This fact is consequential not only for these scientists but for the prize-winners themselves.

More detailed examination of the Nobel prize-winning investigations shows a variety of forms of joint research. The great majority involve only two or three co-workers (although there is some evidence that larger groups are becoming more numerous). The groups run the gamut from teams assembled for one time and then disbanded to pairs of scientists who have collaborated intensively for long periods of time, such as Cournand and Richards, cardiopulmonary physiologists and laureates in medicine, or Moore and Stein, the biochemist laureates who have worked so closely and published together for so long that "their names have become inseparable in the minds of most biochemists" (Richards, 1972, p. 492). They have been comprised of scientist peers such as Yang and Lee and of scientists of distinctly unequal experience such as John Bardeen, twice a physicist laureate, and his young co-workers, Leon Cooper and J. R. Schrieffer.

In the absence of systematic investigation of the social arrangements making for effective scientific work of various kinds,[3] it is not possible now to draw any conclusions beyond those already indicated. Significant scientific contributions are not always the creations of individual investigators. Collaborative research is not a new phenomenon nor one that is predominantly American. Those who believe that joint efforts trivialize science (and all other creative activities) appear to be out of touch with the actual conditions under which major contributions have been made and, as it happens, under which most current research is done. That this is so raises no small problems for scientists who are institutionally motivated to seek recognition for their individual contributions but who participate in research as collaborators.

## Claims to Prize-Winning Research

The communication network linking the inner circle of the scientific elite generally ensures their knowing about what they often describe as the "interesting work" going forward in their field. Communication about

---

3. For an authoritative account of the status of research on group problem solving, see Kelley and Thibaut (1969). These studies have focused on questions of general interest such as the circumstances under which group performance exceeds and falls short of the level of the most proficient members. However, they typically deal with problems unrelated to science and employ subjects who, unlike research scientists, have little training and motivation to deal with the problems presented to them. See also Janis (1972) for an intriguing analysis of ineffective problem solving in lofty political circles.

who is doing or planning what is largely informal—by word of mouth, telephone, or letter (Menzel, 1958; Garvey and Griffith, 1967; Garvey, Lin, and Nelson, 1970; Griffith, Jahn, and Miller, 1971). Even though elite scientists are concentrated in a few places, such informal communication is imperfect and, in any case, does not substitute for communication in the form of publication when it comes to validating claims to scientific work. The norms of science generally assign priority of discovery, and its attendant rewards, to the scientist who has published it first (Merton, 1957).

When a scientist is the sole author of papers reporting research there is little ambiguity about his role in it, although it is not unusual for junior (and other) associates, to see themselves as indispensable collaborators who have been unfairly done out of the recognition due them. But the question of intellectual property is, of course, far more complex in the case of collaborative investigations and papers with many authors. And, as I have noted, scientific work has become increasingly collaborative since the turn of the century in all the sciences (Zuckerman, 1965). With two-thirds of the investigations honored by Nobel prizes being the outcome of joint research, a clear majority of the future laureates have confronted problems of earmarking their individual contributions to these collective efforts.

It will be remembered that the mentors of the future laureates were comparatively generous in according credit to the young apprentices in their collaborative work. Not that inner and sometimes overt conflict over intellectual property rights was entirely absent. But, by and large, the mentors of the future laureates, typically well established and secure, were more willing than the mentors of others, still trying to make their own mark, to give public recognition to young collaborators. And, as will appear in chapter 7 when we examine the laureates' publications after they have themselves become widely recognized and especially after having received their prizes, they often reenact the same pattern of generosity toward their apprentices just as they reenact other aspects of the scientist's role in which they have been socialized by their masters.

A striking instance of this pattern of *noblesse oblige* was provided by the great Russian physiologist Ivan Pavlov. Many of the publications reporting the classical research on the physiology of digestion that came from his laboratory did not bear his name but that of one, or several, of his associates. Still, his role in that work hardly went unnoticed. As was observed by Tigerstedt, one of the Nobel examiners considering Pavlov for the first award in physiology (which he received three years later, in 1904): "The same basic ideas recur in all of [these papers] . . . they must consequently be assumed to be the intellectual property of a single person" (Nobelstiftelsen, 1962, pp. 260–61).

But at least some prospective laureates assumed that they could not safely count on the powers of inference in the scientific community to assure them of their intellectual property rights. For them, *noblesse oblige* had its limits, especially when it came to the prize-winning work, which in most cases they themselves had identified as being important. They saw to it that their role as chief or exclusive architect of that work would be recognized. One of the institutionalized ways of staking their claims was to adopt an ordering of authors' names in the publications reporting the research that symbolically highlighted their principal role in it. Thus, as one laureate described his use of this symbolic device, in language that suggests some ambivalence toward the touchy matter of allocating credit among collaborators, particularly when the stakes are very high:

> I think that Blake and Schachtmann [my pseudonyms for his two co-workers] regarded me as the originator of the idea and the main promoter of the experiments. I've never felt with either one of them the slightest awkwardness as to whose idea it was or anything like that. And this is why I was first [author of the crucial papers]. But I would also like to say that without any one of us, it would not have been carried through.

Another laureate was just as direct about the norms and practices for signaling his preeminent role in the prize-winning research:

> Why was my name first? I suppose one should publish in alphabetical order, but it was my job that I had been working on and then he came in. So I don't think it was unfair to him. He did supply a key piece of the thing but his time on it was small compared to mine anyway.

But name-order sequences differ in the clarity with which they signal a distinctive role of one or another author. Some sequences, such as the alphabetical (ABCD) and the reverse alphabetical (DCBA), are ambiguous, obscuring any special role in the research. Other sequences are relatively unambiguous, providing high visibility to a particular author. In papers with three or more authors, for example, this is true for the first author out of alphabetical sequence (ZABC), the last author out of sequence (XYZA), and the first author in apparently random sequence (ZALMB). How often, then, do these various patterns turn up in the collaborative papers of the laureates, and are there patterns of visibility in the papers that eventually brought them the prize different from those in other papers, of presumably less significance?

Figure 6–1 shows that the collaborative papers of the laureates (in-

volving three or more authors)[4] do not generally follow the norm, referred to by one of the laureates, prescribing alphabetical sequences,
which place all authors on a par. Instead, 69 percent of these collaborative papers adopt one or another of the name-order patterns that unambiguously signal a distinctive role for one of the co-authors. The same
figure also shows these unambiguously distinctive sequences occur in
82 percent of the prize-winning papers compared with 63 percent of the
rest. This suggests that the future laureates were especially concerned to
have the record clear for their more significant work, and particularly in
their prize-winning research papers. We note from figure 6–1 that the
laureates tend to take the highly visible positions in 44 percent of these
papers compared with 16 percent of all the other scientific papers they
have published with the same unambiguous sequences. Thus, although
they continue to exhibit *noblesse oblige* in the sense of according prime
position to collaborators much of the time, they are less likely to do so
when it comes to the papers reporting the research that will later bring
them the prize.

This inferred resultant of conflict between the pattern of *noblesse
oblige* and the high stakes of recognition for their more significant work

**Figure 6-1   Percent of Collaborative Papers on Which Laureates are Highly Visible Authors**

| AUTHOR SEQUENCES | UNAMBIGUOUS SEQUENCES | | AMBIGUOUS SEQUENCES | |
|---|---|---|---|---|
| Papers Reporting Nobel Prize Research (468) | 44% | 38% | 9% | 9% |
| Other Research Papers (1113) | 16% | 47% | 19% | 18% |
| Total Papers (1581) | 24% | 45% | 16% | 15% |

Positions of high visibility       Positions of low visibility

---

4. The data reported here are drawn from the bibliographies of the forty-one laureates
who were interviewed. Collaborative papers having just two authors are omitted since
they can be signed only in alphabetical order or in the reverse.

may be reflected in the data reported in table 6–5. The three relatively unambiguous sequences do not confer the same degree of visibility upon a particular author. Thus, the laureates appear as first-author-out-of-alphabetical-sequence (ZABC), occupying a position of great visibility, more than three times as often on the prize-winning papers: 62 percent versus 19 percent. This difference of 43 percent compares with a difference of 31 percent between the two classes of papers in the second most visible sequence (XYZA) and a difference of only 16 percent in the last sequence.

In deciding on authorship and name order as symbolic of the extent of contribution in collaborative papers, it is generally the senior investigator who has the authority (see Zuckerman and Merton, 1972, p. 344). But the exercise of that authority is limited both by social norms and by the need to maintain cooperation in the research group. It appears from the data we have just examined that when the laureates-to-be reach a stage where they take control of these decisions—as they presumably have done in most of the prize-winning work—they do not simply exercise raw power to put themselves uniformly in the forefront so far

**Table 6–5. Percent of Collaborative Papers [a] on Which Laureates Are Highly Visible Authors**

| | PERCENT OF PAPERS ON WHICH LAUREATES OCCUPY MOST VISIBLE POSITION | | |
| --- | --- | --- | --- |
| Unambiguous Author Sequences | Prize-Winning Papers | Other Papers | All Papers |
| First Author Out of Alphabetical Sequence (ZABC) | 62% (120) | 19% (233) | 34% (353) |
| Last Author Out of Alphabetical Sequence (XYZA) | 79 (107) | 48 (185) | 60 (292) |
| First Author in Random Sequence (ZALMB) | 31 (156) | 15 (288) | 20 (444) |
| Total | 54 (383) | 25 (706) | 35 (1089) |
| Ambiguous Author Sequences | | | |
| Alphabetical and Reverse Alphabetical (ABCD and DCBA) | 51 (85) | 52 (407) | 52 (492) |
| Total—All Author Sequences | 53 (468) | 35 (1113) | 41 (1581) |

a. Papers with three or more co-authors.

as the symbolism of authorship is concerned. Instead, they engage in a double, somewhat ambivalent pattern of behavior: they exhibit *noblesse oblige* by taking positions of high visibility in only 41 percent of all collaborative papers *and* they temper this pattern of generosity by providing for their own visibility somewhat more often in reporting the research they thought was especially important and which eventually brought them the prize.

Nevertheless, as we have noted, the laureates adopted this latter pattern in only 54 percent of the prize-winning papers that unambiguously highlight one or another author. This seems to reflect a fair amount of self-restraint. Since the laureates were singled out for an award after intensive investigation by the Nobel examiners, it can be assumed, even allowing for a margin of error in those investigations, that laureates were the chief architects of the prize-winning research in considerably more than half of these cases. Still, the pattern of *noblesse oblige* in granting prime authorship to their collaborators may be relatively cost-free. For the primary mode of communication in science is informal, especially at the research front where these members of the scientific elite are typically at work. As in the case of Pavlov, almost everyone who matters knows about the prime movers in collaborative research of Nobel prize caliber. Thus the social contexts of knowledge support the observed patterns of generosity.

## The Impact of Prize-Winning Research

As their name-ordering decisions suggest, few laureates-to-be had doubts about the scientific significance of the research that would ultimately win the Nobel prize. One physicist remembers telling a group of friends that evening that "I had taken part in the most important experiment that I had ever done in my life. . . . I had to swear them to secrecy since I knew how important it was and how much more work there was to do." Response from others was not always enthusiastic, but it was so much of the time and quite obviously for investigations that provided solutions to problems widely recognized as important.

The scientific implications of the Watson-Crick model of DNA were instantly perceived, as the authors tried to ensure by using unusually portentous language in the first paper on the model in *Nature:* "It has not escaped our notice that the specific pairing we have postulated immediately suggests a possible copying mechanism for the genetic material" (Watson and Crick, 1953, p. 737). Upon reading the manuscript, the usually reserved Max Delbrück wrote Watson, "I have a feeling that if your structure is true and if its suggestions concerning the nature of replication have any validity at all, then all hell will break loose, and theoretical biology will enter a most tumultuous phase" (quoted in Olby, 1974, p. 423). Linus Pauling, who had reason to be more restrained in his judgment, having published an incorrect structure for nucleic acid a few months before, said, "I feel it is very likely that the Watson-Crick structure may turn out to be the greatest development in the field of molecular genetics in recent years" (quoted in Olby, 1974, p. 422). The interest generated by this work among large numbers of researchers was, as we shall see, atypical even for investigations that later won Nobel awards.

From the perspective of individual laureates-to-be, immediate recognition from members of their reference groups—that is, scientists like Delbrück and Pauling whose opinions mattered—was often more important than creating a scientific stir. (See Merton, 1968b, for the general concept of reference groups.) Maria Goeppert Mayer, laureate in physics, gave the extreme version of this view. She recalled that J. H. D. Jensen (her fellow laureate who independently worked out the shell model of the nucleus) had said, "I have convinced Heisenberg and Bohr. You have convinced Fermi. What do we care about the others? You see, when you have convinced Fermi, you have really accomplished something, and if you also have convinced Bohr and Heisenberg, well then . . . the others don't mean anything."

In principle, it is possible to gauge such qualitative assessments of the impact of contributions by citation analysis. In practice, it is necessary to confine that analysis to the short-term impact of investigations published after 1961 since the Science Citation Index, as I noted earlier, was not established until that year.

*Citation Analysis of Prize-Winning Research.* What can be learned, then, about the immediate reception of contributions ultimately to be honored by Nobel prizes? In all, twenty papers published since 1961 can be identified unambiguously as reporting on prize-winning research. The papers examined here are a subset of all such papers, since early citations to papers that appeared before 1961 could not be tabulated. These twenty papers by eleven laureates-to-be (such research usually appears in a series of papers rather than in just one) appeared between 1961 and 1965. Thus it was possible to trace the citation history of the prize-winning papers for as many as eleven years after publication and before the award of the prize. They include publications by the biochemists Anfinsen, Moore, Stein, Axelrod and Sutherland, the molecular biologists Monod, Jacob, Nirenberg, Holley and Khorana (nearly all with co-authors), and two by the physicist Gell-Mann. Given the crudity of citation analysis (even when it focuses on papers rather than authors) and the fact that these publications span a number of sciences (each having a different rate of citation), it is wise to view these data more in terms of orders of magnitude than as precise measures.

To the extent that citations mirror scientists' interest in a particular investigation, the fact that all the papers but one were cited as often as ten times each year after publication, and that this minimum rate continued for eight to nine years, suggests that the papers were initially provocative and continued to be so far longer than the average scientific publication. By way of crude comparison, only 150 articles published in 1975 were cited as often as ten times in the year after they appeared (Garfield, 1976)—about 0.05 percent of all such publications that year. Moreover, the representative paper cited in a given year—a large share of papers are not cited at all—is mentioned about 1.6 times (Institute for Scientific Information, Science Citation Index 1964–1970, Comparative Statistical Summary). There was especially marked interest in a few of these publications, one paper by Monod and Jacob being cited 160 times the year after it appeared, and one by Nirenberg and Matthaei more than eighty times. Since the investigations reported in these papers were expeditiously recognized by Nobel awards, the promptness of their incorporation into the literature is probably atypical even for prize-winning contributions.

Comparatively high rates of citation to these papers continue for about nine years and then begin to trail off. This may indicate exhaustion of their potential for elaboration or, equally likely, such thorough incorporation into the literature that they, as the originating papers, are no longer cited and their visible influence is obliterated. (On obliteration by incorporation, see Merton, 1968g, pp. 28, 35, 38; Zuckerman and Merton, 1973.) They become, as Joshua Lederberg has put it, "household words," and no one "cites" household words. Citation to five of the papers, however, does not flag and continues at a high rate up to the time of the award itself. As noted in chapter 2, a few of these (five out of eighteen) qualify as landmark publications in the sense of being heavily cited year after year rather than following the modal pattern of declining citation after the first few years of use (see Garfield and Malin, 1968, on landmark papers). In the absence of comparable data on the citation histories of other important papers published in the same years in the same fields, it is not possible to say whether the pattern of citation to this sample of Nobel prize-winning research is representative of all such papers.

The class of scientific investigations known as premature discoveries obviously would not be included among the small number of prize-winning investigations that could be subjected to citation analysis. By definition, premature discoveries are initially ignored or rejected only to be recognized much later for their significance. Thus, premature contributions do not form the base for further research until long after initial publication (see Barber, 1961; Stent, 1972). Although a number of Nobel prize-winning contributions were controversial and encountered resistance when they first appeared, only a few have been premature. They include Landsteiner's early research on the effects of human blood transfusions, Staudinger's work on the formation of macromolecules, Onsager's investigations of irreversible thermodynamics, Alfvén's studies of wave movements in a plasma, and Rous's demonstration that one form of sarcoma is caused by a virus. As I noted earlier, it took many years for the Rous work to be accepted. Some flavor of the resistance Rous encountered is conveyed by S. E. Luria, a founder of phage genetics and laureate in medicine. Rous told Luria that when he "discovered that a chicken sarcoma was caused by a virus . . . an eminent pathologist stated that Rous sarcoma could not be a cancer since Rous had discovered its cause—and it was well known that the cause of cancers was unknown" (Luria, 1973, p. 1338). By contrast, there was no skepticism about Lars Onsager's work, few even noticed it. In presenting Onsager for the 1968 Nobel Prize in Chemistry, Professor Claesson of the Royal Academy of Sciences observed:

> Onsager presented his fundamental discovery . . . at a meeting in 1929. It was published in final form in 1931 in the well known journal, *Physical Review,* in two parts. . . . The elegant presentation meant that the size of the two papers was no more than 22 and 15 pages respectively. . . . One could have expected that the importance of this work would have been immediately obvious to the scientific community. Instead it turned out that Onsager was far ahead of his time. [His papers] . . . attracted for a long time almost no attention whatsoever. It was first after the second world war that they became more widely known. During the last decade they have played a dominant role in the rapid development of irreversible thermodynamics with numerous applications not only in physics and chemistry but also in biology and technology [Nobelstiftelsen, 1968, p. 42; see Groenevelt, 1971, for a different view of the reception of Onsager's early papers].

Hannes Alfvén had an even more difficult time, being ignored by some and rejected by others:

> For much of [his] career, his ideas were dismissed or treated with condescension; he was often forced to publish his papers in obscure journals; and he was continually disputed by the most renowned senior scientists working in the field of space physics. Even today there is a rather pervasive unawareness of Alfvén's multifaceted contributions to the field of physics where his ideas

are used with apparently little appreciation of who originated them [Dessler, 1970, p. 604].

Stephen Cole's (1970) research indicates that delayed recognition is rare in the scientific literature. However, delayed recognition may be disproportionately frequent among highly significant research papers for two quite different reasons: first, because their relevance for ongoing research may not be as easy to identify as it is for the run of papers, and second, because the implications of these papers for research in other sciences may be recognized only long after publication. Ordinary scientific papers, by contrast, are closely linked to current research in a narrow specialty and rarely have implications that spill over the specialty's boundaries.

Being ignored (for a short time) is not entirely dysfunctional for researchers pursuing a fruitful line of investigation, as Gerald Edelman reports. Edelman, whose Nobel was awarded for his studies of the structure of antibodies, remembers

> being terribly confused that people wouldn't believe my results on the chains. . . . It was, I would think, lucky for me . . . given my very sparse facilities for really going into the problem. If they [his ideas] had been believed, they could have been snapped up by laboratories with much more expert people than myself and I think it [the structure] would have become much clearer much earlier, had they paid attention. But they didn't and so I had a chance in a leisurely way to explore this whole business [interview recorded for American Association for the Advancement of Science, pp. 9–10].

In the aggregate, however, the work of few laureates-to-be was dismissed or ignored; most had competitors in working their "gold mines," as Carl Cori described the line of research on sugar metabolism that he and Gerty Cori had developed. By the time the awards were made, their research was widely known and used. As citations to their publications suggest, it would also appear to have had considerable staying power in spite of tendencies toward obliteration by incorporation in the scientific literature.

*Citations Before the Award.* As we have seen, laureates-to-be are heavily cited. This is the case because of their high productivity and because of the influential character of a number of their publications— not just those eventually recognized by the Nobel committees. Each year before the award, between the years 1961 and 1971, prospective laureates are cited 222 times on the average. This is more than twice the average of ninety-nine citations for a random sample of American scientists about to be elected to the National Academy of Sciences during the same years and almost forty times the average of 6.1 citations to a representative author in the Science Citation Index (Institute for Scientific Information, Science Citation Index 1964–1970, Comparative Statistical Summary; Garfield, 1975, private communication). Although heavy incidence of citation is by no means a criterion for distinguishing the ultra-elite of Nobel prize-winners and occupants of the forty-first chair from the more extended elite of National Academy members, it does symbolize the fact that members of the ultra-elite in science have generally made more consequential contributions to knowledge. Further, since the Index lists citations according to the first-named author of papers only and since, as we have seen, laureates-to-be increasingly follow the principle of *noblesse oblige* by putting their names last in the author sequence, citations to some of their papers are not recorded under their names. To the extent that first authorship and scientific standing are inversely related, then, crude citation counts underestimate the impact of the work of eminent scientists and the difference in number of citations between the elite and others.

An extreme but nonetheless instructive case in this regard is H. G. Khorana, who received his award in 1968 for synthesizing molecules containing all possible combinations of the four nucleic bases. Khorana published 50 papers between 1965 and 1967 that were cited 383 times during 1967 (Garfield, 1972, p. 7). But, since he was first author on only five of these, just thirty-nine citations appear under his name in that year's Index.

There are also marked differences in the number of different publications cited for the three classes of authors, with an average of sixty-seven for prospective laureates, thirty-seven for prospective Academicians, and just under four for "representative authors." Although prize-winning contributions are often published as a series of papers, these series rarely if ever involve 67 items, suggesting, as I have indicated, that other work by the laureates-to-be is also being cited.

Citation analysis of the work of prospective laureates on the eve of their awards confirms what any scientist knows: Nobel prizes do not go to unknowns. It also reinforces our sense that the prizes do not typically go to "has beens," to scientists whose research once mattered but is no longer significant to working investigators.

A laureate in physics told of inventing (as it turns out, independently inventing) his own measure of the extent to which his papers were used.[5]

---

5. In 1955, Frederick Mosteller, the Harvard statistician and inventor of ingenious unobtrusive indicators, developed what he called the "dirty edge index" for estimating the use of articles in sets of the *Encyclopedia of the Social Sciences* located in various libraries at Harvard. The articles used most heavily were not only dirty at the edges but smudged, inked, and honorably disreputable in appearance.

As a frequent visiting lecturer at universities around the world, he often found himself with unscheduled time that he used by going to the library, searching out the volumes of the *Physical Review* in which his prime papers were published, and examining the amount of grime accumulated at the edges of the pages where his papers appeared. His acid test came when, having determined that the edges were dirty, he tested whether the volumes in question would open of their own accord to the "right" pages.

## EXTENT AND KINDS OF RECOGNITION
## FOLLOWING PRIZE-WINNING RESEARCH

Having made contributions that will ultimately be judged prize-worthy, prospective laureates receive further recognition and acquire new authority. The reward system of science, however, operates with differential effectiveness—spotting the potentials of some scientists early and providing recognition and resources for their work, but being less prompt in identifying and rewarding others. In this section, we shall examine varieties of response by the reward system to research of prize-winning caliber in the form of promotions, honorific awards, and positions of authority. We shall thus be able to identify those laureates-to-be who were long-term but obviously temporary occupants of the forty-first chair before their awards and those whom the committees at Stockholm rewarded more expeditiously.

### Promotions Following Prize-Winning Research

The upward mobility of future laureates through the academic system has already been examined in the context of age-specific rates of promotion. There I noted that since the great majority of American Nobelists (eighty-five out of ninety-two) were academics affiliated with universities or research institutions of like kind, it would be possible to confine the analysis of promotion to their movement up the professorial ladder. How, then, do universities and research institutes respond to prize-winning research?

The data on age-specific rates of promotion intimated what can be reported straightforwardly here: thirty (35 percent) of the eighty-five laureates-to-be who became full professors were promoted some time before they began to publish their prize-winning research.[6]

Not surprisingly, this subgroup of laureates-to-be were promoted when they were comparatively young; they reached the top of the academic ladder at an average age of 34.9, about two-and-a-half years younger than is the case for American laureates as a whole. It is also not surprising that they did their prize research comparatively later than the rest: at 43.2 years as against an average of 39.2 for all the laureates. These two facts taken together make it appear that the system spotted some scientists particularly early and that they were moved ahead on the basis of potential rather than achievement. It turns out that this is not generally so, although there are several cases that approximate this image. Arthur Compton was made a full professor and chairman of the department of physics at Washington University in St. Louis when he was 28—three years before his work on the Compton Effect. Similarly, Vincent du Vigneaud was 31 when he became a full professor of biochemistry at Illinois, but he did not begin to publish the research on sulphur-containing compounds until a year later. But by the time they were promoted, both of these scientists had done as much work as the average professor does for promotion; it simply took them much less time to do it. The majority of the laureates-to-be who were promoted before beginning to publish their prize-winning research had established themselves even more solidly, having done significant work when they were quite young and before they were promoted. E. C. Kendall, for example, isolated thyroxin (the active constituent of the thyroid hormone) when he was 28, was promoted to a professorship at the Mayo Foundation and University of Minnesota at 35, and partly determined the structure of thyroxin a year later. But his prize-winning research on the biochemistry of cortisone and its therapeutic use did not get fully under way for several years later.

In addition to the thirty laureates-to-be who were promoted early in their research careers, sixteen more were made full professors in the course of doing their prize research. Apparently, the evidence at hand was considered sufficient for them to be raised to top rank before the work was complete. Like the others promoted before beginning to publish their prize research, these scientists were younger than the mean for all laureates when they were made full professors, but not by much: on the average, they were 36.8 years at the time, or only a half year younger than the average for the rest. Altogether, 54 percent of the laureates-

6. Here, as earlier in the analysis, the time of prize-winning work is taken as the date of publication of the research cited for the prize or, when a series of publications was cited, the midpoint in that series. The beginning of prize-winning research is the date of the first publication on research cited for an award.

to-be were full professors at the time they were doing their prize-winning research. For a slim majority of future laureates, then, the reward system worked effectively by providing swift recognition for scientific accomplishment. That this is so obviously says nothing about the overall effectiveness of the reward system for the great majority of scientists who never become members of the ultra-elite.

We can see, therefore, that there is a fair amount of slippage in identifying and promoting scientists who will ultimately do consequential research. One indicator of possible errors of this kind is chronologically late promotion. Along these lines, I noted earlier that "as many as" 30 percent of the laureates had waited until after they were 40 to become full professors and that this could be taken as evidence of some malfunction in the system since they were well past the mean age at which future laureates and Academicians were promoted. In respect to the time of prize-winning research, the post-40 full professors were somewhat slower than the other future laureates, having done their prize-winning research more than four years after the mean, when they were 43.6 years old. But it is also the case that they were promoted at the age of 45.6, on the average almost nine years after the more "precocious" fellow laureates. Thus, these future laureates moved up very slowly indeed, more slowly than the rank-and-file scientists whose mobility was examined earlier in this chapter. In gauging the timeliness of academic rewards, both relative timeliness—the pace of according the same rewards to various individuals—as well as absolute timeliness—the gap or interval between the contribution an individual makes and his reward—need to be taken into account. As we shall see, this distinction will be relevant when we examine the relations between prize-winning research and the timing of other rewards.

Chronologically late promotion is a crude indicator of delayed rewards for Nobelists. A more exacting one is the absence of promotion during the five years following what will become prize-winning research, however old the laureates were at the time of that research. Of the thirty-nine prospective laureates yet to be promoted, nine (almost a fourth) were not made full professors in the five years that followed their prize research. Three of these nine were promoted reasonably soon, but not to the rank of full professor, and the six others waited, on the average, more than a decade to be promoted at all.

The three scientists who were promoted but not to full professorships share one important biographical fact: all did their prize research early. E. M. McMillan was 33 when, with Phillip Abelson, he discovered plutonium, the first transuranium element, and Peyton Rous, as repeatedly noted, was also 33 when he did his work on chicken sarcoma. J. D. Watson was the most youthful of the three, being just 25 when he and Crick developed the double-helix model of DNA. In McMillan's case, World War II appears to have been responsible for his slow ascent up the academic ladder. He was mobilized first to work at the Navy's Radar and Sound Laboratory and then moved to Los Alamos to work on the atom bomb. When he returned to Berkeley he was promptly promoted, illustrating how such delays can, on occasion, have little to do with the reward system of science.

Rous's work, as I have noted, was premature and encountered resistance for a long time. This resistance may have contributed to his having to wait until he was 41 to become a full member of the Rockefeller Institute. Although the Rockefeller was not given to early promotions then, Rous waited longer than other prospective laureates affiliated with it, who on the average were promoted to full membership at just over 36. Watson's work was, of course, received enthusiastically, but Harvard waited until he was a seemly 33 to make him a full professor, eight years after the Crick-Watson work.

Although five of the remaining six laureates were foreign born, this fact seems to have had little to do with the age at which they and other foreign-born laureates became full professors. Rather, they represent at least three classes of scientists whose promotions come late. First, as with Rous, the significance of Onsager's work was not immediately perceived by others. Not surprisingly, promotions come slowly to contributors of premature investigations. Second, as I have noted earlier, clinical research in the biomedical sciences is generally recognized more slowly than are experimental investigations. The late promotions of André Cournand and another Frenchman, Alexis Carrel, remind us of the more general pattern. (Carrel, it turns out, was not made a full member of the Rockefeller Institute—the equivalent of a professor—until the year he got his Nobel prize, that delay being a dubious honor which he shares with laureates Harold Urey and Gerty Cori.)

Third are the two women, the only ones among the ninety-two American laureates. Maria Goeppert Mayer was a "voluntary associate" in physics at the Johns Hopkins when she first came to the United States, a lecturer at Columbia while also on the staff of the secret SAM Laboratory during World War II, and moved to the University of Chicago with her chemist husband, Joseph Mayer, after the war. She served as a research physicist at the Argonne Laboratory and as a "voluntary professor" at Chicago. In short, she never had a regular academic appointment until 1960, when she was 53 years of age, nine years after she had done her prize-winning work.

Gerty Cori's career does not look very different except that she and her husband, Carl Cori, worked together and shared the Nobel Prize in Medicine. Both Coris received their medical degrees at 24. When they emigrated two years later, they came to the State Institute for the Study of Malignant Diseases in Buffalo, he as a "biochemist" and she as an "assistant biochemist." They stayed for nine years, until Carl Cori was offered a full professorship in pharmacology and biochemistry by Washington University in St. Louis, thus appearing in the composite statistics as a full professor at 35. Gerty Cori accepted the post of "research associate" and remained in that position for sixteen years. When the Coris were 40, they isolated glucose-1-phosphate, the "Cori ester," and continued their collaborative studies of metabolism. Gerty Cori was promoted to the rank of full professor when she was 51, the very same year she and her husband shared the Nobel prize. Again, the particular cases represent the more general pattern of late promotion among women. This is so for women elected to the National Academy of Sciences as well as for Nobel laureates (Zuckerman and Cole, 1975).

Finally, there is the unique case of John Enders. Enders is not unique in being a Nobel prize-winning associate professor; Carl Anderson also went to Stockholm before his university, Cal Tech, saw fit to promote him. But Anderson was just 31 when he became a Nobelist and thus might still have been considered "young" for a full professorship. Enders was 59.

As we have related earlier, Enders got his Ph.D. at Harvard when he was 33, thus qualifying as a late starter. He stayed on, moving slowly through the ranks from instructor to faculty instructor to assistant professor and finally to associate professor when he was 45. He stayed an associate professor for fourteen years. In the meantime, he continued his research on infectious diseases and, in his early fifties, began work with two young collaborators, Frederick C. Robbins and Thomas H. Weller, on the cultivation of polio virus. Their work culminated in the demonstration that polio virus was not neurotropic as had commonly been believed and could be grown *in vitro,* thus providing the basis for producing vaccine on a large scale. In 1954, when Enders was 57, he and his co-workers received the Nobel Prize in Medicine. The same year, Enders, still an associate professor, earned the Lasker award. Harvard waited until 1956, when Enders was 59, to promote him to the rank of full professor, and that same year gave him an honorary degree —which must be a rare type of double honors. It is said that there was just one chair in each departmental section at Harvard and that Enders simply had to wait for "his" chair to be vacated. Two years before, when the prize was awarded, Weller had been given a name chair in tropical public health at Harvard at the age of 39.

Delay is one thing, outright error another. There is one such case not registered in the data on American laureates reported here. Born and educated in England, Geoffrey Wilkinson worked in nuclear chemistry at Berkeley, shifted to M.I.T. and inorganic chemistry, and was made an assistant professor at Harvard in 1951. Wilkinson published his first results showing the novel "sandwich" structure of the compound ferrocene a year later. In 1956 Harvard decided, as he put it, that it "could do without me," and he was let go (*New York Times,* 24 November 1973, p. 26). Wilkinson returned to England and in 1973 received the Nobel Prize in Chemistry for his development of the chemistry of sandwich compounds.

## Organizational Mobility Following Prize-Winning Research

If prospective laureates are upwardly mobile following their prize-winning research, they are otherwise quite sessile. As table 6–6 indicates, 84 percent of the laureates did not change their affiliations in the five years following their prize research, and presumably this was not entirely for lack of opportunity to do so. Having failed to raise the appropriate questions in interviewing the laureates, I am unable to report on the frequency, speed, or lavishness of the job offers (and counteroffers) that came to them after they had completed their prize research. The biographical data summarized in table 6–6 tell only what the laureates actually did, not what their opportunities were.

Of the forty-six laureates-to-be who were already full professors

Table 6–6. **Organizational Mobility and Promotion of Laureates (1901–72)** [a] **Five Years After Prize-Winning Research**

| Academic Rank [b] of Prospective Laureate at Time of Prize-Winning Research | ORGANIZATIONAL MOBILITY IN THE FIVE YEARS AFTER PRIZE-WINNING RESEARCH | | |
| --- | --- | --- | --- |
| | Stayed | Moved | Number |
| Full Professor | 91% | 9% | (46) |
| Promoted to Full Professor Within Five Years | 70 | 30 | (30) |
| Promoted to Rank Below Full Professor Within Five Years | 67 | 33 | ( 3) |
| Not Promoted Within Five Years | 100 | — | ( 6) |
| All Laureates | 84 | 16 | (85) [c] |

a. Prospective laureates who were not affiliated with universities or research institutes before their awards are excluded from these data.
b. The equivalent of the full professorship was used when a system other than professorial ranking was in force in an institution.
c. Von Békèsy, although affiliated with Harvard until his retirement, was a Senior Research Fellow and is therefore excluded from this computation.

when doing their prize research, just 9 percent moved in the five years that followed that research. This finding is much the same as the 7 percent reported by Brown (1967, p. 39) for all full professors in all fields of science and scholarship in 1962. If prospective laureates had more and better opportunities to change jobs at this time in their careers, as is likely, they apparently also had good reasons to stay put.

Among the thirty-nine prospective laureates who had not yet attained the rank of full professor, moving was more frequent. About a quarter of them (10) went on to greener pastures during the five-year postlude to their prize-winning research, all but one of them having been promoted to full professorial rank in their new institutions at the time of the move. For prospective laureates as for other academics, upward mobility often requires geographical mobility, but since the numbers are so small, not much weight should be given to them. Put in comparative terms, prospective laureates who were not full professors by this time moved two and a half times as often as did professors of the same rank whom Brown (1967, p. 32) studied. Although part of the mobility observed among laureates-to-be seems to occur as a matter of course among academics of roughly the same rank, the residual mobility they exhibited following their prize research probably results from their greater visibility and their having received more and better job offers than the run of scientists.

Half of the fourteen prospective laureates who moved—four full professors and ten of lower rank—went to elite universities or research institutes and half to institutions of lesser prestige.[7] Prestige in and of itself apparently had little to do with where the laureates-to-be chose to go. Most of them, by this time in their careers, were in a position to shape their research environments and had reached the point of conferring prestige on their universities rather than deriving it from the affiliation. As a consequence of this pattern of movement, 77 percent of the academic laureates were working in the elite universities and research institutes five years after their prize-winning research, just the same proportion as five years earlier. Although the individuals associated with particular universities changed, the overall distribution remained the same.

In all, then, these data suggest that the reward system (as it is reflected in promotions) operates with a degree of efficacy by recognizing both scientific potential and scientific achievement. But since both relative and absolute timeliness of promotion varies greatly, promotion is not a sensitive barometer of achievement.

---

7. No academic left a university post for one in government or industry, but John Bardeen transferred from the Bell Laboratories to the University of Illinois.

## Election to the National Academy of Sciences and Other Honorific Awards

The reward system of science also recognizes achievement by conferring a vast array of honors, prizes, honorary degrees, and election to societies of all sorts. Such honors symbolize to the recipients and others that the work the recipient has done matters.

National eminence in science is generally recognized by election to one's national academy: in England, for example, to the Royal Society of London; in France to the Académie des sciences; and in the United States to the National Academy of Sciences.

As of 1976, nearly all (96 percent) of the ninety-two laureates whose careers are examined here were members of the National Academy. About three fourths (78 percent) were elected before their prizes and thus became members of the extended elite before being raised into the more exclusive ultra-elite.[8] An additional sixteen Nobelists (18 percent) were elected after their prizes. That leaves three laureates who will never be admitted to the Academy and one who is well into his eighties and probably never will be: Alexis Carrel, Philip Hench, Max Theiler, and William Parry Murphy, all laureates in medicine, all physicians, and all but Murphy now dead. Since the Academy does not permit posthumous election, its verdicts are unalterable in three cases. Carrel, who later gained a certain notoriety for associating with the Vichy government, was elected to the Académie des sciences in 1927, fifteen years after his prize. The others, however, were not elected to any major academy. Given the fact that more than a third of the physician laureates ever elected to the Academy got in only after their Nobel awards as compared to 5 percent of the laureates in medicine who hold Ph.D.s, it would appear that the Academy has been reluctant to accept physician researchers. In fact, of course, it is not the degree that counts but the greater proclivity of physicians for clinical research, which some scientists think is less rigorous and fundamental than the experimental research typically done by Ph.D.s. That only one of the M.D.s elected to the Academy before receiving his award did what might be considered clinical research is at least consistent with this reading.

Such cases of neglect constitute just one kind of oversight by the Academy if one accepts the premise that Nobel laureates are likely to have contributed at least as much to their fields as the least distinguished Academicians. Another kind of oversight would seem to be exemplified in the sixteen post-Nobel admissions to the Academy. As table 6–7

---

8. That so many laureates were admitted to the Academy before their Nobels may say more about delays in awarding prizes than about the perspicacity of the Academy.

Table 6-7. Mean Age of Laureate (1901–72) at Time of Prize-Winning Research, at Time of Election to National Academy of Sciences, and at Prize

| Time of Laureates' Election to National Academy of Sciences | x̄ Age at Prize-Winning Research | x̄ Age at Election to NAS | Interval between Prize-Winning Research and Election to NAS | x̄ Age at Nobel Prize | N |
|---|---|---|---|---|---|
| Elected Before Nobel Prize | 39.7 | 42.9 | 3.2 | 52.9 | (72) |
| Elected After Nobel Prize | 38.5 | 48.2 | 9.7 | 44.2 | (16) |
| Both groups | 39.3 | 43.8 | 4.5 | 51.3 | (88) |
| Never Elected to National Academy of Science | 39.0 | — | — | 46.7 | ( 4) |

shows, laureates in this category were much younger (by almost nine years) when they got their prizes than those admitted to the Academy beforehand. On the face of it, it would seem that the Academy simply did not get around to electing these scientists before the Nobel committees so expeditiously earmarked them for membership in the ultra-elite. But this is not so. Taking the interval between the prize-winning research and election to the Academy as a crude measure of the timeliness of the Academy's elections, it is evident that the laureates admitted after their prizes waited considerably longer for election; almost ten years on the average went by before their admission as against the brisk three years for the other laureates.[9]

This finding illustrates the general pattern of delaying rewards to physician researchers mentioned earlier. Although a larger number of physicists than of physicians were elected to the Academy after their prizes, the interval between prize research and election for these physicists was 5.7 years—slightly longer than for their physicist colleagues elected before their prize, but far shorter than the average of 15.4 years for the laureates in medicine elected after the award.

Quite apart from field differences, delays of this kind can result from reluctance on the part of groups like the Academy to endorse the Nobel selections. Some Academicians feel that Nobel prizes have all too much influence and that automatic election of laureates would make that influence even greater.

Turning from the minority pattern of delayed rewards to the majority pattern of prompt recognition, the data in table 6–7 also show that Nobelists elected before their prizes were awarded were admitted on the average when they were just under 43 years old. This means that they were elected considerably earlier than the average age of 50.5 for all Academicians admitted between 1900 and 1967. In the absence of information on age-specific scientific production by members of the Academy, we cannot say whether this difference represents more rapid recognition of the laureates' contributions, or whether it takes other members of the academy longer to accumulate the research credentials generally required for election.

*Other Scientific Awards.* Admission to national academies is only one among many formal symbols of esteem accorded to scientists. Since

9. Four laureates—Alvarez, Langmuir, Morgan, and Segrè—were elected to the Academy before they began the research that brought them their awards. Other laureates may also have been admitted on the basis of contributions made before the prize-winning research, and this may account for the short mean interval of 3.2 years between the prize-winning research and election to the Academy indicated in table 6–7.

esteem from knowledgeable peers is the principal (although not the only) reward scientists receive for significant contributions, the reiteration here of the prospective laureates' honors, prizes, and medals is intended to convey a sense of their standing on the eve of their winning the Nobel prize. That each of them had earned, on the average, six of these varied honors before their selection as laureates—twice as many as a matched sample of age peers in the National Academy of Sciences—once again reminds us that a Nobel prize rarely goes to unknowns. And since some distinguished scientists pare the less significant awards from their official biographies as their standing increases (comparison of earlier and later listings for the same scientist in *American Men of Science* indicates that this is so), the actual number of awards prospective laureates received is likely to be underestimated here.

Most of the prospective laureates have won multiple honors. Typically, they also have at least one that carries great prestige, such as the Enrico Fermi Award or the Lasker Award. This pattern holds for laureates in all fields and is consistent with the Coles's finding that the number of awards physicists hold is correlated with the prestige of their highest ranking award (1973, p. 48). Consider the array, from foreign as well as domestic sources, that Eugene Wigner accumulated before he became a laureate in physics in 1963. He had been decorated with the Presidential Medal for Merit, the Max Planck Medal of the German Physical Society, the Atoms for Peace Award, the Franklin Medal of the Franklin Institute, and the Fermi Award. Six colleges and universities had given him honorary degrees, and he had been a member of the National Academy of Sciences for eighteen years. His fellow physicist Hans Bethe, another long-term though temporary occupant of the forty-first chair, also had the Medal of Merit, the Planck Medal, and the Fermi Award. In addition, he held the Draper Medal of the National Academy and the Morrison Prize of the New York Academy. He had five honorary degrees, including one from Harvard (Bethe has spent his entire career in the United States at Cornell), and was also a foreign member of the Royal Society of London—not commonplace even among Nobel laureates.

Wigner and Bethe are atypical in that their Nobels came late in their careers. But the delay in their prizes is less in point than the way in which their collection of awards illustrates the uneven distribution of rewards in science. Things are quite different for the run of scientists. Cole and Cole (1973, p. 48) report that 72 percent of the 2500 academic physicists they studied had never won an award and another 15 percent had just one, typically a postdoctoral fellowship. To put this in another way, 11 percent of the physicists held 70 percent of the honors —another outcome of processes of accumulation of advantage that have marked the careers of most prospective laureates and continue to be evident after they win the prize.

Although most prospective laureates were frequently honored, there were considerable differences among them. Each of the ten most honored had an average of twenty-one awards, honorary degrees, medals, prizes, and the like before their Nobels as compared with the average of about one for those least recognized. The former group is comprised, for example, of Peyton Rous, Linus Pauling, R. B. Woodward, and, as I have indicated, Wigner and Bethe. Pauling had picked up seventeen scientific awards and ten honorary degrees before his Nobel Prize in Chemistry (his Peace prize came eight years later), and Rous had accumulated eighteen awards and ten honorary degrees. As we shall see, the differences among prospective laureates in scientific standing just before they get their prizes affects the impact the Nobel has upon them.

## Influence in Science: Offices, Gatekeeping, and the Exercise of Authority

Most prospective laureates elect to transform the esteem in which they are held—symbolized by the awards and prizes so many of them received—into positions of influence and responsibility. They accept offices in scientific societies, editorships, and a variety of other gatekeeping posts that require them to decide how resources are to be allocated and to whom.

Individual decisions to take on these tasks reflect personal preference and personal views of individual responsibility to the scientific community. Since these weigh differently for different scientists, there is marked variation among prospective laureates in officeholding. Among the ninety-two, there are nineteen presidents of national scientific societies and four vice-presidents, or about one office for every four laureates. This is approximately the same rate of officeholding as in the sample of prospective laureates' age peers in the National Academy of Sciences. Although these posts confer honor on their occupants (or did in the days before scientific societies became politicized), they entail work and consume time. It is evident from the skewed distribution of offices held by prospective laureates that a good many were unwilling to do the work and spend the time such activities require. For these scientists, research seems to have been not their highest priority, but their only priority.

The same variability is reflected in the number of prospective laureates who took on editorships and positions on selection panels of granting agencies and fellowship boards, that array of positions in which scientific authority is exercised. As Michael Polanyi (1963a, 1963b) and a variety of others have pointed out, the exercise of scientific authority is critical in the operation of science. The validity of new ideas and new empirical research must be continually assessed, and assessment is usually the prerogative and responsibility of scientists who have demonstrated expertise through their own contributions. Their judgments therefore have authority. Donald Fleming, the historian of science, succinctly describes the "responsibilities . . . [of] an Influential of the first magnitude":

> The bestowal of the imprimatur of science upon some but not all observations and theories; the redistribution of emphases, as one field or mode of attack plays out and another is taken up; and the admission of newcomers to the scientific community, with a *prima facie* claim to be heard with respect, and thereafter the grading upward of successively smaller groups until only the Influentials of the next generation are left [1954, p. 132].

That prospective laureates would be prime candidates for such tasks is evident. It is a matter of record that they served as editors and on editorial boards of thirty-six journals. Since such other gatekeeping activities as the refereeing of scientific papers and proposals are typically confidential, it is not possible to gauge the extent to which they also took on these jobs. But there is evidence that many of the physicists did so assiduously. In a study (Zuckerman and Merton, 1971) of the *Physical Review,* now the prime journal in physics, Merton and I found that the greatest share of refereeing was carried out by future members of the scientific elite, physicists who were not yet formally recognized but who could be so identified after the fact. Among them were most but not all of the prospective laureates then at work. These younger scientists were expert in their fields and, as the referees' reports indicated, committed to maintaining the quality of published work. Although their participation in what one laureate called "keeping the literature clean" may vary from field to field, the evidence here implies that future laureates were active in this role.

About two thirds of the prospective laureates energetically assumed responsibilities associated with gatekeeping and the politics of science. The remaining one third avoided these roles altogether. Not surprisingly, the willingness to take on a large complement of such positions and to spend the time they require is associated with age and having an established scientific reputation as gauged by the number of awards held. Eminence leads to opportunities for extraresearch activity and may even reinforce a scientist's sense of obligation to take it on. But the fact remains that some laureates-to-be chose not to occupy themselves with these tasks or were not asked to do so. They confined themselves to bench and blackboard.


## PROSPECTIVE LAUREATES AND OCCUPANTS OF THE FORTY-FIRST CHAIR: MULTIPLE INDEPENDENT DISCOVERIES IN PRIZE-WINNING RESEARCH

On the eve of their prizes, most prospective laureates are internationally eminent. Most of them are persuaded that the work they have done is or will be scientifically consequential. This does not mean that most laureates-to-be believe that they are the *only* ones who could have done the research they carried through. At least a few know otherwise from the fact that their prize-winning work was duplicated by others. The existence of multiple independent and simultaneous discoveries by Nobelists suggests that science would have developed much as it did even if one or another laureate had not done the work for which he was honored. This assertion does not denigrate the achievements of Nobelists (the fact is that *they* did the research) but it does imply that even distinguished scientists are likely to affect the pace of scientific development more than its direction.

In all, just five Nobel prizes have been shared by scientists whose investigations are universally recognized to have been independent and simultaneous.[10] (For multiple discoveries as strategic research sites in the sociology of science, see Merton, 1973, chapters 14–17.) These five are:

— The discovery in 1927 of the "interference phenomenon," which provided the first empirical proof of the wave theory of electrons, by Davisson and Germer at the Bell Laboratories and by G. P. Thomson, then at the University of Aberdeen. The two experiments employed very different procedures but were more or less simultaneous and had

10. These are not, of course, the only occasions for multiple independent and simultaneous discoveries involving Nobel laureates—with one another as well as with other scientists. Merton's analysis of the conditions making for multiples suggests that Nobelists, insofar as they are more apt than other scientists to be at the forefront of research and at work on problems generally thought to be important, would be likely participants in such discoveries (Merton, 1973, chapters 16 and 17). Along the same lines, since partners in multiple discoveries have demonstrably similar qualities of mind, we would expect that Nobelists would share multiple discoveries with other Nobelists and occupants of the forty-first chair. In the absence of systematic data on the frequency of these episodes it is not possible to determine whether this is so; approximately two dozen such occasions have been identified so far.

the same physical implications (Thomson, 1967, p. 61). Davisson and Thomson shared the prize in physics in 1937.

— The observation and measurement of nuclear magnetic resonance, by E. M. Purcell, H. C. Torrey, and R. V. Pound at Harvard and by F. Bloch, W. W. Hansen, and M. E. Packard at Stanford, in 1946. Purcell has observed that physicists "have come to look at the two experiments as practically identical but . . . [that] at the time . . . when Hansen first showed up [at our lab] and started talking about it, it was about an hour before either of us understood how the other was trying to explain it." Bloch and Purcell shared an award in 1957.

— The theoretical development of the shell model of the atomic nucleus, by Maria Goeppert Mayer, at the University of Chicago, and by J. H. D. Jensen at work in Hanover (Germany) in 1948. A prize was awarded to them in 1963. Mayer remembered that, upon reading Jensen's paper, she was "at first dismayed—for about five minutes. But then Weisskopf [who had showed it to her] said, 'You know, now I believe it, now that Jensen has done it too.' Jensen later wrote me an awfully nice letter . . . we always talked to one another about 'your theory.' " Mayer and Jensen shared the physics prize with Wigner in 1963.

— The discovery and development of the maser and laser by Charles H. Townes at Columbia, starting in 1951, and by the Russians Nicolai Basov and Alexsandr Prochorov, who presented their work at a meeting in 1952. Although the three shared the physics prize in 1964, the precise chronology of this work is complex and has been the subject of heated discussions about priority. (See the account by J. P. Gordon [1964], one of Townes's collaborators on the production of the first workable maser.)

— The isolation and structural determination of cortisone by E. C. Kendall at the Mayo Clinic and T. Reichstein at the University of Basle, for which they shared the prize in medicine in 1950. (Researchers at two other laboratories apparently did the same work at the same time.) Kendall and Reichstein were also responsible for isolating other cortical substances and Kendall, with P. S. Hench, first used cortisone in the treatment of chronic rheumatoid arthritis.

That there have been just five such cases in all Nobel prize-winning investigations does not of course indicate their frequency in science or even among scientific contributions of this caliber. As noted in chapter 2, there is reason to believe that the award committees ultimately prefer candidates whose work does not pose problems of credit allocation (as multiple independent and simultaneous discoveries inevitably do), and thus the actual frequency of multiples of the first rank is probably greater than their appearance among prize-winning discoveries would suggest.

The significance of multiple independent discoveries for our purposes does not lie in their numbers but, rather, as indicated earlier, in the evidence they provide for the interchangeability and "nonuniqueness" of Nobelists.

Along the same lines, there is evidence that other Nobel prize-winning investigations would have qualified as multiples if only their contributors had not announced them as early as they did. These early announcements, in Robert Merton's words, "forestalled" multiple discoveries (1973, pp. 360–63) by others at work on the same problems at the same time. Since the laureates themselves presumably have less motivation than others to assert that their work would have been quickly duplicated had they themselves not published it, their testimony on the subject would seem to be persuasive.

A physicist laureate observed:

> It was of course a great experiment but it is also my estimation that you do not do these things until a background knowledge is built up to a place where it's almost impossible not to see the new thing and it often happens that the new step is done contemporaneously in two different places in the world, independently. . . . I don't think you could have delayed [that] experiment by more than five years no matter what you'd done.

He went on to indicate that this was almost but not completely so in the case of his work:

> There were other people who were close to it, who were trying to understand this group of facts. One of them gave a paper at the annual meeting of the Physical Society after [we] had gotten [evidence of] the effect. Or course we couldn't say anything since we weren't ready to publish and this was too important to hash around. He gave this paper and I listened to him. He's a friend of mine and I know the effect. But I just listened to him. Later on, I ran into him in the hall and I had to say hello to him and he wanted to talk about his experiment and I listened. Finally he said, "You know, I think that if one did it this way and measured the potentials—maybe from that experiment one could understand it." And I said, "Yes, I think that would be a good experiment." And I walked away, having already done it and knowing the result.

It is a matter of conjecture whether this multiple discovery was truly "forestalled" or whether in fact it was made independently. By this laureate's account, his friend was very close to the discovery indeed.

Nor is this episode unique among Nobel prize-winning investigations.

The growing documentation on the discovery of the structure of DNA indicates that it too probably would have been made even if Watson and Crick had not done so. In Francis Crick's opinion, either Linus Pauling, the chemist (who, it will be remembered, won the Nobel prize for his elucidation of the nature of the chemical bond) or Rosalind Franklin, the crystallographer at King's College, would have made the discovery. "Don't you see," Crick told Anne Sayre, Franklin's biographer, "if I hadn't done something about it [taken on the DNA problem even though it was a King's College project] Pauling would have got it out first. I know Linus was wrong in his first guess, but Linus isn't stupid. . . . He'd have done it" (quoted in Sayre, 1975, p. 212, fn. 21).

Crick seems equally certain that Rosalind Franklin would have solved the problem. According to Sayre, Crick said, "Of course Rosalind would have solved it. . . . With Rosalind it was only a matter of time . . . perhaps three weeks. Three months is likelier. I'd say certainly in three months, but of course that's a guess" (1975, p. 214, fn. 21). As Crick intimates, judgments of this kind always remain in the realm of conjecture. But for our purposes the point is clear enough. Such forestalled multiple discoveries persuade the laureates and others in the scientific community that their contribution to the development of science could have been and probably would have been made without them. It should be evident, then, how thin is the line between laureates and others who have not won Nobel awards.

## Occupants of the Forty-first Chair: Temporary and Permanent

As we have seen, having made important scientific contributions in midcareer or earlier, the majority of future Nobelists experienced gradual but steady increments in prestige as indicated by their upward mobility in rank, their authority, and their honorific awards. A few whose work was not immediately recognized were longer in acquiring the esteem of their peers and the symbols of high status that go along with it but, ultimately, they too made their way into the elite. It comes as no surprise when they are tapped for the Nobel prize. These laureates-to-be stand in distinct contrast to the relatively small number of prospective laureates whom the Nobel prize abruptly raised from relative obscurity to fame. Their ascendance was so rapid that their reputations had not diffused widely enough for them to be recognized by election to academies or to acquire the many prizes and awards conferred on occupants of the forty-first chair. Amounting to about a fifth of the prospective laureates, these "meteors" were rarely active in the politics of science and even less often played a visible role in the exercising of scientific authority.

If most prospective laureates are temporary occupants of the forty-first chair—that is, on a par with prize-winners in the judgment of the scientific community—a small subset was designated as such by Nobel officials themselves. Although deliberations of the Nobel committees are, as I have noted, confidential by statute, Goran Liljestrand, official historian of the prize in medicine had access to the Nobel archives. He broke precedent and revealed the names of sixty-nine scientists who, in the judgment of the committee, had made contributions of the quality and significance necessary for Nobel prizes but had not won them (Nobelstiftelsen, 1962, p. 196ff.). Liljestrand thereby made official what was widely known among scientists anyway—that is, that officially designated laureates have their counterparts who have done just as good work but who have not gotten prizes. Thirty-three of these scientists (who can be considered "honorable mentions") were still living at the time the essay was published and hence were still candidates for a prize if nominated. (See appendix D for a listing of these scientists.) Since 1962, fourteen of the thirty-three Liljestrand revealed as having done research of the caliber required were, in fact, awarded Nobel prizes. Liljestrand had described them as follows:

**H. K. Hartline and Ragnar Granit.** Their investigations of vision were "found to be of the high value required for a Prize" (*Ibid.*, p. 325).

**John Eccles, Alan L. Hodgkin, A. F. Huxley, and Bernard Katz.** They were reported to "have been seriously considered for the Nobel Prize" for their work on nerve cells (*Ibid.*, pp. 307–308).

**The molecular biologists Max Delbrück, A. D. Hershey, S. E. Luria, and André Lwoff.** All were declared to have made "prizeworthy contributions," although "the number of contributors made it difficult to make a selection for the prize" (*Ibid.,* pp. 206, 211).

**Peyton Rous.** His work was said to merit a Nobel award in spite of the fact that the committee had "doubted" its importance when it was first proposed in 1926. (*Ibid.,* pp. 206–207).[11]

**Karl von Frisch, Konrad Lorenz, and Nikolaas Tinbergen.** Their studies of animal behavior were said to "have thrown new light on the old conception of 'instincts.' It is not improbable that these findings will in the future become of importance for our understanding of the behaviour of man, but it was felt that the work . . . in spite of its high

---

11. Rous's co-winner in 1966, C. B. Huggins, was mentioned only as having demonstrated short-term improvements in prostatic cancer using hormone therapy (Nobelstiftelsen, 1962, p. 249).

standard did hardly fall within the domain of physiology or medicine in the sense that Nobel had meant" (*Ibid.*, p. 314). So it was in 1962, when it seemed that von Frisch, Lorenz, and Tinbergen were to be permanent occupants of the forty-first chair. In 1973 the committee apparently changed its collective mind about Nobel's intentions (there was apparently no question about the significance of the research) and the boundaries of the award were redrawn to accommodate these pioneer ethologists.

In all fourteen cases, the scientists who later got Nobel awards were considered earlier and lost out even though their research was meritorious enough in the committee's judgment. That they later became laureates underscores the fact that both eventual laureates and occupants of the forty-first chair stand in something like a Nobel queue.

Like that of other Nobelists, the work of these fourteen temporary occupants of the forty-first chair was heavily used before the Nobel awards. They differ from the rest, however, in that they received an average of one and a half times as many citations as other laureates in medicine named in the same period.

The careers of prospective laureates are marked by early and copious production of published work, by early identification of promise, and, through a series of interlocking selective processes, by affiliations with elite universities and research institutes. The majority of them remained in elite institutions throughout their work life. Few had to deal with neglect of or resistance to their work; it diffused rapidly, and its scientific significance was quickly recognized.

Having done the research for which they would ultimately receive Nobel awards, future laureates increasingly exhibited *noblesse oblige* toward their own young collaborators and thus reenacted the roles they observed their own masters assuming toward their apprentices. In the wider world of science, the majority came to exercise authority and were increasingly honored. As the reward system operated with a degree of efficacy for most of them, prospective laureates gradually became differentiated from the more extended elite. As they moved through the scientific career we can observe the processes by which prestige begets prestige, making them, relative to other scientists, increasingly "rich" in resources, opportunities, and esteem.

Let us look now to the consequences of these "success stories" of modern science, and see what it means to be a laureate.

# Chapter 7

# AFTER THE PRIZE

The announcement of Nobel prize selections rarely comes as a complete surprise to the new laureates or their colleagues. As the Nobelist in chemistry Derek Barton tersely observed, "Scientists usually know where they stand in the international pecking order" (*New York Times,* 31 October 1969, p. 20).

The question in the minds of leading candidates is not whether they will get the prize but when. In many cases, the signs of a coming prize are unmistakable. Einstein was so confident that his anticipated Nobel honorarium was actually included in the divorce settlement he agreed to in 1919 (Hoffman and Dukas, 1972, p. 134) although the long-overdue prize was not his until two years later. Similarly, the prize to Gell-Mann in 1969 was almost a sure thing. The Princeton physicist Marvin Goldberger observed in *Science* on the occasion of Gell-Mann's award, "To the surprise of no one in the international physics community, the Nobel prize . . . was awarded to Murray Gell-Mann. . . . Standard physics cocktail party conversation for the past six years in late October was always, 'I wonder if Murray will get it this year'" (1969, p. 720).

Still, the shift from expectation to reality is often unsettling if only because the timing of the award is never certain. T. D. Lee, only 31 when he got the news, did not find it entirely cheering. " 'My God,' he said, 'what happens now to the rest of my life?' " (quoted in Wilson, 1969, p. 72). New laureates find that winning the Nobel prize is altogether different from receiving other scientific honors. It immediately transforms them into celebrities, outside science as well as within it. Werner Forssman, a German laureate in medicine, felt like "a village priest raised to a cardinalate" (quoted in Gray, 1961, p. 78). And the American physicist C. J. Davisson found that he had changed "overnight from an exceedingly private citizen to something in the nature of a semi-public institution" (1937, p. 63). Even at the outset, it is clear to most Nobelists that their social identities have been altered once and for all by having won an award. They are marked men and women. For most, the initial transformation is not an unmixed blessing, and it presages further changes in their work as well as in their social relations with other scientists and with the wider society.

## CONTINUING AMBIVALENCE TOWARD THE PRIZE

The ethos of science has it that scientists should derive their prime gratification from doing research and contributing new knowledge. Scientific honors should not be sought but should come as a byproduct. At the same time, the drive for recognition is also built into the ethos of science since recognition by knowledgeable peers is valued as the chief symbol that scientists have done their jobs well (Merton, 1957).

This ambivalence in the normative structure of science is readily apparent in interviews with the laureates. Most insist that they were motivated by the intrinsic satisfaction of doing science and not by the prospect of rewards, including the Nobel prize. They also report having been wary beforehand of the danger of wanting the prize too much and yet feeling enormous gratification in having won it. A biochemist's reminiscences express some of these complex attitudes:

> [Baker, a pseudonym] was just over seventy when I went to his laboratory. A whole group went to his home and Mrs. Baker showed us all of his medals and there was something she said that made me realize that she was disappointed. It was undoubtedly a reflection of her husband's own feelings of disappointment that he had not been recognized by a Nobel award. Driving home with my wife, we got to talking about this and

> I said, "I am never going to worry or have a goal in mind of any prize, even a Nobel award. I refuse to die disappointed if I don't get it." You put your happiness into the hands of some committee, which can be capricious. You've got to work for the fun of it. Men of equal accomplishment don't get it and then they have to rationalize for the rest of their lives. But don't get me wrong, I'm not sorry I got it.

Reports like this one convey the laureates' tendency to deny the importance of the prize at the same time that they express their respect for it. The institutionalized ambivalence in science toward collegial recognition accounts in part for the laureates' insistence that it is only their work that matters.

Although nearly all the laureates who were interviewed expressed the belief that they had deserved their prizes, most were ambivalent about it for one reason or another. One source of ambivalence was the conviction that the research cited for their award was not the best they had done. Another was the conviction that other scientists who had not been so honored were just as deserving. A third reason, for some, was the lateness of the prize. Last and most important of all, many laureates concluded that having won the prize had thoroughly and unalterably disrupted their lives and their work.

## The Selection of Contributions

Nearly half of the laureates who were interviewed, while conceding the scientific significance of their research, were convinced that it was not their best work. One after another made the same observation:

> The experimental work cited in the Prize award was...not necessarily the work I'm most proud of. There were other things that didn't get the notoriety that...were better done or more inspired [Biochemist].

> It's a very good piece of work but I know better things that I've done [Physicist].

> Other papers give me more pleasure when I read them [Physicist].

> It was good work. I wouldn't say the best work [Chemist].

What accounts for their dissatisfaction with the research chosen for the prize? In part, it reflects the priority that many laureates give to contributions that deepen scientists' understanding of large problems. A biochemist observed:

> I think what I got the Nobel Prize for is good but I think the influence I have had on biochemistry comes really from the understanding of the general mechanism of biosynthetic processes that I have provided. One of the greater contributions I made was in a paper in which I mapped out the manner in which an enzyme is used in metabolism. If I had to give myself the Nobel Prize, it is for that it would have come to me. But the Nobel Prize is not given for contributions like that. It has to be for a discovery.

The laureates' dissatisfaction with the research cited for their awards also derives from the value they place on intuition, ingenuity, and success in solving intractable problems. Serendipitous discoveries, however significant scientifically, do not usually meet these criteria. A laureate in physics was quite candid about his own situation.

> Sure, I think the Nobel Prize brought undue rewards. I got it for a purely accidental discovery. Anybody could have done that. This is often true in experimental physics. I think you can happen to be in a position where an important discovery is right there. Not your fault or to your credit, especially. I think there was a real difference between the [Nobel] work and the work we did on [particles]. [The latter] had very much more to it. We were working with clues and paradoxes, trying to resolve a situation that was completely unclear. As I have said many times, the [Nobel] discovery was largely accidental.

Another beneficiary of serendipity, however, was willing to take pride in his work.

> I didn't know what I was after but I was after it, so it wasn't pure chance. I think it was bound to happen. Had I not done it, I don't know when it would have happened but it was just there, waiting for somebody [to do it]. But I did it.

Like serendipitous discoveries, those resulting from having unique access to equipment are frequently disparaged. They could have been "done by anybody," as a laureate in physics observed when commenting on Segrè's prize for the antiproton.

> I am very sorry that he got the Nobel Prize for that. It was very, very good but there are many lovely things that he has done which were better. You see, anyone who had access to the machine could have done that experiment. Segrè is an awfully good

physicist and there are other very nice things he has done. . . .
I am happy he got the Prize; he thoroughly deserves it, but I
would have liked to see him get it for something else.

Apart from their feeling that chance discoveries do not merit substantial rewards, the tentative and incomplete character of many scientific contributions adds to the laureates' complex attitudes toward their
prize-winning research. Richard Feynman conveyed his reservations in
his Nobel address:

> I don't think we have a completely satisfactory relativistic quan-
> tum model, even one that doesn't agree with nature, but at least
> agrees with the logic that the sum of probability of all alternatives
> has to be 100%. Therefore I think that the renormalization
> theory is simply a way to sweep the difficulties of . . . electro-
> dynamics under the rug. I am, of course, not sure of that [1966,
> p. 189].

The incompleteness of science is a fact of the scientific life, and most
scientists understand that their ideas and findings will be superseded.
Some laureates are nonetheless ambivalent about the revision or refor-
mulation by others of the work they began and would prefer to have
been responsible for getting it right in the first place.[1] The case of the
chemist H. O. Wieland is apt in this regard. Wieland, who got his prize
for deducing the structure of the sterol skeleton in 1927, found later that
the structure needed drastic revision and, with Rosenheim and King in
1932, worked out the structure that is now accepted. In one sense, then,
Wieland's prize can be considered one of the Nobel "errors" but, in another sense, since he was responsible for revising and correcting the
work, the choice by the chemistry committee was no error at all. Even
so, Wieland obviously had reason to feel ambivalent about the work for
which he won his prize.

So too for Enrico Fermi, who is considered almost unique among
modern physicists for the power of both his theoretical and his experimental work. He received his award in 1938 in part for having "demon-

---

1. Arthur Kornberg also has an intriguing reason for ambivalence in this regard. When
Kornberg received his Nobel prize in 1959, biochemists believed that he had identified
the mechanism of DNA replication and the enzyme (DNA polymerase) required for its
synthesis (Nobelstiftelsen, 1962, pp. 281–82). In 1970 and 1971, however, two additional
enzymes (Pol II and Pol III) were discovered and shown to be essential for DNA
replication. Pol I, Kornberg's enzyme, turned out to be most active in DNA repair
rather than in its replication (Kornberg, 1974).

Such modifications are of course standard fare in the history of science. What is not
standard here is the fact that Thomas Kornberg, who was partly responsible for identifying Pol II and Pol III and thus for redefining the role of Pol I, is Arthur Kornberg's
son.

strated the existence of new radioactive elements produced by neutron irradiation" (Nobelstiftelsen, 1964–67, vol. 2, p. 408). In his Nobel address, Fermi signaled that he knew even then that something was amiss in the research cited for his prize. In a footnote to his discussion of the discovery of the first transuranium elements by his group he remarked:

> The discovery by Hahn and Strassmann of barium among the disintegration products of bombarded uranium . . . makes it necessary to reexamine all the problems of the transuranic elements, as many of them might be found to be the products of a splitting of uranium [Nobelstiftelsen, 1964–67, vol. 2, p. 417].

So it was. Fission was what Fermi and his group had observed, but they did not know it at the time. It was not until 1940 that element 93 (neptunium) was actually discovered by McMillan and Abelson, work for which McMillan shared the 1951 award in chemistry. That the standing of the physics prize was unimpaired by this "error" was due in part to the citation of Fermi's work on nuclear reactions along with the research on new elements and of course to his great distinction as a physicist. It hardly mattered what he got his prize for as long as he got one.

The(odor)Svedberg's case is similar. He received the Prize in Chemistry in 1926, primarily for studies of Brownian motion, not for his invention of the ultracentrifuge which later had such great impact on the development of molecular biology. Svedberg claimed he had demonstrated the correctness of Einstein's theory of Brownian motion and, in fact, that he had embarked on his experiments before having read Einstein's work. Einstein however rejected Svedberg's claim and later it was decisively challenged by Jean Perrin who, ironically, would win the prize in physics the same year Svedberg won his award. Svedberg's work on centrifugation and its biological applications has been central and qualifies him for a place on the Nobel roster but the fact is that the Chemistry Committee made a mistake in selecting the research if not the man. Svedberg did not even mention the Brownian motion research in his Nobel address. (See Kerker, 1976 from which this account is drawn.) [2] Yet these episodes are the stuff which ambivalence toward the Nobel is

---

2. It seems remarkable to an outsider that the Chemistry Committee could have chosen Svedberg the same year that the Physics Committee chose Perrin. Surely one of the physicists was aware that Perrin had shown that Svedberg was wrong in insisting that Brownian motion was oscillatory. As early as 1909, Perrin wrote that Svedberg was the "victim of an illusion." It should be noted, however, that Svedberg's work continued to appear in standard texts on colloids until 1926 and that disciplinary boundaries are often rigid and impermeable. (See Kerker, 1976 for a thorough analysis of this case.)

made of, especially since the prize makes the prize-winning research so
visible.

## Laureates vs. Occupants of the Forty-first Chair

Many laureates are troubled about the inequities created by awarding Nobel prizes to some scientists but not to others who have done research of much the same caliber. That many are called and few chosen
makes them uneasy. A physiologist put it this way:

> Awards involve a choice. In a choice, there are many people who
> could very well have the award . . . you think . . . maybe someone
> else should have had it. Since you have the award, it is impossible
> to say this because you would seem insolent or immodest. Try-
> ing to be too modest is immodest.

A physicist emphasized the inequities in the reward system in general: "I know people who have produced many more things scientifically
than I have and somehow have gotten less recognition. It doesn't seem
right. If you are smarter than I, work harder, do more things, and so on.
I just happened to hit on a couple of really nice things." He continued,
stressing the disproportionate rewards conferred by Nobel prizes in
particular, "How do you get so much fame for such a small amount of
work? Why so much for so little?" Ramon y Cajal, the 1906 laureate in
medicine, expressed the same feelings: "How could I justify the preference of the Carolinian Institute in the eyes of so many outstanding
investigators who had been passed over and whose superior deserts I
take pleasure in acknowledging?" (1937, p. 549).

These comments and others on the occupants of the forty-first chair
indicate that it is the disproportionate honor that comes with the Nobel
prize that is so disturbing. This is so because the normative structure of
science has it that comparable rewards should accrue to those who do
comparably good work. The existence of occupants of the forty-first
chair reminds laureates that the system in which they achieved so much
success does not operate with as much justice as they would wish.

## Delayed Awards

After several decades of awards, scientists began to develop a sense
of how long the committees would take to make up their minds. Thus
some prizes came to be considered as unusually prompt relative to the

publication of the research that was honored and others as unusually delayed. Among physicists, for example, the award in 1957 to Lee and Yang came swiftly—within two years of publication of their work on parity nonconservation. And Hans Bethe's came unusually late; he had to wait for twenty-nine years for the Physics Committee to conclude what all physicists already knew—namely, that his work on energy production in stars was distinguished and that he was among the most authoritative physicists of the time.

Long delays can be occasions for resentment once the "appropriate" time passes and for ambivalence toward the award when it finally arrives. Max Born was quite candid about his feelings.

> The fact that I did not receive the Nobel Prize in 1932 together with Heisenberg hurt me very much at the time, in spite of a kind letter from Heisenberg. I got over it, because I was conscious of Heisenberg's superiority [Born, 1971, p. 229].

Although Born went on to tell of his "surprise and joy" at being named a laureate twenty-two years later for the statistical interpretation of the wave function, another prize-winner conveyed his disappointment in having his prize come so late:

> More than twenty-five years before I got the prize, I received a letter from Stockholm which said that I was to receive the Nobel Prize that year. It was not written by the Committee but by someone I knew who happened to be in Stockholm. It was the consensus of opinion that it would be awarded to me. Had it been, my life would have been different. I don't know how different but it would have been different. As it is now, it was awarded just a few months before I retired.

The false alarm from Stockholm no doubt made things worse for this laureate. But he is not alone in having the pleasure of winning the prize dampened by having had to wait so long for it.

Taken together, the pervasive ambivalence in science toward recognition and rewards, the laureates' reservations about their prize-winning work and their awareness that others might just as well have won the prize leave many of them uneasy. Yet no scientist has actually turned down a Nobel voluntarily [3] although at least two have considered doing

---

3. In 1936, three years after he had been imprisoned by the Nazis, Carl von Ossietzky was awarded the Nobel Peace Prize. In retaliation, Hitler decreed that no German be permitted to accept a Nobel prize. The decree stood and ultimately prevented the German winners of the 1938 prizes in chemistry and medicine, Richard Kuhn and Gerhard Domagk, and the winner of the 1939 prize in chemistry, Adolph Butenandt, from accepting their awards. In accord with the statutes of the Foundation, their prize money re-

so. Max Delbrück, founding father of molecular biology and importer of physics into biology, has openly expressed his contempt for the "personality cult" associated with Nobel prizes. He was reportedly reluctant to accept his award but finally agreed to do so in order to show his physicist friends that he had not been "talking nonsense" all this time.

In a letter to Arne Tiselius, president of the Foundation, another laureate wrote of his hesitance about accepting the prize:

> I must admit that I found myself with a serious trial of conscience. Should I accept the prize or not? The main reasons for thinking I should say no were that I truly doubted that I truly merited the distinction, that I was afraid that it would put my research at risk, and that it recognizes individual performance which is contrary to the interdependence that characterizes all current research. But I realized that I did not have the courage to say no. I did not wish to mislead my friends and colleagues who would not perhaps see these problems so simply. I did not wish to interfere with the joy of the other laureates whom I deeply admire and esteem. And of course declining the honor would perhaps increase the publicity that would come to me. . . . The period which followed the prize partly justified, partly confirmed my apprehensions [quoted by Tiselius in Nobelstiftelsen, 1964, pp. 16–17; translated from the French].

So much for the laureates' immediate ambivalence toward their awards. Their responses in the long run and the impact of the prize on their careers depend in some measure on how old they were when they got it.

## THE TIMING OF THE PRIZES

On the average, laureates have been about fifty-two years of age at the time of the awards; those who worked in the United States were about a year younger, as table 7–1 shows. W. L. Bragg (later Sir Lawrence Bragg) won his prize in physics in 1915 at 25 and thus holds the record for being the youngest laureate. He is followed by a quartet of

verted to the general fund since no one appeared in Stockholm within a year of the awards to collect the honoraria. After the war, all three received their Nobel medals and the diplomas that accompany them, but the money remained in Stockholm (Nobelstiftelsen, 1962, pp. 173–80, 426fn.). Jean-Paul Sartre refused to accept the literature award in 1964. The Russians Boris Pasternak and Aleksandr Solzhenitsyn were not permitted by their government to collect the 1958 and 1970 awards in literature although Solzhenitsyn picked up his medal later. Andrei Sakharov did not leave Russia in 1975 to accept his peace award (it is said that he feared not being able to return once he left) but his wife, who was in Europe for medical treatment, went to Oslo and accepted the award in his stead.

**Table 7–1.  Mean Age of Laureates (1901–72) at Time of Prize, According to Field and Year of Prize**

| Year of Award [a] | Physics | Chemistry | Physiology or Medicine | All Fields |
|---|---|---|---|---|
| 1901–25 | 47.8 | 50.4 | 51.1 | 49.6 |
|  | (30) | (22) | (23) | (75) |
| 1926–50 | 44.6 | 51.1 | 56.9 | 51.6 |
|  | (24) | (29) | (36) | (89) |
| 1951–72 | 51.0 | 58.9 | 53.5 | 54.2 |
|  | (42) | (32) | (48) | (122) |
| All Years | 48.4 | 53.9 | 54.2 | 52.2 |
| (All laureates) | (96) | (83) | (107) | (286) |
| All Years | 47.0 | 53.1 | 53.6 | 51.2 |
| (American laureates) | (32) | (19) | (42) | (93) |

a. Data for the first fifty years of the award were taken from Manniche and Falk (1957, p. 304ff.).

physicists all of whom won their awards at 31: Carl Anderson, P. A. M. Dirac, Werner Heisenberg, and T. D. Lee. The oldest winners have been the venerable F. P. Rous at 87, Karl von Frisch at 86, and C. S. Sherrington and Albert Claude, both at 75. Some fifteen scientists were in their seventies when they became laureates, in spite of official reluctance to have the prizes become "pensions" (Nobelstiftelsen, 1962, pp. 301–302).

Table 7–1 also shows a moderate increase in the mean age of prize-winners, from 49.6 in the early decades to 54.2 in the later ones. This trend, which has continued since then (laureates named between 1973 and 1976 were 55 on the average), largely results from recent increases in the interval between publication of prize-winning research and the conferring of the awards, as is evident in table 7–2.

It would appear that the backlog of scientific giants who were recipients of the earliest prizes and who did their research in the nineteenth century accounts for some delays in the early awards. The appearance of contemporary candidates and the exhaustion of the older group by the 1920s later reduced the interval between prize-winning research and a Nobel prize but not by much. Recent increases in elapsed time between research and the prizes in every field suggest that the queue of occupants of the forty-first chair waiting their respective turns for a prize has been lengthening.

In a small number of cases, long delays between the prize-winning research and the award resulted more from active resistance or neglect by strategic sectors of the scientific community than from committee procrastination. This was apparently so for Max Born, who observed that although the overwhelming majority of physicists accepted [my statistical interpretation of the wave function] there were always some who did not, among them such great figures as Planck, Einstein, de Broglie and Schrödinger. . . . This may explain why it was twenty-eight years before I was awarded the Nobel Prize for my work [1968, pp. 36–37].

**Table 7–2. Mean Interval between Prize-Winning Research and Prize, According to Field and Year of Prize (1901–72)** [a]

| Year of Award | Physics | Chemistry | Physiology or Medicine | All Fields |
|---|---|---|---|---|
| 1901–25 | 11.1 | 14.8 | 12.0 | 12.5 |
|  | (30) | (22) | (23) | (75) |
| 1926–50 | 10.7 | 10.8 | 13.1 | 11.6 |
|  | (24) | (29) | (36) | (89) |
| 1951–72 | 14.0 | 19.2 | 14.1 | 15.5 |
|  | (42) | (32) | (48) | (122) |
| All Years | 12.3 | 15.1 | 13.4 | 13.5 |
| (All laureates) | (96) | (83) | (107) | (286) |
| All Years | 10.2 | 13.7 | 12.5 | 12.0 |
| (American laureates) | (32) | (19) | (42) | (93) |

a. Data on age at prize-winning research for 1901–50 are taken from Manniche and Falk (1957, p. 304ff.).

Controversy also marked the reception of Hermann Staudinger's work in macromolecular chemistry—his detractors called it *Schmierenchemie,* greasy chemistry—and delayed his prize (Hess, 1970, p. 664). Similarly, J. B. Sumner's efforts to isolate the enzyme urease met with resistance, as did Hannes Alfvén's studies in cosmic physics, with the consequence that they waited a long time for vindication and the prize. But neglect is, as we have noted, a better description of the reception of Lars Onsager's theoretical studies of reciprocal relations in thermodynamics and Karl Landsteiner's discovery of human blood groups, which was to have enormous medical impact some years later. It is not surprising, then, that the gap between the prize-winning research and the award was very long indeed in these instances.

## SCIENTIFIC WORK AFTER THE PRIZE

As originally conceived, the Nobel prize was to serve as both reward for past contributions to science and incentive for future ones. But it appears to have unintended consequences, in the long run and in the short. As one laureate remarked, "The year I won the prize was horrible.

Well, it was wonderful, but I didn't do any work whatsoever." Almost all the laureates reported substantial changes in their work practices following receipt of the prize. Although they addressed themselves far more often to the difficulties of getting work done at all than to maintaining its quality, it is to the latter that we turn first.

## The Significance of Post-Prize Scientific Contributions

Just as baseball players rarely play well enough to be named most valuable player several seasons in a row and authors rarely write novel after novel of the highest distinction, so it is with laureates. Few of them continue to do research of Nobel caliber after the prize. In part, this results from the circumstance that the prize often comes late in the career, after many scientists have made several important contributions. In part, it results from the diverting consequences of the award itself. The effects of the prize on the quantity of laureates' scientific publications will be examined later; here their inability to produce truly first-rate research on a continuing basis is under discussion. That difficulty apparently stems in large measure from the improbability of reproducing any performance of the highest level whether one has a Nobel prize or not. Over and above the impact of the prize, then, the inevitable regression toward the mean that operates in many spheres of human performance seems also to affect Nobel laureates and their research following the prize.[4]

Still, there are a small number of Nobelists whose contributions after their awards were roughly comparable to those they made beforehand.[5]

---

4. When compared to other scientists, the laureates continue to do very good research after their prizes, but they do not match their *own* past achievements. Since these regression effects are apt to be small and since the number of scientists involved is also small, such effects probably would not be visible in aggregate data on scientists' performance. It is not surprising, then, that Allison and Stewart (1974) find no evidence of such regression effects in a large sample of scientists; they report that older scientists are cited more often than younger scientists (a crude indicator of the caliber of their research) and that citation inequality increases with age.

5. The technical difficulties of gauging the impact of post-prize publications by citation analysis are formidable. Not only is it difficult to discriminate between real changes in rates of citation and artifactual changes due to increases in the file of journals covered by the Index but also the laureates' changing authorship patterns, noted earlier, make comparisons of citations to their research before and after the prize difficult to interpret. The difficulties are compounded further by the pervasive pattern of obliteration by incorporation of major contributions to science that is often expressed in implicit textual citation rather than in standard footnotes. Inhaber and Przednowek (1976) compare total citation to Nobelists before and after the prize and find that citations to laureates in physics and chemistry increase after their awards but not to laureates in medicine. Since they do not take into account the interval between prize-winning research and the prize, the laureates' authorship patterns, or the size of the SCI file, "there are," as the authors note, "a number of possible explanations" for their findings.

This is true of the physicists Niels Bohr, Ernest Rutherford, Enrico Fermi, and Frédéric Joliot, the chemists Heinrich Wieland and Hans Fischer, and the medical researchers Emil von Behring and Karl Landsteiner. There is no published evidence that these scientists were reconsidered for Nobel awards, but as noted in chapter 2 five others were. In addition to the double winners, Marie Curie and John Bardeen, those who decisively demonstrated that they could continue to do prizeworthy research are:

— Emil Fischer, who won his Nobel prize in Chemistry in 1902, and in 1914 was the first to synthesize a nucleotide. Fischer was understandably renominated, but this time for an award in medicine. The committee concluded that it was "hardly proper to award him the prize for Physiology or Medicine too" (Nobelstiftelsen, 1962, p. 282).

— Thirteen years after his prize in 1931, Otto Warburg was renominated for his work (with W. Christian) on the respiratory or yellow enzyme and was "found to deserve the honour, but for reasons that can easily be understood he had to give way to others" (Nobelstiftelsen, 1962, p. 293). One paper by Warburg and Christian is recognized as something of a scientific "classic," and though published more than thirty years ago, remains among the twenty papers cited most often in the current literature covered by the Science Citation Index (Garfield, 1974, p. 7).

— Paul Ehrlich began his studies of the trypanosome-specific compounds before his prize in 1908. Salvarsan, the famous No. 606, was not announced until 1910 and was not used to treat human syphilis until 1911. Its success led to Ehrlich's renomination for the Nobel prize in 1912 and again in 1913, but the committee thought that its therapeutic effects had not been decisively demonstrated. Ehrlich died in 1915, conveniently putting the matter to rest, but the official verdict now is that "this contribution [is] a landmark in the development of chemotherapy" (Nobelstiftelsen, 1962, p. 213).

Among living laureates, a few are generally believed to have done very good science indeed since their prizes, so much so that they might be contenders for the prize again.[6] The fact that some laureates continue to do science of the first class after their awards shows that the prize need not upset research careers, although it frequently does. But it leaves

6. The chemist R. B. Woodward is one of those mentioned most often for a second award. His wide-ranging contributions on electrocyclic reactions and orbital symmetry with Ronald Hoffman (1965 and 1969) and his synthesis in 1973, with Albert Eschenmoser and others, of the complex Vitamin $B_{12}$—all post-prize research—are considered to be of prize caliber.

the question open whether most laureates, in the absence of the prize, would have gone on to do further high-caliber work.

If the majority of laureates do not continue to produce at the same level of excellence after the prize, are they productive scientists none-theless? Or is it, as physicist I. I. Rabi suggests, that "unless you are very competitive you aren't likely to function with the same vigor after-ward. You know, it's like the lady from Boston who said, 'Why should I travel when I'm already here?' " (quoted in Bernstein, 1975, II, p. 54).

## Short-Run Changes in Productivity

The productivity of laureates, as indicated by the number of papers published, declines sharply in the five years following receipt of the prize.[7] Having published an average of 5.9 papers a year in the five years before an award, they declined in output by a third, to an annual average of 4.0 papers. However, scientific productivity is not constant over the life cycle; it peaks in the middle years and then tapers off (see appendix E for laureates' age-specific rates of productivity). To assess the impact of the prize independent of this tendency toward declining productivity with aging, we need to compare the laureates with a sample of other scientists whose productivity at the same ages—in the absence of the prize—scarcely decreased over a comparable period, from an annual average of 1.9 to 1.8 papers. This suggests that the new eminence of Nobelists is associated with declining productivity.

Ideally, the effects of the prize would be gauged best by comparing laureates with a sample of equally productive scientists of the same ages in the same fields who are occupants of the forty-first chair. This is so because age-related changes in productivity may vary among scientists who exhibit different levels of initial productivity. It is also possible that regression effects are partly responsible for the observed reduction in publication much as they would seem to affect the scientific significance of the laureates' research after an award. In other words, changes in the laureates' productivity after the prize may be a function of the high rates they maintained beforehand. The difficulties of identifying the

---

7. Enumerations of published productivity are confined to papers published by the forty-one laureates who were interviewed and whose bibliographies were studied intensively. Only papers that could be construed as "scientific"—that is, as contributions to the research literature—were counted. Public addresses, papers in the popular press, and encyclopedia articles were excluded. Since the demands for these extrascientific products increases sharply after an award, their inclusion would have factitiously inflated the laureates' productivity after the prize. The data reported in this chapter differ slightly from those I published earlier (Zuckerman, 1967a). Sufficient time has now elapsed for productivity data to be available for the full five years of publication after the award for all laureates studied.

members of a control group who would meet these criteria need not be spelled out here. Instead, the matched sample of relatively productive scientists used for comparison in earlier chapters is adopted as a less than perfect substitute.[8]

The finding of declining productivity should not be taken to mean that awards for scientific accomplishment *generally* lead to reductions in productivity. The Nobel prize is after all, in a social category by itself. It confers high social visibility upon the recipients, among the general public as well as within the community of scientists, and it carries with it a complex of social demands that may account for the immediate reduction in productivity. The newly crowned laureate is socially defined not only as a great scientist but also as a celebrity and a sage. His new status expands his role set and intensifies role obligations beyond those associated with scientific research.

By the laureates' own testimony, the new status is not an unmixed blessing. It releases new and greater demands by fellow scientists, university and government officials, journalists, organized groups of laymen, and, as E. L. Tatum, for example, reminded me, visiting sociologists (see appendix A on interviewing the laureates). All the laureates testify to this system-induced outbreak of requests for advice, speeches, review articles, greater participation in policy decisions, and other public services. The impact of the prize was described by André Lwoff, the French biochemist, who spoke for himself and his co-winners when he said:

> We have gone from zero to the condition of movie stars. We have been submitted to what may be called an ordeal. We are not used to this sort of public life which has made it impossible for us to go on with our work, . . . Our lives are completely upset. . . . When you have organized your life for your work and then such a thing happens to you, you discover that you are faced with fantastic new responsibilities, new duties [*Medical Tribune,* 5 January 1966, p. 4].

Even before this age of mass media and electronic publicity, becoming a Nobel laureate seems to have had similar, and perhaps even more

---

8. This discussion draws upon conversations with Daniel Kahnemann and Amos Tversky, who have studied the impact of regression effects on performance of many kinds. It should also be noted that the lower rate of productivity exhibited by the matched sample is not altogether disadvantageous for the analysis. Even a small absolute reduction in their productivity would register as a large percentage change. The laureates, by contrast, having established far higher levels of productivity before the prize, would have to have published far fewer papers absolutely to produce the same proportional reduction in productivity. It can be argued that such large absolute changes in the laureates' productivity are not likely to have resulted from random numerical fluctuation alone.

intense, effects since the prizes were still new and few shared expedients had evolved for coping with them. Here is Marie Curie's report to her brother just after the announcement of her first Nobel prize.

> We are inundated with letters and with visits from photographers and journalists. One would like to dig into the ground somewhere to find a little peace.... With much effort we have avoided the banquets people wanted to organize in our honor. We refuse with the energy of despair [quoted in Raven, 1967, p. 114].

Several months later, the impact of her award still had not diminished: "Our life has been altogether spoiled by honors and fame.... I hardly reply to these letters, but I lose time by reading them." Pierre Curie was equally unhappy and wrote of longing for "a quiet place where lectures will be forbidden and newspapers persecuted" (quoted in Raven, 1967, p. 114).

Ramón y Cajal, laureate in Medicine for 1906, also found the experience excruciating.

> In a short time the tattling press broadcast [the event] to the four winds and there was no remedy but to get up onto a pedestal and make myself the focus of the gaze of everyone. Methodically and inexorably the dreaded programme of attentions unrolled itself: telegrams of felicitations; letters and messages of congratulation; acts of homage of students and professors; commemorative diplomas; honorary elections to scientific and literary bodies; streets baptized with my name in cities and even in small villages; chocolates, cordials, and other potions of doubtful hygienic value, marked with my surname; offers of profitable participation in risky or chimerical enterprises, urgent requests for inscriptions ... petitions for appointments.... there was some of everything and to all I had to resign myself, at the same time grateful for it and deploring it.... In a word ... months were squandered in acknowledging felicitations, in pressing friendly or indifferent hands, concocting commonplace toasts, recovering from attacks of indigestion and making grimaces of simulated satisfaction. And to think ... I had chosen the most obscure, recondite, and unpopular of the sciences [Ramón y Cajal, 1937, pp. 546–47].

Few laureates encounter anything like the cascade of demands the Curies and Ramón y Cajal endured. Most Nobelists are not, except in the smallest countries, unique national heroes, and few are as retiring as Ramón y Cajal, who made "heroic efforts during the first few days after hearing about the prize to conceal the event" (1937, p. 546). Nevertheless, many laureates are moved to develop efficient and routinized procedures for dealing with the flood of requests, invitations, and demands. Francis Crick devised a standardized check list, which reads:

> Dr. Crick thanks you for your letter but regrets that he is unable to accept your kind invitation to:

| | |
|---|---|
| send an autograph | help you in your project |
| provide a photograph | read your manuscript |
| cure your disease | deliver a lecture |
| be interviewed | attend a conference |
| talk on the radio | act as chairman |
| appear on TV | become an editor |
| speak after dinner | write a book |
| give a testimonial | accept an honorary degree [9] |

Following the implications of Durkheim's analysis of the consequences of abrupt upward mobility (1951, p. 243ff., first published 1897), we would assume that the impact of the Nobel prize upon productivity would differ among the laureates according to the size of status increment it represented. For many of them, the prize is merely the capstone of a long and distinguished career; for others, it represents a sudden and very great increment in prestige. Contrast the standing of Linus Pauling, for example, with that of some of the other prize-winners. Before his Nobel award, Pauling was already a member of eight academies and had received eleven honorary degrees and eight scientific awards, among them the Davy Medal of the Royal Society and the Langmuir Prize of the American Chemical Society. Other laureates had practically no institutionalized recognition of eminence before their Nobel award; they were members neither of their own national academies nor of foreign ones, had won no major prizes, and possessed no honorary degrees.

We want, then, to compare changes in productivity following the prize among the newly eminent and among those who had previously achieved high rank.[10] Since age and productivity are related, we must also take age into account. Table 7–3 indicates that the greater the increment in rank represented by the Nobel prize, the greater the im-

---

9. Edmund Wilson the literary critic and Marianne Moore the poet—status equivalents of the laureates and equally subject to demands on their time—also used these "all-purpose" refusals with suitable additions and deletions. (Gill, 1976, p. 279).
10. Twenty-eight laureates were classified as eminent before receiving their prizes. As members of the National Academy of Sciences or a foreign equivalent, they had more than three times as many awards on the average as the thirteen laureates who were not Academicians before receiving their prizes. The previously eminent had a total of seventy-three degrees among them compared with just one among the remaining thirteen.

mediate decline in the production of scientific papers.[11] Among both the younger and the older laureates, the previously eminent experience less of a decline than those who have been catapulted into eminence and celebrity. The older laureates have a greater decline in productivity, presumably the joint effect of both the prize and aging. When the laureates are compared with their age peers among scientists, the effect of the prize seems equally disruptive for both the younger and the older winners who had not had a previous period of eminence during which to work out some accommodation to the demands of high status.

These aggregated data on the effects of the prize do not imply that every laureate experienced a decline in productivity. It turns out, in fact, that thirteen of the forty-one whose bibliographies were studied in detail dealt effectively with the demands that accompany the prize and maintained about the same level of productivity immediately afterward as they had before. Consistent with the Durkheimian interpretation, eleven of these laureates were already eminent when the prize was con-

**Table 7–3. Percent Change in Publications Five Years Before and Five Years After Prize, According to Age and Eminence at Prize**

| Age at Receipt of Prize | LAUREATES | | | |
|---|---|---|---|---|
| | Eminent Before Prize [a] | Not Eminent Before Prize | Total | Matched Sample |
| 49 or before | −3% | −29% | −18% | +4% |
| | (159–155) | (219–156) | (378–311) | (128–133) |
| 50 or later | −30 | −52 | −33 | −14 |
| | (582–405) | ( 87–42) | (669–447) | (240–207) |
| Total | −24 | −35 | −28 | −8 |
| | (741–560) | (306–198) | (1047–758) | (368–340) |

NOTE: The numbers upon which each percentage change is based are shown in parentheses.
a. These data exclude the three laureates who assumed administrative positions after their awards. See footnote thirteen for discussion of changes in their productivity.

11. Harry Alpert suggested that laureates who were not eminent before the Nobel prize may have received it sooner after completing their award-winning research and that, as a consequence, reductions in their productivity reflect the need for a period of planning before new research could be published. This may be so, since those who were eminent before winning the prize waited almost 10.9 years for their award compared to the others, who waited only 6.7 years for theirs. When the laureate's age at winning the prize is taken into account, this difference is concentrated largely among younger laureates. Still, it seems unlikely that new laureates would need as much "turnaround time" as these data suggest they had. Moreover, not one laureate mentioned difficulty in getting new work started, but practically all spoke of their problems in pursuing research in progress.

ferred and, not surprisingly, they were also somewhat younger than the rest.

A more detailed deviant case of *a fortiori* analysis confirms the hypothesis that the reduction in scientific productivity after the prize results primarily from increased social demands upon the laureates. Not included in table 7–3 are three laureates who took on major administrative posts shortly after receiving the prize and thus experienced the most thoroughgoing changes in roles. Their productivity fell by 66 percent. As three of the most productive laureates-to-be, these men averaged 10.6 papers annually for each of the five years before the prize. And, despite marked decline in productivity after assuming their new administrative posts, they nonetheless continued to publish an average of 3.9 papers each year.

These statistical indications of the differential effects of the Nobel prize are confirmed by the personal reports of the laureates. One laureate who had not been eminent before his award contrasted his own problems involving adaptation to the new demands with those of an older laureate more practiced in dealing with them:

> After not doing much work since I got the Prize one gets to a kind of crossroads and has to make a decision. I want to get rid of a lot of the honors and get back to work. But how do you do it? You have to discharge a certain number of obligations and fight off new ones. That's easier said than done. Professor Jones knows how to do it. He knew how to do it from the beginning. He's a little older than I and smarter anyway. He knew how to do it from the start.

By contrast, a laureate who had been eminent long before receiving the prize turns out to be selective and hard-boiled about the many requests that reach him:

> I've not found the prize a burden. It can be one very easily if you start accepting all the invitations. But I've never done that. I've tried to limit this business only to those cases where I have a special interest or a special friendship. So I haven't let it be a burden.

And, he continues, "Once you begin to say yes, people are more likely to ask, so I didn't in the first place." In other words, it appears that scientists who achieved eminence before receiving a Nobel prize had gradual anticipatory socialization to the demands that divert highly placed scientists from their work. Other laureates, being elevated suddenly into the ultra-elite, had no such preparation for the management of fame, and it is from these scientists that the prize appears to take its greatest toll.

Nonetheless, the percentage changes in the laureates' productivity after the prize, however marked in comparison with their age peers' in the matched sample, conceal important differences in their rates of absolute productivity. Although the laureates' productivity decreases at a rapid rate, they are on the average still more productive after the award than are the matched scientists who continue to publish. In the five years after their prizes, laureates published, as we noted, *twice* as many papers annually as the matched sample (4.0 papers versus 1.8). Thus, although the prize is associated with declining productivity, the laureates' research scientific careers by no means come to a halt.

## Authentic and Spurious Changes in Productivity

Until now, the sheer quantity of published papers has been taken as an index of productivity among Nobel laureates. Here we ask whether the prize has consequences that affect the correlation between the number of papers published and what, in the abstract, may be thought of as real scientific productivity. The issue at hand is not specifically whether the number of published papers is an adequate measure of scientific productivity but whether the prize brings about changes in laureates' patterns of publication that make *comparisons* of pre- and post-prize productivity factitious. The beginnings of an answer to this question can be pieced together by drawing upon the interviews with the laureates. We begin by distinguishing between *authentic* changes and *pseudo* changes in productivity induced by the social and psychological consequences of a prize.

Just as the Nobel prize brings about changes in the laureates' status and thus makes for reductions in their productivity, it can also enhance their productivity. Such authentic changes following the prize contrast with pseudochanges, which are shifts in published output attributable to having won the prize but unrelated to changes in the laureates' scientific activity.

Although the prize leads some winners to believe along with the lady from Boston that they need travel no longer since they have already arrived, many laureates, as noted earlier, are persuaded that others were as qualified for the prize as they were, and this motivates them to prove to themselves and their colleagues that it was deserved. Rather than rest on their oars, they feel, as one of them put it, "that you want to prove the award was justified. Why, that's the pressure you're under. You don't want to have it said that a mistake was made." Others think that the evaluation system requires scientists, Nobel prize-winners or not, to prove their worth continually if they are to have access to the resources they need to carry on. Older laureates particularly believe it is important to demonstrate that their future does not lie in the past. A biochemist observed:

> Even if you are an older person, you have to publish something or you won't get grants or what-not. If you are regarded as completely inactive and people merely say at the end of their papers that you have encouraged them, then it doesn't say very much. So you are obliged professionally to come up with something.

The award also enhances laureates' opportunities for publication in the sense that after the award they are more often asked (and expected) to contribute to books, special issues of journals, symposia, and the like. In these and kindred respects, the award reinforces motivation and increases opportunity for continued and significant productivity.

Pseudochanges in productivity are increases or decreases in numbers of published papers that do not register the extent or significance of scientific work. The award contributes to such pseudochanges in several ways. The practice among laureates of sometimes omitting their names from collaborative papers is one factor making for apparent but not actual decreases in productivity. Laureates do not exclude their names in all cases of joint work, but they do so with greater frequency after receiving a Nobel, to counteract the operation of the Matthew Effect (Merton, 1968f) and thereby improve their co-workers' chances of being credited with the work. A chemist remarked:

> Especially during recent years, it's been evident that if my name was on a paper, people would remember it and not remember who else was involved. I have felt and I think the people working with me have felt that they would get more credit if my name were not attached to these papers.

Having received the Nobel prize, many laureates are not only less concerned about recognition than they were earlier but also can afford to do without the increment of credit that one more paper will bring. They exercise *noblesse oblige* in this respect as well as in arranging co-authorship on collaborative papers, thereby reenacting the role behavior of their masters toward younger colleagues (see Zuckerman and Merton, 1972, pp. 342–45). The practice of foregoing the privilege associated with high rank reappears over and over again in the laureates' role behavior. Although it reduces their apparent productivity, the exercise of *noblesse oblige* contributes to the equitable allocation of recognition.

Some laureates said that having won the Nobel prize made them more hesitant to publish work that might be judged mediocre. Their standards for publication (and those they believe others demand of them) are raised after the award, resulting in fewer publications than might otherwise have appeared. A physicist described his experience:

> After you've done something good and received such high recognition for it, it's hard to publish anything without feeling it's below the stature you've gained. It becomes very hard to do anything that you might call pedestrian, and a good many people just quit. At the present time, it's difficult for me to keep going because of all of this extraneous honor.

## Long-Run Changes in Productivity

So far, the short-run effects of the Nobel prize have been at issue. Are the changes seen in the laureates' productivity temporary or more enduring? Following the procedure used earlier, laureates and their age peers in the matched sample of scientists are compared so as to gauge the extent to which reductions in productivity initially attributed to the prize persist and, if they do, the extent to which aging may also contribute to them.

Laureates' published productivity declined even further in the sixth to tenth years after their awards, dropping by 27 percent in addition to the reduction of a third that occurred immediately after the award. The laureates published an annual average of 2.9 papers as compared with their earlier rate of 4.0 papers.[12] Once again, the productivity of the matched sample of scientists who did not receive the award declined far less, about the same rate as in the preceding five years; their annual rate of productivity stood at 1.7 papers, down from 1.8. Nevertheless, even with their marked relative reductions, the laureates, as before, were more productive than the matched sample of their age peers—in fact, nearly twice as productive.

These aggregate figures do not indicate whether reductions in the laureates' productivity in the long run are concentrated, as they were earlier, among older scientists and those who experienced abrupt changes in status. If the Durkheimian analysis holds, age-related reductions may be expected to persist but those associated with changing rank should

---

12. These data are drawn from the bibliographies of all but two of the laureates whose productivity following the prize is recorded in table 7–2. The two excluded did not have their prize for a full decade.

not. Scientists who were not eminent when they received their awards should in time adjust to their changed status and come to resemble their eminent colleagues. This is precisely what occurs. Those thrust into eminence apparently learn how to cope with the demands and obligations attendant on a Nobel prize. That is, after a time, they cope about as well as other laureates who had achieved eminence earlier and more gradually. Reductions in productivity in the sixth to tenth years after the prize are about the same for both groups of laureates, 24 percent and 28 percent, respectively.[13] The experience of having a Nobel prize eradicates the differences between these groups that existed before the award. Further, decreases in laureates' productivity associated with aging also diminish. Those who won their prizes before they were 50 years old show a 24 percent reduction in papers published in this period as against a similar reduction of 30 percent for those who won the prize after reaching the age of 50. Much more needs to be known about age-related patterns of productivity to account for these findings.

In the long run as in the short, the Nobel prize and the vast celebrity it brings make it difficult for laureates to continue doing the very work that brought them their award in the first place. Whether the prize and its sequelae slow down or otherwise impair the development of science is not self-evident. In fact, the prize may have very little effect indeed. The laureate in physics I. I. Rabi made the point in his characteristically earthy fashion:

> Some of the science that I might have done later [after the prize], I have to admit was done by others, perhaps not with the same personal quality I always tried to inject in all my scientific work. But anyway it was done. So I can't say that the progress of science was impeded at all by my getting the Nobel Prize [quoted in Bernstein, 1975, II, p. 56].

As Rabi suggests, the heroic view of the history of science does not square very well with empirical fact. The Nobel prizes, in spite of their frequently damaging effects on individual productivity, probably do not greatly interfere with scientific growth. Whether on balance they turn out to be functional for science is a question to be examined later on.

---

13. A more detailed look at the laureates who took on administrative posts after their award confirms the analysis. Abrupt changes in their status were associated with marked immediate reductions in productivity. After the initial period of adjustment, these reductions diminished. As noted earlier, the three eminent scientists who became administrators after their prize showed an initial decline of 66 percent in productivity but, in the second five years after the award, their productivity dropped only 12 percent. They continued to be among the most productive laureates, averaging 3.4 papers annually 6 to 10 years after the prize.

## CHANGES IN SOCIAL RELATIONS AND
## WORK PRACTICES

Along with its diverse consequences for the quantity and significance of their research, the prize often changes laureates' relations with colleagues and collaborators, leads them to alter long-accustomed work practices, and is an occasion for redefining their scientific roles.

Several laureates observed that the prize erects barriers of deference between them and their colleagues. One physicist mused: "These prizes separate people emotionally. Other scientists say, 'He has received the Nobel prize. He is not quite my brother any more; he is the first-born.' The distance between them and me is much greater now." Distance is sometimes transformed into envy and the inclination to remove the hero from his pedestal. Thus younger and less well known scientists sometimes attempt to make their reputations by demolishing the work of a Nobelist—what one chemist laureate called the "David and Goliath syndrome." But malice is obviously not limited to the young. Ramón y Cajal wrote:

> A few histologists and naturalists who always distinguished me with their disdain or their unfriendliness rose violently against me. It was high time, according to my pious confrères, to crush the neuron doctrine for good, burying at the same time its most fervent supporter. There was in their invectives so much injustice, and they were . . . so disproportionate to the insignificance of my polite observations of earlier times, that it would be ingenuous to believe that there was not a certain etiological connection between them and the award of the Nobel Prize [1937, p. 560].

Even when malice was absent, laureates often found that they no longer had sufficient access to candid evaluation of their work. A biochemist mused:

> It's funny. If you find something really new, people don't believe you. But if I come out with something, people tend to believe it immediately and I become sad. I worry that it is because of who I am and that I am getting old, that it is not really new stuff. Good science is what people reject initially. Of course now I am a great authority. It is not pleasant. People are more willing to accept what I say or at least not to criticize. There is no corroboration for what I am doing now. Before they would have said, "It's rubbish, nobody can repeat it." Now I don't hear a word except "we wait for the evidence."

Receiving less response to their work than in the past, laureates felt even more exposed to status-seeking criticism. This, along with their much enhanced visibility, makes some of them hesitant to speculate. A physicist remarked that "it isn't fun to be a notorious person," and another mused:

> It's a little difficult at conferences, for instance. When I ask a question it attracts attention, and that's not good. When I ask a question I want to know the answer. It is simpler if it is not noted that I am prominent. One wants to explore ideas and some-times to talk nonsense because if you always talk responsibly you become a statue.

Although the sense of being more visible after the Nobel prize seems to be shared by all laureates, an enhanced sense of vulnerability is not. Rather, the prize gives some more self-assurance than they had earlier. The immunologist and laureate in medicine, Macfarlane Burnet, reported that "having been to Stockholm gave me confidence to say what I thought, even if it was unpopular with some of my scientific colleagues" (quoted in Sherwood, 1974, p. 17). Since self-assurance can be easily interpreted as arrogance, it is not surprising that some laureates are thought to become more arrogant after the prize. As laureate A. V. Hill put it, "The Prize makes some people absolutely crazy. I know of one winner whose ego was so inflated by it that he put the initials N.L. [Nobel Laureate] after his name" (quoted in Stuckey, 1975, p. 34). Such changes rarely escape notice by other scientists, who are on the lookout for them. As a consequence, Nobelists' relationships with their scientific colleagues are often strained.

It is the laureates' relations with their co-workers that change most decisively for the Nobel award raises questions of distributive justice. The award of a Nobel prize deepens the ever present curiosity about the relative contributions of collaborators (an important source of strain in joint research in any case) since it forces attention to delicate matters of credit that most co-workers prefer to keep undefined. Thus, the heightened visibility that comes with the Nobel prize accentuates one of the major stresses in group research.

These stresses are, as we shall see, consequential when the laureate is still collaborating with those who shared in the prize-winning research but, continuing collaboration or not, the award creates difficulties. As noted in chapter 6, approximately two-thirds of all prize-winners received their awards for collaborative work, a proportion that has risen steadily from 41 percent when the prizes were first awarded to the current level of 79 percent. The proportion of laureates sharing awards with at least one co-worker has also been rising, but the gap between the two figures has not been closing. About half of all laureates who have won awards for joint research have not shared them, and this has been so from the beginning.

The source of strain among joint winners and in groups with only one winner is much the same—the feeling that recognition has not been allocated justly. For collaborators sharing the prize, the issue takes the form: Has equal recognition been given for distinctly unequal contributions? And, for the research groups in which only one scientist has received an award, the question is: Has unequal credit been given for roughly equivalent contributions? One laureate who shared the prize described the tension it created among other members of the research team: "The group couldn't have survived my getting the prize. Questions of credit and who did what started coming up."

As this laureate implies, the Nobel prize may be even more disruptive to groups in which only one person is honored, particularly when his collaborators do not regard him as the principal figure in the research. A physicist laureate described his own experience:

> It's unfortunate that Jones [a pseudonym] wasn't recognized somehow in this. While his contribution was not necessarily central theoretically, he was indispensable to the experiment. It would have been more sensible if we had shared the prize, particularly because the two of us were in the habit of working together here and it doesn't help any in our relationship to have this separation drawn.

Assuming that scientists feel more deprived when they receive no formal credit at all than when they share it with others, research groups with only one prize-winner should be under greater strain than groups in which several shared the award. A crude indicator of the extent to which the prize disrupts research groups is the duration of collaboration following the award. For the pairs of joint winners who were still working together at the time of the Nobel prize, the average duration of collaboration afterward was 5.4 years, as compared with an average of 3.6 years among single winners and their prime co-workers.[14] These averages obscure the fact that one of the five pairs of scientists who shared the prize worked together until one partner died. André Cournand and Dickinson Richards, who shared the award in medicine in 1956, hold the record among American laureates for the longest collaboration, having begun their joint work in 1932 and continued it until Richards's

---

14. Prime collaborators were those whom the laureates judged as having contributed most to their work and whose names appeared most frequently on papers reporting prize-winning research. Since some laureates had more than one prime collaborator and the duration of joint research after the prize was not the same for all of them, these were treated as separate pairs.

death in 1973, even though their interests turned from pulmonary physiology to the history of medicine and medical education.

Some prize-winning collaborations end soon after the Nobel prize for quite other reasons. These terminate not because of strains deriving from differential recognition but because the younger co-winners want to establish themselves as capable of working independently of their senior and more distinguished colleagues. Far from feeling that the senior scientists have received undue credit for the work, some of these younger laureates believe that they did not deserve the prize as much as the seniors did. As one of them put it: "In part, I think, this Nobel prize was for recognition of past things he has done. He has done many more outstanding things, real innovations, than I have." Having received so much recognition, these younger laureates feel a need to prove to themselves and others that they have not been "riding on the coattails" of their elders, as one of them described it.

> I suppose there comes a time when you feel you'd better get out on your own. I'd been working with him four and a half years. It was getting to a point where I wasn't sure what ideas were his and which were mine, and I knew perfectly well that I couldn't stay indefinitely.

And so they embark on new collaborations or upon individual work.

The Nobel brings about other changes in patterns of research. Some laureates are sought out by students more frequently than before—although a biochemist observed. "I suppose more students have come. But people aren't so dumb. They come if they think you have something to offer. They don't come just because you got the Nobel Prize." The intensified demands that follow the prize reduce the time available for research. To cope with these demands and still continue their research, the laureates often collaborate with students, delegating some tasks to them that, before the prize, they would have looked after themselves. This seems to be so even for those who like doing the "routine" work. A biochemist said, with some sadness:

> I'd enjoy doing more of the technical side myself. That's the thing I've not been able to do in recent years. I just haven't the time. When I can't do it, I don't get the same kick out of research. I feel I'm missing something when someone else does it for me.

These accommodations made in the research role itself are accompanied by marked redefinitions of the laureates' scientific and social roles generally. As eminent scientists and Nobelists, they command scientific resources in the form of jobs, research money, and access to publication outlets. This puts them in a position to support the work of other scientists more effectively than they could have earlier. Some therefore curtail research and turn their energies to facilitating the work of others. Contributions to science of this kind are not readily measured by publication counts. As a case in point, one of J. D. Watson's admirers observed that if Watson's publications were totted up after his prize, it would look as if he had not been doing much. But, he continued, Watson's laboratory at Harvard and the Cold Spring Harbor Laboratory (where he is director) still have "the best students, the best postdocs, and do the best work" in their field. Whether this judgment of Watson's laboratories is sound is less in point than Watson's redefinition of his role and the evident inappropriateness of publication counts as a measure of this kind of scientific activity.

Other laureates choose to make use of their enhanced opportunities in other ways. The older ones especially find the philosophy and history of science congenial and turn their attention in these directions. Consider only the recent publications along these lines by laureates Jacques Monod, François Jacob, Peter Medawar, John Eccles, Eugene Wigner, and Werner Heisenberg. It may be that these interests are no more pronounced among Nobelists and other eminent scientists than among their age peers in science generally. The apparent proclivity of Nobelists for history and philosophy may result only from their having greater opportunity than do other scientists to write and speak on these questions. But it is also likely that, having done important scientific work, they are sensitized to epistemological issues and to questions concerning the process of scientific discovery. Since Nobelists in any case are frequently cast in the role of philosopher and social sage, this turn of mind fits in with expectations others have of them.

Still others make use of the widespread tendency to define laureates as wise men by advocating various political and social causes. In doing so, they transform scientific authority into political and social influence. Having the prize seems to confer extra legitimacy and visibility on their views—even those having little to do with science. Thus, in a study of science writers' assessments of the most visible American scientists, five of the top ten were Nobelists: Linus Pauling, Joshua Lederberg, George Wald, Glenn Seaborg, and J. D. Watson (Goodell, 1977). Except when advocacy is in the cause of science—when, for example, the teaching of creationism along with Darwinian evolution was publicly denounced by a number of laureates—such activities are apt to be criticized by scientific colleagues as publicity seeking and unscientific. (Similar charges are of course made against other scientists who have become public figures.) This is so because the competence of the advocates to speak out on issues is often not established. This makes the majority of laureates leery of making public statements on issues not directly connected with their scientific expertise, and a small fraction refuse to lend their names to political causes of any kind, even those they privately support. Whatever their response, the Nobel prize greatly increases the opportunities laureates have for public activity, and their support is often sought as validation for all manner of political and social positions.

Most laureates continue to give prime attention to scientific work. Their staying power in this regard is substantial. By way of example, of the forty-one laureates whose bibliographies were studied in detail, all nine who had reached the age of 70 continued to publish even though the oldest one, at 80, felt no obligation to do so any more. "After all," he said, "enough is enough," but his papers appeared in the journals nonetheless. Among the matched scientists, however, just three of those who were past 70 continued to be productive. This difference results from the laureates' being consistently subjected to greater expectations, their own and others', for continuing productivity, from their having established routines which facilitate it, and from their greater opportunities to publish their work. Not only do the laureates exhibit greater staying power in the sense of remaining longer in the research role, they also continue to be more productive. Laureates who lived past the age of 60 published about four papers annually, more than three times the annual rate of 1.3 papers for their age peers. (See appendix E on age-specific rates of productivity.) Thus the gap in productivity between the two groups increased as they aged.[15]

## NEW REWARDS AND RECOGNITION

The award of the Nobel prize releases a flood of new benefits, monetary and honorific, as well as liabilities. All apart from the honorarium attached to a prize, laureates are in a position to transform their considerable prestige into hard cash. Some serve as directors of corporations and levy substantial fees for consultations with science-based industries. No detailed information is available on just how much a Nobel prize is worth in this sense, but Harold Urey has estimated that its value

---

15. This finding is consistent with Allison and Stewart's report of growing inequality in productivity and citations with professional age in a synthetic cohort of scientists. This inequality seems to result from research-role attrition among scientists less esteemed by their peers. But the more scientists are reinforced by recognition (in the form of citations to their work), the more likely they are to remain active and continue to publish (Allison and Stewart, 1974).

is "four to five times as much as the [prize] money I received, through increased salaries from universities, expense accounts of all kinds [and so on]" (quoted in Sherwood, 1974, p. 16). Some are no doubt more successful in exploiting the prize, others less so. But in a way, although not quite in the form that Alfred Nobel intended, the prize makes it possible for most laureates to be free of financial worries.[16]

As laureates, they also become prime candidates for other honors, since association with the Nobel prize, as we have noted, seems to enhance the prestige of other awards and the standing of the organizations that confer them. Choosing laureates has advantages; those responsible for selecting recipients obviously do not wish to make mistakes and so they protect themselves by giving awards where the Nobel has already committed itself. Award-giving organizations want to recognize significant scientific contributions and, even if the work of occupants of the forty-first chair is taken into account, there are not that many truly significant contributions to honor.

It turns out that the number of awards Nobelists receive is largely (but not entirely) a function of the length of time they survive their prizes. Those who live a long time and are widely esteemed as scientists-statesmen are the most copiously honored. A few special cases make the point. At last count, Harold Urey, a laureate for more than forty years, has been given twenty-three honorary degrees and twenty-one prizes of various sorts, and is still being honored. Next in seniority among the Americans, Arthur H. Compton, who had his prize for thirty-five years, acquired twenty-four honorary degrees and ten other awards after his Nobel. And R. A. Millikan, a power in political as well as scientific circles, was awarded twenty honorary degrees and sixteen prizes during the three decades he was a laureate. And we know that this is not strictly an American phenomenon. The British physicist W. L. Bragg, a laureate for fifty-six years and thus the longest-lived Nobelist of all, was honored by memberships in the Royal Society, the Académie des sciences, and the National Academy of Sciences along with several others, held sixteen prizes and honorary degrees, and finally was knighted at age 51. The Argentinian Bernardo Houssay, a laureate for only twenty-eight years, has a record twenty-seven degrees, is a

---

16. Some laureates have made a point of using their prize money for special purposes. Max Delbrück gave his share of the 1969 award to Amnesty International, an organization devoted to arranging for the release of political prisoners in all nations. His co-winner, Salvador Luria, gave part of his share to antiwar organizations (*New York Times*, 17 October 1969, p. 24). John Bardeen used money from his second prize to establish the Fritz London awards and to support the Fritz London Memorial Lectures in honor of the distinguished physicist (Gross, 1973). And finally, in an interesting turnabout, Georg von Békèsy named the Nobel Foundation heir to his estate of $400,000 in art objects—about ten times the amount he received from his prize (*Medical Tribune*, 17 January 1973).

fellow in eleven academies, and has received fifteen decorations of various sorts, many of them after his prize. But this is not to imply that, once a laureate, other scientific honors are there for the asking. Marie Curie's candidacy for the Académie des sciences was rejected after her first and second prizes (Meadows, 1972, p. 231) and that of her daughter, laureate Irène Joliot-Curie, was rejected four times (*New York Times,* 14 February 1971, p. 76). In these cases, the disadvantage of their gender apparently outweighed their scientific distinction.

The heaping of awards on prize-winners is one process by which honor in science comes to be concentrated among a comparatively small number of investigators and contributes to the accumulation of advantage (Zuckerman, 1971). Although there is evidence that a small number of scientists contribute disproportionately to science (Price, 1963; J. Cole, 1970; Garfield, 1970b; Cole and Cole, 1972), the concentration of honor, as registered in the distribution of formal awards, may become disproportionate to their later research contribution.

In response to this accretion of honor, some laureates choose not to display all their symbolic riches. As we have noted, they do not stake claims in published biographies to all their awards but prefer to be selective—once again following the principle of *noblesse oblige*—and because many intuitively recognize that scientific awards tend to be arranged according to prestige, in a Guttman scale (see Merton, 1957). After the prize, about a fourth of the laureates trim their listings in biographical dictionaries such as *American Men and Women of Science* by excluding awards from local societies and sections of national ones. Linus Pauling, for example, omits his multiple honorary degrees and some local prizes. Georg von Békèsy and Joshua Lederberg managed to achieve the height of inconspicuous eminence by choosing not to list their Nobel prizes.[17]

## Formal and Informal Standing in Science

Although the Nobel award puts all prize-winners at the top of the stratification pyramid of science, they differ greatly in the esteem accorded them by fellow scientists. Current reputation or informal standing seems largely to depend on current research achievement. Although a lifetime of important work evokes respect, the scientific com-

---

17. Such reticence is the bane of editors of biographical dictionaries. Lederberg's and von Békèsy's Nobel prizes were restored to their listings in the twelfth edition of *American Men and Women of Science,* apparently whether they authorized it or not. Somewhat atypically, Harold Urey draws special attention to his having won the Willard Gibbs Medal of the Chicago section of the American Chemical Society "because I received it before I received the Nobel prize" (Sherwood, 1974, p. 16).

munity is acutely sensitive to signs of its members' being "over the hill."

As the laureates' research is superseded by newer contributions, many experience a growing gap between their public prestige as laureates and their prestige among scientific colleagues. Some observe, with irritation, that young scientists are unaware of early contributors. A physicist remarked:

> You can't pick up an issue of the *Physical Review* without seeing this material referred to in a half dozen articles. This recent work [a new major contribution] was made with it and I don't think one physicist in a thousand knows now that I developed it.

The emphasis on moving ahead in science and the concomitant obliteration of past contributions through incorporation into new ones literally erases the bases of prestige of older scientists (see Zuckerman and Merton, 1972). As a consequence, reputation must be replenished periodically by new contributions, except for the few, like Einstein, whose early work was so monumental that nothing more would have been required of him. In science, most heroes have short lives, and those who outlive the usefulness of their research are likely to find the experience disconcerting, at the very least.

Still, many laureates whose standing among active researchers has dimmed with time continue to occupy positions of authority. As gatekeepers on foundation boards, fellowship committees, and review panels, they have a say in the allocation of resources and rewards to an extent that younger scientists may not recognize. In combination with their public standing, these activities give them considerable influence in the scientific community. As Arne Tiselius put it, "I have my prestige to spend" (1967, p. 7).

## INSTITUTIONAL MOBILITY AFTER THE PRIZE

Given the propensity of universities and research institutes to acquire Nobelists as symbols of organizational excellence, laureates should have ample opportunity to move if they choose. Few do so. As we saw in chapter 6, the laureates are not given to moving from one institution to another. Just ten of the ninety-two laureates permanently based in the United States when they received their awards changed their affiliations in the five years following the prize. Two of the ten, Georg von Békèsy and Edward C. Kendall, did so only upon reaching the retirement age of 65. An eleventh, John H. Northrop, arranged a joint appointment at the

Rockefeller Institute (now University) and Berkeley, thereby establishing two home bases and making it possible for both institutions to claim him.

The laureates' low rate of mobility after the prize in part reflects their age distribution. In general, institutional mobility diminishes after academics reach the age of 50 or so (Brown, 1967, p. 38). Laureates were no more mobile in the five years after their prizes than their age peers in the matched sample; 12 percent of the laureates moved as against 17 percent of the others. Contrary to the widespread impression that competitive bidding after the prize leads laureates to sever old ties and to establish new ones, the laureates turn out to be more sessile afterward than they were before. But the impression lingers as a result of the publicity efforts of a small number of institutions to capitalize on the reputations of their "acquisitions."

Not only do laureates tend to stay put after they get their prizes, but many remained at the same institutions for the greater part of their careers. As noted earlier, three of them never left the universities where they took their bachelor's degrees. It turns out that nine more stayed on where they took their doctorates and an additional fifteen, where they had their first jobs. These data do not square very well with the image of laureates as itinerant careerists.

Table 7–4 records the comparative stability of laureates' affiliations after they get their Nobel prizes.[18] It also provides further evidence for the clumping phenomenon noted earlier in the laureates' educational origins, the places that first employed and promoted them, and the sites of their prize-winning research. More than half the laureates were associated with just five universities when they got their prizes and the same is true afterward. Harvard, California at Berkeley, the Rockefeller, Cal Tech, and Columbia maintained their dominance as institutional homes of Nobelists.

Table 7–4 also shows that the laureates who moved were no more likely to go to elite universities than elsewhere, suggesting that at this time in their careers improvements in their immediate work situations matter more than prestige and ambiance. However, if the laureates who moved following the prize did so to better their work environments, their productivity afterward shows no signs of it. It drops by 37 percent—somewhat more than the reduction of 27 percent for all laureates—perhaps reflecting the costs inevitably associated with moving and settling in.

---

18. American universities and research institutes also gained eleven laureates through permanent immigration after they had done their prize-winning research: Peter Debye, John Eccles, Albert Einstein, Enrico Fermi, James Franck, Viktor Hess, Karl Landsteiner, Otto Loewi, Otto Meyerhof, Otto Stern, and Albert Szent-Gyorgyi.

**Table 7-4. Affiliations of Laureates (1901–72) Five Years After Prize**

| Laureates' Affiliations | TIME OF AWARD | | INTERIM | | FIVE YEARS AFTER AWARD | |
| --- | --- | --- | --- | --- | --- | --- |
| | N | Cumulative Percent | Gains | Losses | N | Cumulative Percent |
| *Harvard | 14 | 15 | | −2 | 12 | 13 |
| *California, Berkeley | 9 | 25 | +2.5 | | 11.5 | 26 |
| *Rockefeller Univ. & Foundation | 11 | 37 | +1 | −1.5 | 10.5 | 37 |
| *Cal Tech | 8 | 46 | | −1 | 7 | 44 |
| *Columbia | 6 | 52 | | | 6 | 51 |
| *Chicago | 4 | 56 | +1 | | 5 | 56 |
| Stanford | 4 | 61 | +1 | | 4 | 61 |
| Bell Laboratories | 3 | 64 | | −1 | 2 | 63 |
| *Cornell | 3 | 67 | | −1 | 3 | 66 |
| National Institutes of Health | 3 | 71 | | | 3 | 70 |
| Washington Univ. (St. Louis) | 3 | 74 | | | 3 | 73 |
| *Mayo Clinic | 2 | 76 | | −1 | 1 | 74 |
| *M.I.T. | 2 | 78 | +1 | −1 | 2 | 76 |
| *Wisconsin | 2 | 81 | | −2 | 0 | 76 |
| *Princeton | 1 | 82 | +1 | | 2 | 78 |
| *Other elite universities and research institutes [a] | 4 | 86 | +1 | | 5 | 83 |
| Other universities, research institutes, and laboratories [b] | 13 | 100 | +3 | −1 | 15 | 100 |
| | (92) | | (+11.5) | (−11.5) | (92) [c] | |

a. At the time of their awards, laureates were associated with Illinois, the Institute for Advanced Study, the Carnegie Institution, and Yale, all elite institutions. After the prize, Willis Lamb emigrated to England and was a member of New College, Oxford. This transition seemed appropriate to include here.

b. At the time of their awards, laureates were associated with Brown, Indiana, New York University, Rochester, Rutgers, St. Louis, California at Los Angeles, California at San Diego, Vanderbilt, Western Reserve, Salk Institute, and General Electric. After the award, William Shockley moved from Bell Laboratories to found the Shockley Laboratory, Clevite Corporation. Earl Sutherland moved to Miami and Georg von Békésy to Hawaii.

c. Earl Sutherland and F. P. Rous died soon after their prizes. Their post-prize affiliations are included nonetheless.

* Elite institution.

The Nobel prize—with its psychological and social consequences for the recipients—is associated with a variety of changes in their work. Both immediately after the prize and in subsequent years, the significance and the quantity of their research decline, hardly the outcome Alfred Nobel hoped for when he established his prizes. The social demands of the prize are more immediately disruptive for laureates who experienced comparatively large gains in prestige than for those who had already learned how to deal with eminence and fame. But the initial effects of new-found eminence pass, and laureates who were unaccustomed to fame learned how to deal with it at least as well as others who were more accustomed to it. Having the prize interferes with their long-term productivity no more than it does with the others'. After the prize, most laureates continue to publish at a far greater rate than other productive scientists of the same ages. Since the disruptive effects of the prize may depend on new laureates' awareness of and efforts to accommodate to them, these findings may turn out to be self-disconfirming.

Changes in social roles and relations with colleagues and co-workers ensue. Prize-winning collaborations in particular terminate soon after the award. At the same time, the formal honors bestowed on the laureates multiply, typically in recognition of old rather than new scientific achievements. Most laureates find, however, that the multiplication of awards has little to do with the esteem in which they are held by active researchers, which depends more on the present than the past.

Long after the occasion for the Nobel prize is only vaguely remembered (or wholly forgotten), the cachet of having acquired one lingers on. Being a laureate indelibly certifies the superiority of one's achievements. But permanent elevation into the ultra-elite of science is not without its costs, as many laureates observed. A physicist elected to speak for all of them:

> There are many things about the Nobel Prize that remind me of the Lord's Prayer. "Lead me not into temptation." Very few people have survived it with a whole skin. It leads to difficulties . . . and to opportunities, but there are always the temptations.

# Chapter 8

# THE NOBEL PRIZE
# AND THE ACCUMULATION OF
# ADVANTAGE IN SCIENCE

## THE EFFICACY OF THE PRIZE

The Nobel prize involves more than the bestowing of honors for meritorious contributions to scientific knowledge. It has become an institution with effects that ramify throughout the social system of science. Few leaders in the fields covered by the prizes have been altogether free of Nobel aspirations and even scientists in the ineligible fields have strong opinions about the prize. Laureates as well as scientists who have been passed by have their own brands of ambivalence toward it. How well does the prize do the job of selecting the prime scientific achievements? And, can we say whether, on balance, it is good for science? These are, of course, generic questions that can be raised about any institutional arrangement, and the Nobel prize, with all its luster, is obviously not exempt from such appraisal.

How well the prize works depends, first, on how effectively it does what it was manifestly intended to do—advance the growth of scientific knowledge—and, second, on whether its unanticipated consequences are more or less consistent with that purpose.

As we have seen, Nobel originally intended the prizes to honor the most significant contributions to physics, chemistry, and the biological sciences. He apparently wanted to direct attention to science of the first class at a time when it was less publicly visible than it is now, and he wanted to make it possible for his laureates to continue their research unhampered by financial constraints.

By the most obvious criteria, the importance of research that has been honored and the scientific excellence of the laureates, one must conclude that the Nobel committees have chosen well. Few scientists complain about the overall quality of the list of Nobel prize contributions or about the merits of most laureates; when they do complain, it is generally about sins of omission rather than commission. It might be argued that this is no small achievement to maintain for three quarters of a century.

But certain limitations in the selection procedures raise questions about the legitimacy of the institution itself. As we have seen over and over again, the phenomenon of the forty-first chair testifies to some of these limitations. Occupants of that chair have done research as consequential for science as the laureates yet have not received the prize. The phenomenon of the forty-first chair is of course the outcome of exponential growth in the population of scientists and, along with it, growth in the subset of scientists qualified for the prize. Even if that subset has grown more slowly than the scientific population (as it likely has), it has far outstripped the number of available prizes. The Nobel committees must choose among growing numbers of scientists who have greatly advanced their sciences.

This growing number of potential candidates introduces a further strain on the equitable and effective operation of the Nobel prize as an institution. How are choices to be made among the number of more or less equally deserving candidates? Chapter 2 sketched a model of successive approximations in the decision making process involving both universalism and particularism in sequence. According to this model, the ideal of universalism in the cultural structure of science is more or less realized in the first phase of winnowing. The model also allows secondary criteria, both universalistic and particularistic, to enter as the roster of candidates is winnowed further. In the process, new occupants of the forty-first chair are generated and form a cadre whose contributions to scientific knowledge are on first approximation just as consequential as the laureates'.

Occupants of the forty-first chair are also brought into being by statutory limitations on the fields covered by the prizes. As we have noted, these limitations exclude a considerable share of those now at work in astronomy, the marine and earth sciences, zoology, and mathematics, as well as all the social sciences except for economics—fields in which major advances have been made in the past seventy-five years. Such limitations might seem to be innocuous, especially since other respected awards are available to investigators who work in the excluded fields. But, as I have indicated, the Nobel prize has a special meaning outside of science. Scientists who are ineligible for prizes are inadvertently demoted to second-class status without anyone's intent. It is almost as though a discipline lacking its own Nobel cannot be a discipline contributing greatly to scientific knowledge.

There are signs that the Nobel establishment is beginning to modify the rules laid down generations ago by extending the fields eligible for the prize. Along with accepting the new Nobel prize in economics under the aegis of the scientific awards, the committees have stretched the traditional territory covered by the prizes: once in 1973, by conferring the prize in physiology on von Frisch, Lorenz, and Tinbergen for research on animal behavior rather than physiology, narrowly conceived, and again in 1974, by conferring the prize in physics on Ryle and Hewish for research that was clearly in astronomy, not physics. (While this volume was in press, a detailed sociological analysis of the Ryle and Hewish research was published. See Edge and Mulkay, 1976.) These decisions may set precedent for greater leeway in future prize selections.

It would not of course solve the other problems of equity associated with the phenomenon of the forty-first chair simply to increase the number of awards given each year. In the absence of great discontinuities in the quality of scientific contributions, there would remain the phenomenon of the also-rans whatever the cutting point. Moreover, the multiplication of the prizes by dividing them into smaller awards would dilute their symbolic value as well as their monetary value. In short, the problems that go with greatly increasing the number of prizes would scarcely make for greater distributive justice.

In the course of their seventy-five-year-long history, the Nobel prizes have, in some quarters, become such extravagant symbols of scientific excellence as to be turned into institutionalized exaggerations. Chapter 2 examined the ways in which the success of the awards has led to their use for purposes quite at odds with their intrinsic aims. The counting and claiming of laureates by regions, institutions, and organizations constitute just one example of the general pattern of efforts to transfer the prestige of the prizes to all manner of organizations and causes. These practices have for a time resulted in lionizing the prize-winners and in accrediting them as the spokesmen for science. As the prestige of the prize has risen, this has led to conditions that tempt some laureates into claiming general wisdom and authority as well as demonstrated expertness of a high order. The social control system in science may work to curb some of these excesses, but it scarcely eliminates them. This leads a good many scientists to express skepticism about the value of the prize, this at the very time that it seems to have achieved its greatest popular esteem. In this sense, at least, the Nobel prize in science has been victimized by its own success, a development that has invited debate among scientists the world over about the prize as an institution.

Scientists contending that the prize has not been good for science generally put forward two reasons for their position: the prize does not advance scientific knowledge, and it ultimately serves to demean the quality of the scientific life. The first argument runs that the prize, with its emphasis on rewards rather than on direct motivation for the advancement of knowledge, tends to divert scientists from research on deep intractable problems to "prizable" ones. The process, it is said, results in collective scientific attention being focused on a limited range of problems, with many others remaining underinvestigated. Although few of these critics go on to make the point, by the same logic, the prizes might be claimed to keep ambitious scientists from research directed to major practical problems and to reinforce their more or less exclusive concern with fundamental problems.

In point of fact, next to nothing is known in any systematically documented way about the effects of the Nobel prizes on problem choice in science. We still need to find out, for example, whether there is greater concentration of research attention in fields covered by Nobel prizes than in the rest, and in the event, how far this can be attributed to the existence of the prizes. Since the reward system in science gives credit for problems solved rather than for brave but unsuccessful efforts and for research on central questions rather than on peripheral ones, it tends in any case to create "bandwagon" effects in the foci of scientific attention. The secondary effects of the prizes along the same lines may be swamped by these broader tendencies.

Critics of the prize have also argued that it actually impedes the advancement of science by fostering an excessive concern with the recognition and rewards that subverts primary interest in the work itself. More good science, it is said, would get done if the prize were not so conspicuously there on the horizon. It is not difficult to diagnose occasional clinical cases of this displacement of interest but it has not been much honored by systematic investigation into its merits. Nor is it clear just how the hypothesis might be systematically investigated. In this connection, it is worth reiterating an observation made before: the institutionalized quest for recognition in science seems to be as old as modern science itself (Merton, 1973, part 4). The existence of the

Nobel, and of other major awards, may have intensified the strong emphasis on collegial recognition for some scientists but it has not introduced an altogether new element into the reward system of science.

Nobel aspirations are one thing, but getting a prize is quite another. Assessments of the impact of the prize on scientific advance must also include some reckoning of its effects on the laureates. The data reported in Chapter 7 suggest that receiving the Nobel prize often has the effect of slowing the continuing flow of laureates' contributions to science. The symbolic success has a potential for subverting their scientific work as they yield to the demands and temptations that come with being celebrities. In many individual cases then, the prize was found to impede rather than to advance scientific work. But it seems safe to conclude that, in the aggregate, the course of the history of science was not much changed by such effects on the laureates. As the phenomenon of multiple independent and simultaneous discoveries (noted in chapter 6) reminds us, the work that this or that laureate might have done but did not do, will tend to get done by others—not perhaps in precisely the same manner but with a more or less equivalent impact on the development of their sciences. The very existence of the prize does much to symbolize the heroic theory of scientific growth by underscoring the contributions of particular individuals and having them all appear to be uniquely indispensable to the actual course of development. But the occurrence of multiples, not infrequently among the laureates themselves, puts that theory into serious question.

Especially since the appearance of *The Double Helix* (Watson, 1968), the prize has been singled out as contributing to competitive practices in science, among them, keeping one's ideas secret until they are ready to be publicly unveiled (Chargaff, 1968: Lwoff, 1968). The prize is said to make for what one laureate has called "the presence of sharp elbows in science." It is suggested that the quality of the scientific life has greatly declined, that sensibility and cooperation are being replaced by rivalry and contention for priority, and that the Nobel prize has contributed to the general deterioration. This complex hypothesis requires better evidence than we have on historical change in the extent of competitiveness and on the further question of whether competitiveness is dysfunctional for the advancement of scientific knowledge. But even now it would seem that the prize is more nearly a symbolic expression of competitiveness in science rather than a prime source of it. The most severe critics have generally maintained that intense competition and the resulting acrimony have worsened only in the last several decades. Yet the Nobel prizes have been around for three-quarters of a century and have operated in much the same way throughout this period.

The disarmingly simple question of whether the Nobel prize is good

for science thus turns out to be disconcertingly complex. Plainly, the
prize is not essential to the effective pursuit of scientific knowledge since
it came comparatively late in the centuries-long history of modern science, yet it appears to have made major contributions in its time. The
exaggerated significance often attached to the prize and its frequent
diversion to purposes having nothing to do with scientific excellence are
perhaps not beyond hope of control by communities of scientists and by
the Nobel establishment itself.

From all this, it appears that the Nobel prize as an institution gets
high marks in its effort to identify good science and to honor good scientists. But it has been far less effective in taking into account the vast
changes in the cognitive map of science that have taken place since the
prize was first instituted at the turn of the century. Far less is known
than we need to know in order to assess the actual effects of the prize
on the advancement of scientific knowledge and on the quality of the
scientific life. We have yet to discover whether, in an ironic turnabout,
the prize has diminished rather than augmented the legitimacy of the
reward system in science.

So much for the prize as an institution and for the engaging question
of whether it is, on the whole, "good for science." We need now to
review its role in the stratification system of science, with special reference to the accumulation of advantage.

## THE ACCUMULATION OF
## ADVANTAGE IN SCIENCE

Chapter 3 outlined a model of the principal social processes that
seem to produce the accumulation of advantage in science. That model
has scientists who show promise early in their careers being given greater
opportunities in the way of research training and facilities. To the extent
that these scientists are as competent as the rest or more so, they ultimately will do far better in terms of both role performance and rewards.
Access to resources and facilities often affects the quality of scientific
role performance, which, in its turn, evokes greater or lesser rewards.
Rewards, in turn, can be transformed into resources for further work;
scientists who are initially advantaged gain even greater opportunities for
further achievement and rewards.

How well do the histories of Nobel laureates fit this model of scientific achievement and the allocation of rewards? As I have repeatedly
indicated, there is evidence of accumulating advantage in the careers of
laureates. Most, although not all, get an early start by going to the academic centers of science (the most distinguished universities) for gradu-

ate work. More than half study under Nobel laureates, and a large share of the rest study with occupants of the forty-first chair and the more extended elite, thus establishing a pattern of eminence begetting eminence. The laureates claim the development of scientific taste, standards, and self-confidence as the most beneficial result of their apprenticeships. None of these is easily taught, usually being acquired by example rather than precept. It also happens that studying under members of the scientific elite located the laureates-to-be in the network of scientific communication, thereby improving their opportunities to publish and to acquire further facilities: good appointments and productive research milieus. Elite training can be considered a scarce resource that is differentially distributed and later contributes to differential scientific achievement. Since elite training is not consumed with use, unlike other resources, it continues to contribute to the accumulation of advantage.

Future Nobelists got their degrees and began to publish earlier and more copiously than other scientists and continued to do so throughout their careers. Fairly early, they also began to distinguish themselves in terms of quality of scientific contribution from their age peers who later joined the extended scientific elite, the other members of the National Academy of Sciences. Some evidence, which is not nearly systematic enough, suggests, just as we would expect, that the career beginnings of occupants of the forty-first chair exhibit the same characteristics: they also studied under distinguished scientists, began research early, were rewarded early, and thus look much the same as the Nobelists.

Future laureates typically do the research that eventually brings them the Nobel prize in their late thirties and early forties, establishing once and for all their capacity to do research of the first class and accelerating their upward mobility in terms of appointments, positions of influence, and honorific awards. With the exception of a few laureates whose work was at first resisted or neglected, laureates were rewarded soon after publishing their research. They are, as I have said, the success stories of science.

At the time of their prizes, most laureates were already eminent investigators, recognized the world over as having made truly consequential contributions to science. Those few laureates who were not yet eminent found themselves rapidly elevated to the highest stratum of science. After the prize, both sets of laureates moved into a new phase of accelerated acquisition of honors and generally found that they could command whatever resources they needed to get on with their work.

Altogether, then, this composite portrait of the laureates' careers seems to square rather well with the model suggested by the accumulation of advantage: the spiraling of augmented achievements and rewards for individuals and a system of stratification that is sharply graded.

Close examination of the processes involved in the accumulation of advantage finds that strict conformity to the merit principle in science contributes to marked inequalities in scientific performance and rewards. Laureates and occupants of the forty-first chair have both contributed disproportionately to science and have received a correspondingly large share of rewards for having done so. There is, however, no way of ascertaining whether other scientists would have done equally well had they been advantaged in the same ways as the laureates and other members of the ultra-elite. This is so because the opportunity structures scientists confront at successive stages of their careers tend to reflect their past achievements. Those who do well when they are young have a better chance to continue to do well as they get older. The processes involved in the accumulation of advantage cast considerable doubt on the conclusion that marked differences in performance between the ultra-elite and other scientists reflect equally marked differences in their initial capacities to do scientific work.

This conclusion is further put in doubt by the intrusion of self-fulfilling prophecies (Merton, 1968a) into the evaluation and reward systems of science. Being earmarked as a promising young scientist often leads to better objective conditions for work. This facilitates research performance, which in turn confirms the initial prediction. Under such circumstances it is not possible to determine whether those who were initially benefited were in fact more promising and whether the evaluation system is as effective as the "results"—the greater achievements of who were selected—would seem to indicate. Since appraisals of the effectiveness of evaluation systems tend to be based on "results," it is not surprising that many conclude that evaluation in science works well and that the indeterminacies created by the operation of self-fulfilling prophecies and the accumulation of advantage are rarely taken into account.

The same pattern of accumulation of advantage observed for individual scientists appears also to characterize scientific organizations, for they, too, are sharply graded in achievements and prestige. The two stratification hierarchies—of individuals and organizations—are in fact tightly interconnected through exchanges of prestige and through self-selection and selective recruitment.

Universities and research organizations derive prestige from having eminent scientists on their staffs. But the prestige of an organization is something more than the aggregated prestige of its members. Distinguished scientists have something like a multiplier effect on the standing of their organizations, not least by attracting other established scientists and needed resources. They also improve opportunities for younger and less well-known colleagues to do good work by providing research facilities or lending their support to efforts to acquire them. In their turn, if younger scientists make their own contributions to science, the recognition they receive redounds to the organization, and thus some of the reputational stock on which they drew earlier is replenished.

Self-selection by high-quality students and staff and the use of rigorous standards in recruitment help to ensure the production of high-quality work, which in turn heightens institutional repute. This makes for further self-selection by competent individuals, increases opportunities to be demanding in recruitment, and reinforces the processes of accumulation of advantage for organizations. Self-selection by individuals of promise and their selective recruitment by a small number of institutions produce concentrations of laureates and members of the National Academy of Sciences in a few universities and research laboratories.

These concentrations make it likely that a small number of organizations will be evocative environments and thus sites of major scientific achievements. It is not merely that numbers of able scientists assembled in one place do good work; it is also that their interaction is likely to lead to significant new scientific developments of the sort that make an institution a leader in science and part of the organizational elite.

If the stratification of individuals and organizations in science seems to derive largely from universalistic judgments of scientific merit, there are also particularistic elements in the evaluation and reward systems of science. There are intimations that particularism plays a role in the laureates' early careers, especially for those apprenticed to Nobelists. At the same time, since all the laureates later made substantial contributions to science, particularistic criteria were plainly not the only ones determining their entry into the ultra-elite. (See Mulkay, 1976, for an independent but similar view.)

Particularism plays a more visible role later in the laureates' careers, as the Matthew Effect confers further authority and influence on those who are already influential and authoritative and brings further honors to those who have already been honored. There is, as I have noted, a distinct tendency for laureates to be honored after the prize, usually for old achievements rather than new ones. Thus, achieved status tends to be transformed into ascribed status quite without intent and often without the recognition of those involved in the process.

## Constraints on the Accumulation of Advantage

On the face of it, it would seem that there is nothing in the evaluation and reward systems of science to constrain the accumulation of advantage. Why, then, has concentration gone no further? What processes, if any, put a brake on the accumulation of advantage? As Robert Merton suggests in his second study of the Matthew Effect (1975), these questions are deeply relevant to distributive justice in science. I do not address myself to this issue here but, rather, try to sketch out some constraints on the accumulation of advantage that can be detected in the careers of Nobel laureates.

First, success in science, at least at the lofty level of the Nobel prize, can undermine the very activities that brought it about. Since the subversive potential of success has been an enduring theme in social and political thought, the phenomenon is plainly not confined to science. (For one classic example, see Weber [1958b, first published 1904–1905] on the ultimate effects of wealth on ascetic Protestant merchants.)

I have noted repeatedly that the Nobel prize can subvert the research activity of laureates by subjecting them to the distractions of being a celebrity. As the laureate in medicine Robert Holley observed, ". . . recognition leads to demands on one's time; it is a way in which success is self-limiting" (quoted in Sherwood, 1974, p. 14).

Success can set in motion social and psychological processes that limit later achievements. Having been moderately successful, some people develop excessive expectations about the quality of their own performance and are subjected to equivalently excessive expectations by others. Not many have the self-confidence to work effectively under these conditions.

Such excessive expectations, their own and others', have limited effects on Nobel laureates. For one thing, they are aware of how infrequent sustained excellence in science really is. For another, other scientists tend to be skeptical that the laureates will continue any serious work at all; they are not given to inflating Nobelists' scientific capacities. Instead of being plagued by others' excessively high expectations, laureates encounter the polar opposite, being subjected to the belief that they have, as one physicist likes to put it, "shot their wad."

Success can also be subversive by seeming empty and meaningless once it is attained; it thereby undercuts motivation to continue among those who once pursued it most avidly. Nobel laureates seem less susceptible to this aspect of the wrecked-by-success syndrome than artists and writers whose success, unlike that of scientists, depends more often on popular acclaim than on judgments by those qualified to make them. Although many laureates were, as chapter 7 indicated, ambivalent about certain aspects of the prize, not one reported that the prize had no savor.

Second, the accumulation of advantage seems to be countered by the presence of a ceiling on scientific standing. Although there is no systematic evidence, it appears that once scientists enter the ultra-elite, new increments of achievement and prestige neither elevate them higher in the stratification system nor enable them to command an increasing share of resources. These later increments of accomplishment serve essentially to confirm their rank.

Third, the ability of individual scientists to use resources is limited. Individual scientists probably differ in their thresholds of resource-saturation but even the most expansive ones among them have their limits. The Coles' (1976) studies of the peer review process suggest that review panels are well aware that scientists cannot effectively use more than a certain amount of support—even if they ask for more—and take these limitations into account in distributing the means of scientific production. All of this imposes its own kind of constraint on the accumulation of advantage.

Fourth, the accumulation of advantage is limited by what we have seen to be the occasional exercise of *noblesse oblige* in assigning resources for research yet to be done and credit for research already completed. Although such gestures express and reinforce the superior-subordinate relationship between donor and recipient, they nevertheless have the objective consequence of reallocating honor and resources that would otherwise be monopolized by the donor. Younger scientists benefit from expressions of *noblesse oblige* and older eminent ones lose little. This transfer of payments is not costly and it provides the participants with a sense that the reward system can be made to work equitably.

Various social and psychological processes thus limit the accumulation of advantage in ways that are still barely understood. The data on the laureates are roughly consistent with the model of accumulation presented here. But they are not detailed enough or buttressed enough by comparative data on other strata of scientists to provide good estimates of the interaction between achievement and ascription at successive stages of their careers. Studies of the accumulation of advantage and counteracting forces thus raise questions of general sociological interest as well as questions specific to the operation of science as an institution.

Along quite different lines, close analysis of the career patterns of laureates suggests that there are differences between physicists, chemists, and biological scientists in, for example, the proportions who had laureates as mentors and in their age at doing prize-winning research. I have noted these differences between the sciences but have not tried to account for them. Another direction for further inquiry, then, is the comparative analysis of the various sciences and the interconnections between their cognitive characteristics and their social structures. It has been observed, for example, that the discovery of fundamental principles in the biological sciences is, in fact, less frequent and, in theory, perhaps less likely than in the physical sciences. The exact significance of new research may be less often clear in the biological sciences (Delbrück, 1965) with less agreement on the contributions and contributors meriting recognition. On the side of social structure, this should make for a system of ranking scientists that is more ambiguous as well as less sharply differentiated. This is just one way in which cognitive features of science may impinge upon the stratification of achievement and rewards.

Finally, in examining the Nobel prize as a social institution and the careers of the Nobel laureates, we noted elements of both achievement and ascription. We also noted that the laureates are not *the* most important contributors to twentieth-century science but rather that there are also occupants of the forty-first chair who have contributed as much to science or more. Plainly, the reward system of science has its imperfections; rewards do not always accord with quality of contribution nor do they always serve to elicit continuing role performance of the highest quality. Even so, the laureates have, on the whole, greatly advanced scientific knowledge and it is the significance of those contributions which now, as at the turn of the century, affirms the preeminence of the Nobel prize.

# Appendix A

# INTERVIEWING
# AN ULTRA-ELITE

The Nobel laureates are not simply members of the scientific elite; they are, as a group, among the most eminent men and women of science. As we have seen, they are distinguished by their small relative and absolute numbers and their high visibility; they are regarded as an elite by other high-ranking individuals and organizations, and they are preeminent among scientists in achievement, authority, and rewards.

Judged by these standards, which are close to those established by Pareto for a genuine elite in a particular sphere of activity, Nobel laureates may be fairly described as an ultra-elite when compared with the elites interviewed in many other sociological studies. They occupy positions in the social structure substantially higher than the large sample of American social scientists interviewed for *The Academic Mind* (Lazarsfeld and Thielens, 1958; Riesman, 1958a) and the small samples of natural and social scientists interviewed in several other studies (Crane, 1964; Gaston, 1969; Hagstrom, 1965; Storer, 1962; Swatez, 1970). In the system of American social stratification, they are a substantial cut above the school superintendents interviewed at length for the study in role conflict (Gross, Mason, and McEachern, 1958) or those occupying high places in the power structures of American communities (Dahl, 1961; Hunter, 1950; Warner and Abegglen, 1955). Perhaps the only roughly comparable elite interviewed by sociologists is found in Rock Caporale's study (1964) of Vatican Councils, in the course of which he interviewed a large sample of cardinals, archbishops, and bishops, comprising the elite of the Roman Catholic Church. Properly so, for it will be remembered from chapter 7 that the laureate Werner Forssmann felt as though he had been raised to the cardinalate of science.

I emphasize the ultra-elite character of the laureates here to explain why I present a fairly detailed account of how the interviews were obtained. This account not only provides a basis for assessing the interview materials on which this study draws heavily but also may have some bearing on the distinctive problems of interviewing members of elites, suggesting some pertinent ideas for interviewing topmost elites in such other institutional realms as government,[1] economy, and religion.

## PROCEDURES OF INTERVIEWING
### The Initial Phase

The first phase of making contact with the laureates involved the sheer mechanics of locating them. Standard scientific directories for the

---

1. Several investigators, among them Matthews (1960), have interviewed members of top political elites. Substantial portions of these elites were not interviewed, however, nor were these data subjected to systematic analysis. See Dexter, 1970, for a guide to interviewing elites.

most part provided the necessary information. Other means were employed to locate about 15 percent of them.[2] The laureates are a scattered elite, and only a few are in regular contact with one another. However, they are not uniformly distributed among academic and research organizations but, as we have demonstrated, are clustered at the most distinguished universities and research institutes, with most of them being located around the major cultural centers—Boston, New York, and San Francisco.

Forty-one of the fifty-six laureates living in the United States in 1963 were interviewed. The remaining fifteen were not interviewed for the following reasons:[3]

|                                          | Number of Laureates |
| ---------------------------------------- | ------------------- |
| Illness                                  | 2                   |
| Absence from United States               | 2                   |
| No reply to letters requesting interview | 5                   |
| Too busy                                 | 6                   |

There is compelling evidence that two laureates were too ill to be interviewed; one died within a few months of receiving the request and the second so soon after he was contacted that no reply was possible. One laureate, tallied in the no-reply category, could not be located, leaving only three laureates who never answered the letters sent to them. Answers were received from the two laureates who were not at the time in the United States.

Of the six laureates who refused to be interviewed on the ground of being too busy, four held appointments at Harvard, one at Berkeley, and one at Yale. A typical letter of refusal read:

> *I am fully occupied now. I am preparing a paper and will sail for Europe in a few weeks.*

> *I have a very rough schedule in the coming month [in reply to the second letter sent three months later]. My schedule has not eased at all. I am involved in so many experiments now that it is not possible to see you.*

---

2. Difficulties encountered in locating the laureates were instructive. They testified to the extraordinary amount of traveling done by eminent scientists.
3. Four of those who were interviewed did not reply at first but were reached by telephone and appointments were arranged. Those who let me know that they were "too busy" to see me were of course using one of the coping mechanisms for dealing with demands on their time described in chapter 7. Even more efficient, if not cooperative, in this regard, were the laureates who did not reply to my letters at all.

The variety and weight of the demands made on the laureates' time were
frequently mentioned in the interviews—on occasion with a meaningful
glance at the interviewer. In view of the many requests directed toward
this top elite, it is surprising that as many as forty-one of them were
willing to devote time to the interview.

About 80 percent of the laureates in physics and chemistry and 70
percent of the laureates in medicine were interviewed. Of the seven
prize-winners in medicine who were not interviewed, four were located
in Boston and associated with Harvard. Altogether, 57 percent of the
Boston laureates refused to be interviewed. The refusal rate for laureates
in other geographical areas never exceeded 10 percent.

Not only did most laureates consent to an interview, but for the most
part they did so promptly. On the average, only eight days elapsed between the time the laureates received the first letter of request and the
date they replied.[4] Five answered the same day the letter was received
and two waited as long as thirty-one days to reply. When no answer to
the first letter was received within three weeks, a second letter was sent.
The average time between the date of receiving the second letter and
answering decreased to four days.

## The Letters

The first six laureates I wrote to received rather lengthy letters explaining the nature of the investigation. It was apparent from their very
short and pointed replies that the laureates themselves did not indulge in
lengthy correspondence. Typical examples are these one-line letters
from a chemist and a biochemist:

> *I should be available in Berkeley after the 29 of August.*

and

> *I will be in town on Friday August 30th, so just call me when
> you get in and we'll talk.*

This fact coupled with the remark by one laureate that he hadn't
time to read my letter through led to substantial editing of its original
contents. Later versions of the letter requesting an interview contained
only four pieces of information. I identified myself as a Fellow of the

---

4. It was not possible to arrange for appointments with eleven laureates at the time the
first letters were sent. A second request was sent about four months later. The time
elapsed between receiving and answering these letters was six days on the average.

Social Science Research Council at Columbia University and a recipient of a grant from the National Science Foundation. These allusions were intended to provide institutional legitimacy for both the request and the investigation. The purpose of the study was then stated succinctly and, to provide legitimation by their peers, the names of some laureates who had already been interviewed were given.[5] Finally, the letter sent to the laureate stated when I would be visiting the area in which he lived.[6] Laureates located in New York and its environs were asked when it would be "least inconvenient" for them to have the interview.

Letters of confirmation were sent at once to all laureates who agreed to be interviewed and, whenever possible, information was supplied on how I might be reached when I arrived in their city. Within a day or two after the interview was completed, a handwritten thank-you note was sent to each laureate. The contents of these notes varied, but some mention was usually made of the laureate's distinctive contribution to the study.

## Making Appointments for the Interviews

About one third of the laureates replied to my initial letter giving specific dates and times that they would be free; the rest asked to be notified when I arrived in their area. None of the laureates who consented to be interviewed was unable to keep the appointments that had been made, although three found that new obligations limited the time set aside for the interview.[7] Four laureates who had not answered the letter were reached by telephone and appointments were made at that time: in two cases on the same day, and in the other two several days later.

---

5. In all cases, permission was asked and granted to mention names of those already interviewed in letters to their fellow prize-winners.
6. The text of the letters varied slightly. The following text was used in arranging for the second round of interviews on the West Coast.

> As a Fellow of the Social Science Research Council at Columbia University, I am engaged in a study of collaboration among scientists. One part of the investigation deals with the collaborative experiences of eminent scientists, particularly Nobel laureates.
> I have been privileged to talk with a good number of your fellow laureates on the East Coast and in the Middle West. A small grant from the National Science Foundation will enable me to visit California. I am writing to ask whether it will be possible to see you at any time during the later part of March or early April. I would greatly appreciate the opportunity of meeting with you.
> Sincerely,

7. One laureate called to say that he had only a half hour on the day appointed for the interview and therefore wished to reschedule it. This was arranged without difficulty.

## Attitudes toward the Prospect of
## Being Interviewed

Long before the interviews took place, the laureates indicated in a variety of ways that they were not unreceptive to the idea of being interviewed. Several offered to make hotel reservations for me, and five letters included invitations to lunch.[8] Rather detailed directions were given for getting to their offices or homes, and in two cases plane and bus schedules were sent. One laureate arranged for transportation to his laboratory and another sent a hand-drawn map giving the route to his home.[9] Several who had delayed their answers wrote letters of apology explaining that they had been away from their offices. Three telephoned (all long-distance calls) to arrange for appointments, and two had their secretaries leave word confirming appointments at the offices of other laureates whom they assumed would be visited. One managed to find time for the interview the day before he was to leave for Europe and another on the day he returned from Russia. Another agreed to be interviewed while on a holiday, after his secretary had made it clear that this was impossible.[10]

The majority of replies were simple acceptances, but the import of the indicators of receptivity cannot be overlooked. The principal factors affecting receptivity presumably were the legitimacy of the interviewer's request, judged by her affiliations with Columbia University, the Social Science Research Council, and the National Science Foundation;[11] the laureates' sense of obligation to other investigators;[12] the self-contained nature of the proposed interview;[13] and not least, as we shall see, the sheer interest of the laureates in the subject of the inquiry. The relative importance of these elements probably varied from case to case, but each of them was made evident by the laureates in some part of the interviews.

---

8. One laureate, visiting New York, even asked to have hotel reservations made for him. By the time all the interviews were completed, the interviewer had been asked to lunch with eight laureates and had had lunch with the associates of three more.

9. Upon receiving information on where the interviewer would be staying, one laureate sent a special delivery letter explaining that no hotel of that name existed in his city. Reservations were then made by the laureate's secretary for the interviewer at the university-owned hotel.

10. The casual reply to the interviewer's apology for requesting an interview on a holiday was, "What self-respecting physicist doesn't work on holidays?"

11. By contrast, Rock Caporale, S.J., has reported that the legitimacy of his credentials was questioned by some of the members of the religious elite—cardinals, archbishops, and bishops—he interviewed for his dissertation in sociology at Columbia University.

12. A number of laureates referred to their heightened sense of responsibility to other researchers and to their communities after receiving their prize.

13. One laureate described his attitude in this way: "What I didn't want was many hours or many sessions. But this kind, I think, covers very pertinent questions."

## Preparation for the Interviews

Every interview was preceded by intensive and detailed preparation by the investigator. This preparation ordinarily consisted of five steps. The standard directory, *American Men of Science,* provided some information on the laureates' education and careers, their scientific awards and major research interests. *Who's Who* supplied data on government and community affiliations. The official publication of the Nobel Foundation, *Nobel: The Man and His Prizes* (1951; 1962), was used as a preliminary source of information about the laureates' scientific work. *Les Prix Nobel,* the Nobel Foundation annual, also provided useful biographical material, particularly on the identity of laureates' coworkers and significant teachers.

These biographical data were used in the initial phase of the interview as well as in analyses of the migration of laureates, changes of jobs, promotion rates, and the like. By piecing together information on the locations of scientists at particular times, quasi-sociometric maps were constructed. These were used to develop specific lines of questioning, particularly on the conception of evocative research environments. These "career maps" of the successive locations of scientists—for example, those at Göttingen in the 1920s and 1930s or at Los Alamos during World War II—enabled me to ask about particular associates who might not have been mentioned by the laureates spontaneously. This preparation gave me a detailed acquaintance with the names of the laureates' major associates and with their work.

The annual volumes published by the Nobel Foundation were the fundamental source for the second phase of preparation. They contain the laureates' Nobel addresses [14] as well as their remarks at the Nobel banquet. The addresses given by the laureates in Stockholm were a particularly rich source of information. The laureates frequently use this occasion to review the work honored by the Nobel Committee, to mention the contributions of significant colleagues, and to describe the idiosyncratic histories of their discoveries. Information from these lectures often suggested new lines for discussion in the interviews. Speeches made at the Nobel banquet provided additional material on contributions by other scientists to prize-winning discoveries. On this ceremonial occasion, laureates tend to give generous (sometimes excessive) recog-

---

14. There is substantial delay in publication of *Les Prix Nobel,* the annuals devoted to each year's awards. Reprints of the addresses in *Science* were, therefore, useful in preparing for interviews with recent winners. In several cases, addresses were published in German. I am indebted to Gerda Lorenz for her patience in helping me translate these papers. Nobel addresses have now been reprinted (Nobelstiftelsen, 1964–67).

nition to other scientists and to minimize their own contributions; their remarks were interpreted in this context.

The next phase of preparation involved locating additional publications, especially those written by the laureates for lay audiences (du Vigneaud, 1952; Waksman, 1954; Rabi, 1960; Szent-Györgyi, 1963). Secondary sources, essays, magazine articles, newspaper reports (Berland, 1962; Bernstein, 1962; Farber, 1953; Heathcote, 1953; Kaplan, 1939; Ludovici, 1957; MacCollum and Taylor, 1938; Stevenson, 1953), and summaries of particular fields of research (Asimov, 1960; Moore, 1961; Williams, 1959) were reviewed to provide the interviewer with information.

*Abstracts* of each field were carefully examined for each period so that preliminary bibliographies could be constructed for each laureate. The names of collaborators, duration of joint authorship and patterns of name ordering in the authorship of collaborative papers as well as the foci of their research were obtained from these indexes. Complete bibliographies were requested from each laureate; these provided the data for the analysis of laureates' publication patterns.[15]

Finally, a one-page summary of each laureate's career and his work was prepared for use in the interview itself. These guides made a great deal of information readily available to me during the course of the interview.

## Some Functions of Preparation for the Interview

Intensive preparation facilitated the process of interviewing in two principal ways. First, it provided testimony to the seriousness of the interviewer and helped to legitimize her demands on the laureate's time. This provided a firm basis for the temporary laureate-interviewer relation. As a laureate in chemistry remarked:

> You've done your homework, haven't you? I'd feel terrible if I didn't do something for you. No, this is something very few people do.... You can charm me but you can't bluff me. You've done well, okay?

And, second, questions based on materials gathered in preparation often called forth responses that would otherwise not have been elicited, par-

---

15. Seven bibliographies were substantially updated for my use. I am grateful to the secretaries of these laureates for freeing me from this task. Their courtesy greatly facilitated my work.

ticularly if an entirely standardized interview guide had been employed.

Members of the ultra-elite are understandably unwilling to devote time to projects they consider trivial. As another chemist admitted at the close of the interview:

> I said to myself before you came, "If she wants to ask me about social things, I will get her out of here fast." I am not interested in social things. You don't think I am unkind to say that? [No.] But you asked me about the real things. You know, science is a human thing, much more than most people realize. You never get that in books. What is written is never quite right. You have to hear it from the people who were there.

Almost all the laureates were acutely concerned with maximizing the use of that inevitably scarce resource, time; as one said, "I am constantly concerned with clock and calendar." In part, their commitment to the intellectually profitable use of their time led them to subject me to an almost continuous series of tests to ascertain the degree of my competence and commitment. The results of these tests affected both the laureates' willingness to continue and the quality of their responses.

Several types of responses indicated that the laureates were continuously evaluating my performance as interviewer just as they subject their own colleagues to incessant evaluation. A prime indicator of this evaluation was, of course, the laureates' direct statements of what they thought of the interviewer and the interview. Sometimes these consisted of simple interjections, unrelated to my prior remarks. A biochemist said:

> I'll say that you've done a great job in preparation for this interview. Do you do as well on all of them? [I try to.] I should say so. The information is accurate and the material very extensive.[16]

More often, the evaluations came in direct response to a question, as can be seen from this interchange:

INTERVIEWER: After you moved to C, did you continue to use samples made at B?

LAUREATE: Some, at the beginning. . . . You've read the history pretty thoroughly, haven't you?

An exchange with another biochemist reflects the same kind of appraisal:

---

16. This laureate later wrote to the sponsor of my dissertation to reiterate this statement.

INT.: X was working in Y's laboratory at the same time. Did you work together at all?

L.: You did your homework very well.

Two aspects of my remarks called forth these evaluations: in the first case, the particular piece of information was not widely known, and in the second, the question involved piecing together quite disparate bits of data. In a few instances, explicit positive evaluations seemed to have been partly motivated by the laureate's wish to be "supportive."

Another indicator of the adequacy of preparation is the "how do you know that?" type of response. Such responses were elicited in two forms: rhetorical appreciative questions indicating surprise, and direct questions regarding my sources of information. An example of the first came from a biochemist:

INT.: How did Dr. Z come to be involved?

L.: That's interesting. How do you come to ask? She worked with . . .

A physicist made a similar remark:

INT.: That was the period when you were working in solid state?

L.: Yes, how do you know that?

The second type of response required an answer from the interviewer and, in effect, constituted one type of test to which I was subjected. Positive evaluation came only when I had dealt with the question to the laureate's satisfaction. For example, a physicist:

INT.: You went on to write a paper with . . . on B potentials.

L.: Where did you find that?

INT.: In *Science Abstracts.*

L.: Oh, that's funny that you found that. It was never published. We prepared the paper and even sent in an abstract. . . . But [we found out] it was wrong.

Another physicist:

INT.: You wrote, of course, that he's played an important role in your career. In the biography in *Les Prix Nobel* . . . you mentioned X and Y and Z. . . .

L.: . . . I think in the text I say there was some collaboration with Z. Yes.

INT.: That's right. Work on crystals.

L.: [Goes to bookcase and gets volume.] Well, that isn't mentioned here. You must have access to some other . . . Yes, here. This is what you mean. Okay, that's true.

A biochemist:

INT.: You encountered some resistance to this work, didn't you? The Nobel citation reports that many people didn't think that . . .

L.: I don't think that the citation states anything about that. Where did you find that?

INT.: This is what was written. "X maintained for several years that Y compound . . . was the active principle, and he defended this idea against a growing skepticism of his colleagues."

L.: Yes, well, the skepticism was justified.

For some laureates, the testing of the interviewer was designed to determine the extent of her knowledge. For example, "Have you read Hadamard? [Yes.] You have? Good." Or:

L.: I don't know whether you know much about W.

INT.: I know that he is at R Laboratory, of course, and I know that you two did a book together.

L.: Very good.

This kind of test enabled the laureates to identify the appropriate level at which to answer the questions and, secondarily, reinforced their sense that the interviewer was not totally ignorant of some aspects of the world in which they live.

Another variety of testing involved interviewing the interviewer not only on the extent and accuracy of her knowledge but on the intent of her questions. One conversation went this way:

INT.: [Collaboration] seems to be a fairly critical issue right now. . . .

L.: In what way?

INT.: Well, for example, you mentioned declining satisfaction when you work with others. . . . And at the same time, the kind of

work you do seems to require it in some cases. And this puts
a certain amount of strain on scientists.

L.: It does and it's not a good thing. It takes a lot of joy out of
doing physics.

A fourth type of testing was more complex. Some laureates, irritated at finding themselves in the situation of being interviewed, seemed to be testing my tenacity. They expressed their irritation and may have hoped that this would lead to a quick termination of the interview. Sometimes this took the form of outright (though temporary) antagonism:

I never have any time. . . . That's because of . . . people like you.

Sometimes by more gentle indirection:

I have tried to assure the maximum of time [for research] but it
always gets cut into by committee meetings and . . . interviews.
That was mean but I couldn't resist. You know I really don't
mean it.

Or by this biochemist's initial expression of resistance to the project
as a whole:

L.: With the arrogance of the scientist, I should say that I don't
think it is possible to make a good study of collaboration
among scientists . . . unless one had had some work in natural
science. . . .

INT.: Well, as you know, we all labor under certain disadvantages.
My own sense is that you have to familiarize yourself as
much as possible with the kind of work you're studying.

L.: You can't do that very well at a distance. . . . Anyway, I guess
we oughtn't argue about that.

Occasionally, the laureates wanted to know if the interviewer would
stand her ground. The same biochemist said:

INT.: You received the prize for a whole career of work.

L.: It wasn't put that way.

INT.: Don't you believe that this was so?

L.: Yes, I guess it was.

In most cases, when I responded with a mixture of sympathy and the determination to continue, the mood of the laureates changed. Their responses became longer and more detailed. Sometimes their very postures changed to one of sprawling ease from one of uncomfortable vigilance, and they laughed more frequently.

Later in this chapter, I shall report how information gathered during the preparatory phase was used to frame questions for the interview. However, the present concern here has been to demonstrate the ways in which intensive preparation facilitated the *process* of interviewing top elites; it legitimated their expenditure of time by giving evidence of the interviewer's serious purpose and, to a degree, of her competence for the task at hand. Appropriate credentials were important for obtaining appointments for the interview but intensive study of relevant documentary materials is needed to give the interviewer some small insurance against the testing to which he or she is subjected.

## The Interview Guide

Even after initial pretesting, the interview guide was revised several times. When the interviewing began, the guide consisted of sets of questions designed to elicit detailed information on patterns of collaboration and individual work. The laureates were treated as interviewees on matters pertaining to their own collaboration and as informants in characterizing their fields and its customs.

During the first interview, it became obvious that the standard pattern of interviewing was inappropriate for this top elite. Answers to any one question tended to be elaborate and to cover a number of matters that appeared later in the guide. The standard and routine quality of the interview seemed to interfere with rapport. Members of this top elite are accustomed to being treated as individuals who have minds of their own, following their own bent. They are not receptive to being treated as just another interviewee, by the use of what they soon detect are standard questions rather than questions tailored to their particular interests and histories. They resent being encased in the straitjacket of entirely standardized questions. Nevertheless, as we shall see, the formal structure of the interview questions was retained in all interviews, even though the actual wording was adapted to the interests of each laureate. Since the interviews were used not as sources of data for quantitative analysis but rather as qualitative sources on the variety of experiences in science, the attempt to make the interviews strictly comparable by the use of fixed questions was abandoned. It seemed more sensible to focus on the issues most relevant for particular laureates, especially when the time available did not allow for intensive questioning on all issues. For example, certain laureates were known to have worked with a single collaborator over a long period. The interview, then, was designed to deal intensively with the character of collaborations of long duration and the ways in which these differ from short-term associations. In addition, reliance on an entirely standardized format did not easily permit the use of information uncovered during the preparation phase.

These considerations as well as great variations in the time available [17] for the interviews led to revisions in the interview guide. Certain questions were earmarked as essential for all laureates and others designed for particular laureates. After the third interview, the new procedure was used exclusively. By that time, the interview consisted of two major parts: the first dealing with detailed and concrete data based on the laureate's biography [18] and the second, of a relatively standard set of questions covering types of collaboration and individual work other than those already discussed, practices of authorship, characterizations of their fields, and responses to having won the prize. This format was not changed in a significant way for the rest of the interviews.

## A Note on the Use of Biographical Summaries

The summary sheets prepared for each interview were invaluable during the first part of the interview. They included information on where the laureate was located during various periods, lists of other scientists at work in the same place, the dates of major researches as well as the co-authors of the papers reporting them, and cases of multiple independent discoveries, if any. A second sheet consisted of a list of particular questions usually prompted by something the laureate had written. Relevant scientific terms and their definitions were also included. Before each interview, these summaries were carefully reviewed and much of their contents committed to memory. They were, however, always used for reference during the interview.

## Further Revisions of the Interview Guide

After twenty interviews were completed, the transcripts were analyzed in an attempt to locate important and recurrent themes so that new questions might be added to the standard portion of the guide.

---

17. On the average, the interviews lasted an hour. The shortest was thirty minutes and the longest three and one half hours.
18. This part of the interview resembled the technique described in Merton, Fiske, and Kendall (1956).

During this preliminary phase of analysis, the operation of the Matthew Effect was identified, particularly in the form of allocation of inordinate credit for joint research to eminent scientists at the expense of their less-well-known co-workers. Questions were added to elicit more detailed information on how this process worked as well as on structural mechanisms promoting autonomy of scientists and on ambivalence toward receiving the prize. Hyman (1964) has discussed the advantages of "successive" designs in which new hypotheses developing out of early stages of data collection are tested in later phases (see also Duijker and Rokkan, 1954). Successive designs do not yield strictly comparable data, but this was not a great problem in this investigation.

Some brief remarks should be made on the cumulative aspects of interviewing individuals who know one another. It was particularly useful for the study of scientific collaboration to have interviewed pairs and trios of co-workers.[19] These interviews provide valuable data on patterns of interaction and on the variety of ways in which particular events were perceived by people who played different roles in them. For that reason, interviews focused on occasions on which laureates had worked with one another. It was inevitable that the laureates would be aware that some of the questions being asked were based on remarks made by other laureates. This presented certain difficulties in maintaining the confidentiality of the interviews but it also appeared to motivate the laureates to report these events in detail so that the record would be complete. In this sense, the content of the interview guide was based on prior interviews as well as on earlier preparation.

The final structure of the guide for interviewing the remaining twenty-one laureates consisted of a biographical section and a relatively standard section comprised of questions designed to cover limited areas that had not been mentioned spontaneously earlier in the interview.

## Use of a Tape Recorder

As noted earlier, all interviews except one were recorded on tape.[20] The advantage of recording is obvious: it provided exact transcriptions

---

19. Pairs and trios included joint prize-winners and other laureates who had worked together on investigations that did not lead to the prize. A more extended investigation, beyond the scope of this one, should of course involve interviews with all those engaged in particular scientific collaborations.

20. One laureate preferred not to have the interview taped. Extensive notes were taken during this interview and were transcribed immediately afterward. He remarked that he "froze up" at the prospect of being recorded and assured the interviewer that he would give a better interview without it. No other laureate objected to the use of the tape recorder.

of what was said. The usual disadvantages of recording are less apt to occur when interviewing top elites, unlike some other types of interviewees. The laureates are accustomed to having their remarks recorded; some use tapes for their own work. They are not reticent and soon ignore the presence of the machine.[21] Tape recording turned out to be beneficial for both interviewer and interviewees. The full attention of the interviewer was focused on what was being said, and the usual pauses required to transcribe the contents of the interview did not intrude. The laureates were probably more articulate and discursive than they would have been if they were pressed to adjust the pace of their remarks to the speed at which the interviewer could take notes. The advantages of complete accuracy for a top-level interviewee have already been noted. The tapes themselves are, in addition, historical documents of some interest.[22]

## TECHNIQUES OF INTERVIEWING

The ideal interview is composed of a sequence of questions that are directly responsive to what the interviewee has already said and, at the same time, covers the pertinent items in the interview guide. In exploratory interviews, there is the additional problem of being alert enough to new themes so that they may be pursued even at the cost of sacrificing additional material on subjects that have already been given some measure of clarification.

Some laureates, of course, were less responsive to being interviewed than others. Their discomfort or irritation is usually evident in their remarks. Even when they did not express themselves explicitly, the cues were unmistakable. Monosyllabic replies were one indicator, as in this retort by a biochemist:

INT.: Do you find that you receive an inordinate amount of credit when you co-author a paper with a student?

L.: What do you think?

---

21. In several senses, the presence of the recorder acted as an ice-breaker. It happened to be a very attractive machine and many laureates asked about its operation and cost. During the interview, some expressed concern as to whether the machine was recording properly. A brief portion of the tape was occasionally played back to reassure them. On one occasion, when the recorder was not operating properly, the interviewee, a physicist, repaired it with dispatch.
22. Transcripts of these interviews are deposited in the Oral History Collection of Columbia University.
Some of the quotations included in this chapter, especially those concerning evaluation of the interviewer, may seem to indicate a lack of modesty on the part of the writer. They are intended to convey a sense of what occurred during the interviews and have been included with some embarrassment. Having the tapes provide some assurance that the transcripts are accurate.

There were also behavioral cues. At the beginning of one interview, the laureate sat about four feet from me in a chair on rollers. He began to retreat so that by the end of the interview he was at least ten feet from his original position. I tried to become sensitive to these signs of faltering interest or of hostility in order to make new efforts to interest or pacify the interviewee. There were, however, two or three cases in which no effort on my part greatly improved the quality of rapport.

## A Note on the Use of Technical Language

A further advantage of intensive preparation is growing familiarity with some of the technical language deployed by the laureates. In the early phase of most interviews, the laureates tried to avoid the use of language I might not understand. When given cues that he would be understood—particularly by my using such terms—the laureate relaxed and his vocabulary more closely approximated his usual one. In one such situation, a physicist remarked:

> I see that you have already acquired the terminology of the physicist.

Two types of "technical" language were used: terms and phrases employed in their scientific work and idiom used by scientists to describe their colleagues, their fields, and the experience of doing research. The latter type includes the terms "operator," "hot" and "quiet" fields, "taste," "theoretikers," "*Geheimrat* professor," and so forth. The scientific language as well as the trade argot was used to make the interviewee feel that he was not talking to a total alien. It was not intended to convey expertise on my part and did not seem to be perceived in that light. Inappropriate or awkward use of such terms only disrupts the interview, and so they were used with discretion.

## Types of Questions

Two techniques were used to give continuity to the interviews: questions were preceded by bridging remarks such as "you said earlier that . . ." and links between apparently discontinuous parts of the interview were clarified; these frequently took the form of comparative questions, for example: "How does . . . differ from . . .?" It is not always possible to provide continuity. On these occasions, this too was signaled, for example, by "I'd like to return to . . ." or "Now I would like to move to another problem." This top elite seemed thoroughly aware of discontinuities in the conversation, and it seemed sensible to let them know that I also was aware of abrupt shifts rather than hope that they would not be noticed.

As has been noted, specific questions based on prior preparation elicited detailed and concrete replies. These questions served several purposes beyond that of testifying to my competence. First, it was sometimes the case that the laureate had written on a pertinent subject in an instructive fashion but that further detail was needed. Only by concrete questions based on published materials could I begin to elaborate the differences—for example, between the collaboration one laureate had with an early co-worker and with his collaboration with his fellow laureate. This kind of questioning would typically begin with "You wrote in . . . that . . ."

Second, observations I had made during the course of my research formed the basis for another sort of question. One interchange with a biochemist exemplifies this:

INT.: I wanted to ask you, did you take any of the people in your lab with you to W? I notice there's an overlap of publication dates with R and T.

L.: . . . I believe when a man moves he should transfer, in a small way, a part of his environment. . . . You teach through the environment you've created.

A thorough knowledge of the career of the laureates enables one to ask questions that elicit replies useful for sociological analysis and tends to diminish the number of rambling responses that have no definite location in time and place.

Another technique used to call forth detailed replies involved phrasing questions or comments in rather extreme form. These usually elicited counterreplies of "No, no." The laureates tend to be unwilling to settle for only approximate versions of what they wish to convey. This sometimes involved rephrasing what they had said: "So one might say that . . ." The risks involved in using this technique are great; it sometimes interferes with the interviewee's sense that he is being understood. Even when it was used in its blandest form, the laureates occasionally expressed irritation and countered with, "I've already explained that."

This technique is, of course, quite different from situations in which the interviewer simply misunderstands or makes a mistake, such as confusing one scientific term for another. Errors of the latter kind seemed to have little effect on the interview; nonspecialists are not expected to have full command of (for them) specialized knowledge. In fact, a great deal of credit is given for even a small degree of scientific literacy but small prices are paid for lack of it. In other cases when I had misunderstood, the laureate immediately clarified the situation. A physicist replied to the following:

INT.: So . . . you . . . focused the attention of physicists on this problem?

L.: Let me say that this is a slightly unfair statement. I am not accusing you but what we did was the following.

Errors of this sort turned out to be fruitful when the interviewer's misapprehensions were corrected.

Even with so self-confident a group as Nobel laureates, the interviewer is occasionally obliged to provide social support. Several types of occasions called for supportive comments. Sometimes a laureate was not sure that he should continue with a particular observation. Especially when it was desirable to have him do so, I signaled this:

INT.: I'm interested [that] you've reflected on this.

L.: I haven't reflected on this for the interview but—this is off the cuff—I've always been conscious of these things. . . . It's very easy . . . somebody shows an interest and seems to understand you and then you get to talking too darned much— which I think I have.

INT.: Not at all.

Some questions were perceived as threatening, especially those that seemed to require the laureates to disclose details of conflicts between co-workers or that reminded them of distressing events in their own careers. It was possible in some cases to reassure the laureate that he was not being asked to reveal confidential information about particular people or events. A biochemist hesitated:

L.: It's not necessary to talk specifically about it.

INT.: Oh no. I'm much more interested in the types of occasions that make for conflict and what happens rather than in the people involved.

There were other occasions, however, when the laureates made it clear that they would not answer a particular question. A laureate in medicine replied:

INT.: You said that Dr. X was not the easiest man in the world to work with. What exactly do you mean?

L.: I don't know that I'd like to talk about that.

Another version of this occurred when the laureate cut himself short after telling only part of a story.

There was a lot more to that story than I've told and want to tell.

Questions on relative contributions of scientists to particular discoveries were sometimes sensitive. One physicist replied:

You'll have to wait until I publish my memoirs. After I die, you'll find out some of the answers to that question.

Further pursuit of that line would have been fruitless. So far as possible, the transition to new lines of questioning was casual and matter of fact. Long pauses to elicit more information were not used when the laureates had clearly indicated that they would not continue. Some sensitive areas for particular laureates were anticipated, and questioning on these was put off until the end of the interview. As it turned out, my fears that these might result in termination of the interview were unfounded. On the whole, the laureates have learned to evade, without much embarrassment, questions that they do not wish to answer.

The phrasing of questions also seemed important. The laureates responded with unusual precision to the wording of questions. For this reason, it is difficult to use the routine procedure of beginning with a general question and moving successively to more concrete issues. The laureates frequently asked for definitions of terms or supplied complex variations on the meaning of particular wording. At times, this served to clarify matters that I should have clarified for myself. For example, when asked if the Nobel Foundation had honored their "most important" work, several of them differentiated between work having the greatest immediate impact on their field, work most relevant for the later development of the field, work giving them the most understanding of the problems that interested them and work that turned out to be most challenging for them. Not only was it illuminating to discover that there might be little overlap between these types but the conceptual clarity that many of the laureates bring to describing experience turned out to be helpful in improving the interview guide.

A final problem involved the coherence of the standardized section of the interview guide. When selected questions had been discussed in detail in the early part of the interview and were not immediately followed by questions bearing on the same issue, attempts were made to bridge such as "I would like to return to this question of. . . ."

These techniques developed in the course of interviewing; the first few times they were used, I was not fully aware of their benefits and costs. Their effectiveness was judged on the basis of evaluated experience. Even now, their effects on the interviews as a whole have not been decisively estimated.

## Images of the Interviewer

Spontaneous comments by the laureates indicate that they saw me from three distinctive perspectives. I was, for some, an expert in my own field; for some, as having more than a layman's average knowledge in their own fields; and many perceived me as a part of the communications system that links the laureates.

Two sorts of indicators suggest that I was seen as an expert in the sociology of science. First were the remarks indicating some assumed differentials in knowledge—of the "you probably know this better than I" variety. And second were the direct questions of fact that they assumed I could answer. An example of the first came from a physicist:

> I don't know if science could exist without some sort of recognition. . . . I think that's a . . . question that you could probably tell much better than I.

The same man asked about the trends of multiauthored papers:

> Well, you must know the statistics. I don't know [them]. You have them? [Yes.] I'm curious to know what they are.

Or more specifically about Nobelists, another physicist asked:

> I don't know whether B is actually much younger than I am. You probably know this better.

On rare occasions, I could report something—in this case, a mere biographical fact—that the laureate did not know.

    L.: Do you know P?

    Int.: No . . . but of course . . . I know that C the chemistry laureate
          worked with him in E.

    L.: Oh, I didn't know that.

For some, seeing me as a social scientist meant that I was also seen as an outsider, someone who could not have had direct experience with doing science. A biochemist remarked:

> I don't think you understand. . . . I don't know why you should. People don't.

Responses of this kind were infrequent but, in part, they were a consequence of perceiving me as "only" a social scientist.

As the interviews proceeded and I gave some indication of knowing something about the laureates' work, it became clear that some of them assumed that I was knowledgeable in their fields. A physicist asked:

> But you seem to know a lot about what I did and what L did in physics. You have been studying it?

On the one hand, they assumed that I had read the relevant materials, "You probably read about that in X's article," although I had not mentioned that paper earlier. Or another said, "If you looked at . . . [a book by this laureate], as I think you did, you'll find that expressed." Another remarked, "You know a little chemistry. I won't go through it." They also assumed a familiarity with the names of other scientists and with their work. "You know about Uhlenbeck." In a less explicit way, this assumption was demonstrated over and over by mentions of names and research without qualifying or descriptive remarks.

All the laureates had been told that I had seen or would eventually be seeing their fellow prize-winners. Although the laureates are scattered around the country and are not in regular contact, many of course know one another. They often asked if I had seen some special friends, and some asked, as a physicist did in response to a question: "Did X tell you that?"

Protecting the anonymity of the interviewees was made difficult because they knew the other individuals involved in the investigation and were understandably curious about what others had said. A physicist asked:

> L.: Have you talked with . . . yet?
>
> INT.: Yes.
>
> L.: . . . Interesting. There's a problem there, isn't there?
>
> INT.: Well —
>
> L.: Yes. Okay. Much more interesting than my problems.

Another physicist saw me as a direct link between the laureates. After describing a particularly complicated episode, he added:

> L.: Don't tell Y about that when you talk to him. I'm not sure I ever told him about it. I don't want to take the chance of you telling him.
>
> INT.: I wouldn't think of it.

Most laureates understood immediately that I was not free to report what others had said and also expected that what they said would be treated confidentially. A biochemist remarked:

> I have some idea of what kind of person you are and I do not think you would do that [quote people without their permission].

The interview is a social situation in which both parties develop a set of images of one another and use these as guides to behavior. Interviews are, however, quite special, for the interviewee is asked to give far more information about himself and has access to fewer explicit cues from the interviewer than would usually be the case. In spite of this, the laureates obviously made attempts to locate me socially, to evaluate my performance by devising tests, and to develop a sense of my personal qualities.

## Interviewing Top Elites and Others

It is prosaic but true that most interviewees appreciate being treated as individuals. Early interviews with the laureates indicate that standardized interviewing is less effective with this top elite than procedures designed with their special competences in mind. Nobel prize-winners in science were, in general, eager to discuss the nonscientific aspects of their work. By contrast, some are irked by requests for opinions on subjects that one described as ranging "from air pollution to population control," subjects they do not feel particularly able to handle or in which they have little interest. Ordinary interviewees are more apt to discuss issues about which they have little information.[23] The laureates' responses indicated extraordinarily detailed perceptions of questions. Many were acutely aware of what they had said previously and clarified the relations between disparate sets of ideas expressed at different times

23. Hyman and Sheatsley (1950) remark on the willingness of Americans to express opinions "on ... every conceivable issue" despite their lack of information.

in the interview. Their replies contained almost endless qualifications and explicit references to constraining conditions. As a group, they were not comfortable with loose generalizations. Once they decided that they did not wish to answer particular questions, standard techniques of eliciting further information were not at all useful. On the other hand, they were sensitive to behavioral cues of appreciation and would usually elaborate with little prodding.

In all cases, the differences in rank between the laureate and the interviewer impressed itself on the situation. This is not to say that rank differences were treated invidiously, although greater difficulties probably would have arisen had I been a graduate student in science instead of being in a wholly unrelated field. However, the frequency of evaluative statements by the laureates indicated that they saw themselves as judges and saw me as the object of judgment. Nevertheless, some of the laureates did ask for evaluations of their performance as interviewees. A physicist queried:

How much of this do you want?

And a chemist:

Are you interested in knowing something about the background of my co-workers?

This type of response seemed to be directed not so much at pleasing me as at eliciting some measure of their effectiveness in meeting the purpose of the interview, with the intent to change their performance if that seemed appropriate. The laureates, in this sense, differed from interviewees who characteristically develop an acquiescence set. The prizewinners were not concerned about impressing me with their own importance. They tend to be aware of their eminence and to be at home with it.


## SUMMARY

By all measures, the Nobel laureates are located at the uppermost level of the prestige hierarchy of science. As a top elite, they are accustomed to expressing themselves publicly. They are anxious to use their time profitably and expect competence and seriousness of purpose from those who make demands on them. Intensive preparation by the interviewer is an effective indicator of commitment and is useful in framing appropriate lines of questioning. Top elites are critical of the interviewer's performance as well as of their own. Their responses tend to be both elaborate and precise. They are sensitive to the interviewer's language and behavioral cues. They wish to perform effectively as interviewees as they do in other roles.

Appendix B

# NOBEL LAUREATES IN SCIENCE, 1901–76

## NOBEL PRIZES FOR PHYSICS

**1901** Wilhelm K. Roentgen (German) for discovering X rays.

**1902** Hendrik Antoon Lorentz and Pieter Zeeman (Dutch) for discovering the Zeeman effect of magnetism on light.

**1903** Antoine Henri Becquerel and Pierre and Marie Curie (French) for discovering radioactivity and studying uranium.

**1904** Baron Rayleigh (British) for studying the density of gases and discovering argon.

**1905** Philipp Lenard (German) for studying the properties of cathode rays.

**1906** Sir Joseph John Thomson (British) for studying electrical discharge through gases.

**1907** Albert A. Michelson (American) for inventing optical instruments and measuring the speed of light.

**1908** Gabriel Lippmann (French) for his method of color photography.

**1909** Guglielmo Marconi (Italian) and Karl Ferdinand Braun (German) for developing the wireless telegraph.

**1910** Johannes D. van der Waals (Dutch) for studying the relationships of liquids and gases.

**1911** Wilhelm Wien (German) for his discoveries on the heat radiated by black objects.

**1912** Nils Dalén (Swedish) for inventing automatic gas regulators for lighthouses.

**1913** Heike Kamerlingh Onnes (Dutch) for experimenting with low temperatures and liquefying helium.

**1914** Max T. F. von Laue (German) for using crystals to measure X rays.

**1915** Sir William Henry Bragg and Sir William L. Bragg (British) for using X rays to study crystal structure.

**1916** No Award

**1917** Charles Barkla (British) for studying the diffusion of light and the radiation of X rays from elements.

**1918** Max Planck (German) for stating the quantum theory of light.

**1919** Johannes Stark (German) for discovering the Stark effect of spectra in electric fields.

**1920** Charles E. Guillaume (French) for discovering nickel-steel alloys with slight expansion, and the alloy invar.

**1921** Albert Einstein (German) for contributing to mathematical physics and stating the law of the photoelectric effect.

**1922** Niels Bohr (Danish) for studying the structure of atoms and their radiations.

**1923** Robert A. Millikan (American) for measuring the charge on electrons and working on the photoelectric effect.

**1924** Karl M. G. Siegbahn (Swedish) for working with the X-ray spectroscope.

**1925** James Franck and Gustav Hertz (German) for stating laws on the collision of an electron with an atom.

**1926** Jean Baptiste Perrin (French) for studying the discontinuous structure of matter and measuring the sizes of atoms.

**1927** Arthur H. Compton (American) for discovering the Compton effect on X rays reflected from atoms, and Charles T. R. Wilson (British) for discovering a method for tracing the paths of ions.

**1928** Owen W. Richardson (British) for studying thermionic effect and electrons sent off by hot metals.

**1929** Louis Victor de Broglie (French) for discovering the wave character of electrons.

**1930** Sir Chandrasekhara 'Venkata

From *The World Book Encyclopedia.* © 1976 Field Enterprises Educational Corporation. Nationality designations are based on citizenship at the time of the award.

Raman (Indian) for discovering a new effect in radiation from elements.

**1931** No Award

**1932** Werner Heisenberg (German) for founding quantum mechanics, which led to discoveries in hydrogen.

**1933** Paul Dirac (British) and Erwin Schrödinger (Austrian) for discovering new forms of atomic theory.

**1934** No Award

**1935** Sir James Chadwick (British) for discovering the neutron.

**1936** Carl David Anderson (American) for discovering the positron, and Viktor F. Hess (Austrian) for discovering cosmic rays.

**1937** Clinton Davisson (American) and George Thomson (British) for discovering the diffraction of electrons by crystals.

**1938** Enrico Fermi (Italian) for discovering new radioactive elements beyond uranium.

**1939** Ernest O. Lawrence (American) for inventing the cyclotron and working on artificial radioactivity.

**1940–1942** No Award

**1943** Otto Stern (American) for discovering the molecular beam method of studying the atom.

**1944** Isidor Isaac Rabi (American) for recording the magnetic properties of atomic nuclei.

**1945** Wolfgang Pauli (Austrian) for discovering the exclusion principle (Pauli principle) of electrons.

**1946** Percy Williams Bridgman (American) for his work in the field of very high pressures.

**1947** Sir Edward V. Appleton (British) for exploring the ionosphere.

**1948** Patrick M. S. Blackett (British) for his discoveries in cosmic radiation.

**1949** Hideki Yukawa (Japanese) for discovering the meson.

**1950** Cecil Frank Powell (British) for his photographic method of studying atomic nuclei and his discoveries concerning mesons.

**1951** Sir John D. Cockcroft (British) and Ernest T. S. Walton (Irish) for working on the transmutation of atomic nuclei by artificially accelerated atomic particles.

**1952** Felix Bloch and Edward Mills Purcell (American) for developing magnetic measurement methods for atomic nuclei.

**1953** Frits Zernike (Dutch) for inventing the phase contrast microscope.

**1954** Max Born (German) for research in quantum mechanics, and Walther Bothe (German) for discoveries he made with his coincidence method.

**1955** Willis E. Lamb, Jr. (American), for discoveries on the structure of the hydrogen spectrum, and Polykarp Kusch (American) for determining the magnetic moment of the electron.

**1956** John Bardeen, Walter H. Brattain, and William Shockley (American) for inventing the transistor.

**1957** Tsung Dao Lee and Chen Ning Yang (Chinese, worked in the United States) for disproving the law of conservation of parity.

**1958** Pavel A. Cherenkov, Ilya M. Frank, and Igor Y. Tamm (Russian) for discovering and interpreting the Cherenkov effect in studying high-energy particles.

**1959** Emilio Segrè and Owen Chamberlain (American) for their work in demonstrating the existence of the antiproton.

**1960** Donald A. Glaser (American) for inventing the bubble chamber to study subatomic particles.

**1961** Robert Hofstadter (American) for his studies of nucleons, and Rudolf L. Mössbauer (German) for his research on gamma rays.

**1962** Lev Davidovich Landau (Russian) for his research on liquid helium gas.

**1963** Eugene Paul Wigner (American) for his contributions to the understanding of atomic nuclei and elementary particles, and Maria Goeppert Mayer (American) and J. Hans Jensen (German) for their work on the structure of atomic nuclei.

**1964** Charles H. Townes (American), Nikolai G. Basov and Alexander M. Prokhorov (Russian) for work in quantum electronics leading to development of the maser and laser.

**1965** Sin-itiro Tomonaga (Japanese) and Julian S. Schwinger and Richard P. Feynman (American) for basic work in quantum electrodynamics.

**1966** Alfred Kastler (French) for his work on the energy level of atoms.

**1967** Hans Albrecht Bethe (American) for his contributions to the theory of nuclear reactions, especially his discoveries on the energy production in stars.

**1968** Luis W. Alvarez (American) for his contributions to the knowledge of subatomic particles.

**1969** Murray Gell-Mann (American) for his discoveries concerning the classification of nuclear particles and their interactions.

**1970** Hannes Olof Gosta Alfvén (Swedish) for his work in *magnetohydrodynamics,* the study of electrical and magnetic effects in fluids that conduct electricity, and Louis Eugène Félix Néel (French) for his

discoveries of magnetic properties that applied to computer memories.

**1971** Dennis Gabor (British) for his developmental work in *holography,* a method of making a three-dimensional photograph with coherent light from a laser.

**1972** John Bardeen, Leon N. Cooper, and John Robert Schrieffer (American) for their work on *superconductivity,* the disappearance of electrical resistance.

**1973** Ivar Giaever (American), Leo Esaki (Japanese), and Brian Josephson (British) for their work concerning the phenomena of electron "tunneling" through semiconductor and superconductor materials.

**1974** Antony Hewish (British) for the discovery of *pulsars,* celestial objects that give off bursts of radio waves, and Sir Martin Ryle (British) for his use of small radio telescopes to "see" into space with great accuracy.

**1975** Aage N. Bohr and Ben R. Mottelson (Danish) and James Rainwater (American) for their research on nonspherical atomic nuclei.

**1976** Burton Richter and Samuel C. C. Ting (American) for independently discovering the J or psi particle, a unique form of subatomic matter.

## NOBEL PRIZES FOR CHEMISTRY

**1901** Jacobus Henricus van't Hoff (Dutch) for discovering laws of chemical dynamics and osmotic pressure.

**1902** Emil Fischer (German) for synthesizing sugars, purine derivatives, and peptides.

**1903** Svante August Arrhenius (Swedish) for his dissociation theory of ionization in electrolytes.

**1904** Sir William Ramsay (British) for discovering helium, neon,

xenon, and krypton, and determining their place in the periodic system.

**1905** Adolph von Baeyer (German) for his work on dyes and organic compounds, and for synthesizing indigo and arsenicals.

**1906** Henri Moissan (French) for preparing pure fluorine and developing the electric furnace.

**1907** Eduard Buchner (German) for his biochemical researches and for discovering cell-less fermentation.

**1908** Ernest Rutherford (British) for discovering that alpha rays break down atoms and studying radioactive substances.

**1909** Wilhelm Ostwald (German) for his work on catalysis, chemical equilibrium, and the rate of chemical reactions.

**1910** Otto Wallach (German) for his work in the field of alicyclic substances.

**1911** Marie Curie (French) for discovering radium and polonium, and for isolating radium and studying its compounds.

**1912** François Auguste Victor Grignard (French) for discovering the Grignard reagent to synthesize organic compounds, and Paul Sabatier (French) for his method of adding hydrogen to organic compounds, using metals as catalysts.

**1913** Alfred Werner (Swiss) for his coordination theory on the arrangement of atoms.

**1914** Theodore W. Richards (American) for determining the atomic weights of many elements.

**1915** Richard Willstätter (German) for his research on chlorophyll and other coloring matter in plants.

**1916–1917** No Award

**1918** Fritz Haber (German) for the Haber-Bosch process of synthesizing ammonia from nitrogen and hydrogen.

**1919** No Award

**1920** Walther Nernst (German) for his discoveries concerning heat changes in chemical reactions.

**1921** Frederick Soddy (British) for studying radioactive substances and isotopes.

**1922** Francis W. Aston (British) for discovering many isotopes by means of the mass spectrograph and discovering the whole number rule on the structure and weight of atoms.

**1923** Fritz Pregl (Austrian) for inventing a method of microanalyzing organic substances.

**1924** No Award

**1925** Richard Zsigmondy (German) for his method of studying colloids.

**1926** Theodor Svedberg (Swedish) for his work on dispersions and on colloid chemistry.

**1927** Heinrich O. Wieland (German) for studying gall acids and related substances.

**1928** Adolf Windaus (German) for studying sterols and their connection with vitamins.

**1929** Sir Arthur Harden (British) and Hans August Simon von Euler-Chelpin (German) for their research on sugar fermentation and enzymes.

**1930** Hans Fischer (German) for studying the coloring matter of blood and leaves and synthesizing hemin.

**1931** Carl Bosch and Friedrich Bergius (German) for inventing high-pressure methods of manufacturing ammonia and liquefying coal.

**1932** Irving Langmuir (American) for his discoveries about molecular films absorbed on surfaces.

**1933** No Award

**1934** Harold Clayton Urey (American) for discovering deuterium (heavy hydrogen).

**1935** Frédéric Joliot and Irène Joliot-Curie (French) for synthesizing new radioactive elements.

**1936** Peter J. W. Debye (Dutch) for his studies on molecules, dipole moments, the diffraction of electrons, and X rays in gases.

**1937** Sir Walter N. Haworth (British) for his research on carbohydrates and vitamin C, and Paul Karrer (Swiss) for studying carotenoids, flavins, and vitamins A and $B_2$.

**1938** Richard Kuhn (German) for his work on carotenoids and vitamins (declined).

**1939** Adolf Butenandt (German) for studying the chemistry of sex hormones (declined), and Leopold Ružička (Swiss) for his work on polymethylenes.

**1940–1942** No Award

**1943** Georg von Hevesy (Hungarian)

for using isotopes as indicators in chemistry.

**1944** Otto Hahn (German) for his discoveries in atomic fission.

**1945** Artturi Virtanen (Finnish) for inventing new methods in agricultural biochemistry.

**1946** James B. Sumner (American) for discovering that enzymes can be crystallized, and Wendell M. Stanley and John H. Northrop (American) for preparing enzymes and virus proteins in pure form.

**1947** Sir Robert Robinson (British) for his research on biologically significant plant substances.

**1948** Arne Tiselius (Swedish) for his discoveries on the nature of the serum proteins.

**1949** William Francis Giauque (American) for studying reactions to extreme cold.

**1950** Otto Diels and Kurt Alder (German) for developing a method of synthesizing organic compounds of the diene group.

**1951** Edwin M. McMillan and Glenn T. Seaborg (American) for discovering plutonium and other elements.

**1952** Archer J. P. Martin and Richard Synge (British) for developing the partition chromatography process, a method of separating compounds.

**1953** Hermann Staudinger (German) for discoveries in macromolecular chemistry.

**1954** Linus Pauling (American) for his work on the forces that hold matter together.

**1955** Vincent Du Vigneaud (American) for discovering a process for making synthetic hormones.

**1956** Sir Cyril Hinshelwood (British) and Nikolai N. Semenov (Russian) for their work on chemical chain reactions.

**1957** Lord Todd (British) for his work on the protein composition of cells.

**1958** Frederick Sanger (British) for discovering the structure of the insulin molecule.

**1959** Jaroslav Heyrovský (Czech) for developing the polaro-

graphic method of analysis.

**1960** Willard F. Libby (American) for developing a method of radiocarbon dating.

**1961** Melvin Calvin (American) for his research on photosynthesis.

**1962** John Cowdery Kendrew and Max Ferdinand Perutz (British) for studies on globular proteins.

**1963** Giulio Natta (Italian) for his contributions to the understanding of *polymers,* and Karl Ziegler (German) for his production of *organometallic compounds.* The work of both men led to the production of improved plastics products.

**1964** Dorothy C. Hodgkin (British) for X-ray studies of compounds such as vitamin $B_{12}$ and penicillin.

**1965** Robert Burns Woodward (American) for his contributions to organic synthesis.

**1966** Robert S. Mulliken (American) for developing the *molecular-orbital* theory of chemical structure.

**1967** Manfred Eigen (German) and Ronald G. W. Norrish and George Porter (British) for developing techniques to measure rapid chemical reactions.

**1968** Lars Onsager (American) for developing the theory of reciprocal relations of various kinds of thermodynamic activity.

**1969** Derek H. R. Barton (British) and Odd Hassel (Norwegian) for their studies relating chemical reactions with the three-dimensional shape of molecules.

**1970** Luis Federico Leloir (Argentine) for his discovery of chemical compounds that affect the storage of chemical energy in living things.

**1971** Gerhard Herzberg (Canadian) for his research in the structure of molecules, particularly the fragments of some molecules called *free radicals.*

**1972** Christian B. Anfinsen, Stanford Moore, and William H. Stein (American) for their fundamental contributions to the chemistry of *enzymes,* basic substances of living things.

**1973** Geoffrey Wilkinson (British) and Ernst Fischer (German) for their work on *organometallic compounds,* substances which consist of organic compounds and metal atoms.

**1974** Paul John Flory (American) for his work in polymer chemistry.

**1975** John Warcup Cornforth (Australian, worked in Britain) for stereochemical investigations of enzyme-catalyzed reactions and Vladimir Prelog (Swiss) for research on the stereochemistry of organic molecules and reactions.

**1976** William N. Lipscomb (American) for his studies of the structure and bonding of boron hydrides and their derivatives.

# NOBEL PRIZES FOR PHYSIOLOGY OR MEDICINE

**1901** Emil von Behring (German) for discovering the diphtheria antitoxin.

**1902** Sir Ronald Ross (British) for work on malaria and discovering how malaria is transmitted.

**1903** Niels Ryberg Finsen (Danish) for treating diseases, especially *lupus vulgaris,* with concentrated light rays.

**1904** Ivan Petrovich Pavlov (Russian) for his work on the physiology of digestion.

**1905** Robert Koch (German) for work on tuberculosis and discovering the tubercule bacillus and tuberculin.

**1906** Camillo Golgi (Italian) and Santiago Ramón y Cajal (Spanish) for their studies of nerve tissue.

**1907** Charles Louis Alphonse Laveran (French) for studying diseases caused by protozoa.

**1908** Paul Ehrlich (German) and Élie Metchnikof (Russian) for their work on immunity.

**1909** Emil Theodor Kocher (Swiss) for his work on the physiology, pathology, and surgery of the thyroid gland.

**1910** Albrecht Kossel (German) for studying cell chemistry, proteins, and nucleic substances.

**1911** Allvar Gullstrand (Swedish) for his work on dioptrics, the refraction of light through the eye.

**1912** Alexis Carrel (French) for suturing blood vessels and grafting vessels and organs.

**1913** Charles Robert Richet (French) for studying allergies caused by foreign substances, as in hay fever.

**1914** Robert Bárány (Austrian) for work on function and diseases of equilibrium organs in the inner ear.

**1915–1918** No Award

**1919** Jules Bordet (Belgian) for discoveries on immunity.

**1920** August Krogh (Danish) for discovering the system of action of blood capillaries.

**1921** No Award

**1922** Archibald V. Hill (British) for his discovery on heat production in the muscles, and Otto Meyerhof (German) for his theory on the production of lactic acid in the muscles.

**1923** Sir Frederick Grant Banting (Canadian) and John J. R. Macleod (Scottish) for discovering insulin.

**1924** Willem Einthoven (Dutch) for inventing the electrocardiograph.

**1925** No Award

**1926** Johannes Fibiger (Danish) for discovering a parasite that causes cancer.

**1927** Julius Wagner von Jauregg (Austrian) for discovering the fever treatment for paralysis.

1928 Charles Nicolle (French) for his work on typhus.
1929 Christiaan Eijkman (Dutch) for discovering vitamins that prevent beriberi, and Sir Frederick G. Hopkins (British) for discovering vitamins that help growth.
1930 Karl Landsteiner (American) for discovering the four main human blood types.
1931 Otto H. Warburg (German) for discovering that enzymes aid in respiration by tissues.
1932 Edgar D. Adrian and Sir Charles S. Sherrington (British) for discoveries on the function of neurons.
1933 Thomas H. Morgan (American) for studying the function of chromosomes in heredity.
1934 George Minot, William P. Murphy, and George H. Whipple (American) for their discoveries on liver treatment for anemia.
1935 Hans Spemann (German) for discovering the organizer-effect in the growth of an embryo.
1936 Sir Henry H. Dale (British) and Otto Loewi (Austrian) for their discoveries on the chemical transmission of nerve impulses.
1937 Albert Szent-Györgyi (Hungarian) for his discoveries in connection with oxidation in tissues, vitamin C, and fumaric acid.
1938 Corneille Heymans (Belgian) for his discoveries concerning the regulation of respiration.
1939 Gerhard Domagk (German) for discovering prontosil, the first sulfa drug (declined).
1940–1942 No Award
1943 C. P. Henrik Dam (Danish) for discovering vitamin K, and Edward Doisy (American) for synthesizing it.
1944 Joseph Erlanger and Herbert Gasser (American) for their work on single nerve fibers.
1945 Sir Alexander Fleming, Howard W. Florey, and Ernst B. Chain (British) for discovering penicillin.
1946 Hermann Joseph Muller (American) for discovering that X rays can produce mutations.
1947 Carl F. and Gerty Cori (American) for their work on glycogen conversion, and Bernardo Houssay (Argentine) for studying the pancreas and the pituitary gland.
1948 Paul Mueller (Swiss) for discovering the insect-killing properties of DDT.
1949 Walter R. Hess (Swiss) for discovering how certain parts of the brain control organs of the body, and Antônio E. Moniz (Portuguese) for originating prefrontal lobotomy.
1950 Philip S. Hench, Edward C. Kendall (American), and Tadeus Reichstein (Swiss) for their discoveries on cortisone and ACTH.
1951 Max Theiler (South African who worked in the United States) for developing the yellow fever vaccine known as 17-D.
1952 Selman A. Waksman (American) for his work in the discovery of streptomycin.
1953 Fritz Albert Lipmann (American) and Hans Adolf Krebs (British) for their discoveries in biosynthesis and metabolism.
1954 John F. Enders, Thomas H. Weller and Frederick C. Robbins (American) for discovering a simple method of growing polio virus in test tubes.
1955 Hugo Theorell (Swedish) for his discoveries on the nature and action of oxidation enzymes.
1956 André F. Cournand, Dickinson W. Richards, Jr. (American), and Werner Forssmann (German) for using a catheter to chart the heart's interior.
1957 Daniel Bovet (Italian) for discovering antihistamines.
1958 George Wells Beadle and Edward Lawrie Tatum (American) for their work in biochemical genetics, and Joshua Lederberg (American) for his studies of genetics in bacteria.
1959 Severo Ochoa and Arthur

Kornberg (American) for producing nucleic acids by artificial means.

**1960** Sir Macfarlane Burnet (Australian) and Peter B. Medawar (British) for research in immunity reactions.

**1961** Georg von Békésy (American) for demonstrating how the ear distinguishes between various sounds.

**1962** James D. Watson (American) and Francis H. Crick and Maurice H. F. Wilkins (British) for their work on DNA structure.

**1963** Sir John Carew Eccles (Australian) for his research on the transmission of nerve impulses, and Alan Lloyd Hodgkin (British) and Andrew Fielding Huxley (British) for their description of the behavior of nerve impulses.

**1964** Konrad E. Bloch (American) and Feodor Lynen (German) for their work on cholesterol and fatty acid metabolism.

**1965** François Jacob, André Lwoff, and Jacques Monod (French) for their discoveries concerning genetic control of enzyme and virus synthesis.

**1966** Francis Peyton Rous (American) for discovering a cancer-producing virus, and Charles B. Huggins (American) for discovering uses of hormones in treating cancer.

**1967** Ragnar Granit (Swedish) and H. Keffer Hartline and George Wald (American) for their work on chemical and physiological processes in the eye.

**1968** Robert W. Holley, H. Gobind Khorana, and Marshall W. Nirenberg (American) for explaining how genes determine the function of cells.

**1969** Max Delbrück, Alfred Hershey, and Salvador Luria (American) for their work with *bacteriophages.*

**1970** Julius Axelrod (American), Bernard Katz (British), and Ulf Svante von Euler (Swedish) for their discoveries of the role played by certain chemicals in the transmission of nerve impulses.

**1971** Earl W. Sutherland, Jr. (American), for his discovery of the ways hormones act, including the discovery of cyclic AMP, a chemical that influences the actions of hormones on body processes.

**1972** Gerald M. Edelman (American) and Rodney R. Porter (British) for their discovery of the chemical structure of antibodies.

**1973** Nikolaas Tinbergen (Dutch-born) and Konrad Z. Lorenz and Karl von Frisch (Austrian) for their studies on animal behavior.

**1974** Christian de Duve (Belgian) and Albert Claude and George E. Palade (American) for their pioneer work in cell biology.

**1975** David Baltimore (American), Renato Dulbecco (Italian, worked in the United States), and Howard M. Temin (American) for their discoveries on the interaction of tumor viruses and the cell's genetic material.

**1976** Baruch S. Blumberg (American) for his investigations on hepatitis B virus, and D. Carlton Gajdusek (American) for his research on slow acting viruses and their role in disease.

## NOBEL PRIZES FOR ECONOMICS*

**1969** Ragnar Frisch (Norwegian) and Jan Tinbergen (Dutch) for their work in *econometrics,* the developing of mathematical models to analyze economic activity.

* The prize in economics was established in 1969.

**1970** Paul A. Samuelson (American) for his efforts to raise the level of scientific analysis in economic theory.

**1971** Simon Kuznets (American) for his interpretation of economic growth and its measurement.

**1972** Kenneth J. Arrow (American) and Sir John Hicks (British) for their pioneering contribution to general equilibrium theory and welfare theory.

**1973** Wassily Leontief (American) for his development of the input-output method of economic analysis.

**1974** Friedrich von Hayek (Austrian) and Gunnar Myrdal (Swedish) for work in the theory of money and economic change and in the analysis of the relationship between economic and social factors.

**1975** Tjalling C. Koopmans (American) and Leonid V. Kantorovich (Russian) for their contributions to the theory of optimum allocation of resources.

**1976** Milton Friedman (American) for his research in consumption analysis, monetary theory, and stabilization policy.

# Appendix C

# PRIZE-WINNING RESEARCH:
# SPECIALTY AND
# YEAR OF AWARD

**Table C–1. Laureates in Physics, According to Specialty and Year of Award (1901–72)**

| Specialty of Prize-Winning Research in Physics [a] | 1901–19 | 1920–39 | 1940–59 | 1960–72 | Total |
|---|---|---|---|---|---|
| Nuclear structure and reactions; experimental methods in nuclear physics | — | 2 | 6 | 6 | 14 |
| Cosmic radiation and elementary particles | — | 3 | 7 | 3 | 13 |
| Radio waves, solid state electronics, quantum electronics | 2 | — | 4 | 7 | 13 |
| X-rays in atomic research | 4 | 4 | 1 | — | 9 |
| Magnetic quantum effects; quantum electrodynamics | — | — | 6 | 3 | 9 |
| Atomic theory applied to gases and liquids; low temperatures and high pressures | 3 | 1 | 1 | 1 | 6 |
| Quantum nature of energy | 2 | 4 | — | — | 6 |
| New quantum theories and the quantized atomic model | — | 4 | 2 | — | 6 |
| Radioactivity | 4 | — | — | — | 4 |
| Effects of magnetic fields on optical spectra | 4 | — | — | — | 4 |
| Physics and technology | 2 | 1 | — | 1 | 4 |
| Waves as quanta, particles as waves | — | 3 | — | — | 3 |
| Electron: charge and mass | 2 | — | — | — | 2 |
| Electrons from metals | — | 1 | — | — | 1 |
| Magneto-hydrodynamics | — | — | — | 1 | 1 |
| Antiferromagnetism | — | — | — | 1 | 1 |
| Totals | (23) | (23) | (27) | (23) | (96) |

a. Classification of specialties represented by prize-winning research from Nobelstiftelsen (1972, pp. 387–475).

**Table C–2. Laureates in Chemistry, According to Specialty and Year of Award (1901–72)**

| Specialty of Prize-Winning Research in Chemistry [a] | 1901–19 | 1920–39 | 1940–59 | 1960–72 | Total |
|---|---|---|---|---|---|
| Constitution of organic compounds and biochemistry | 3 | 10 | 8 | 12 | 33 |
| Radioactivity and atomic chemistry | 2 | 5 | 4 | 1 | 12 |
| Chemical change | 3 | — | 2 | 3 | 8 |
| Chemical thermodynamics and technical applications | 2 | 3 | 1 | 1 | 7 |
| Colloids, chromatography, and surface chemistry | — | 3 | 3 | — | 6 |

**Table C–2.    (Continued)**

| Specialty of Prize-Winning Research in Chemistry[a] | 1901–19 | 1920–39 | 1940–59 | 1960–72 | Total |
|---|---|---|---|---|---|
| Molecular structure, valence forces, X-ray and electron interference by gases and vapors | 1 | 1 | 1 | 2 | 5 |
| Preparative organic chemistry | 2 | — | 2 | 1 | 5 |
| Molecular structure of organic substances | 2 | — | — | — | 2 |
| Inorganic chemistry | 2 | — | — | — | 2 |
| Microchemistry | — | 1 | 1 | — | 2 |
| Agricultural chemistry | — | — | 1 | — | 1 |
| Totals | (17) | (23) | (23) | (20) | (83) |

a. Classification of specialties represented by prize-winning research from Nobelstiftelsen (1972, pp. 279–384).

**Table C–3. Laureates in Physiology or Medicine, According to Specialty and Year of Award (1901–72)**

| Specialty of Prize-Winning Research in Physiology or Medicine[a] | 1901–19 | 1920–39 | 1940–59 | 1960–72 | Total |
|---|---|---|---|---|---|
| Classical genetics and molecular biology | 1 | 1 | 6 | 14 | 22 |
| Microbiology and immunology | 6 | 1 | 4 | 2 | 13 |
| Intermediary metabolism | — | 4 | 5 | 2 | 11 |
| Digestion, respiration, and circulation | 2 | 3 | 3 | — | 8 |
| Hormones | 1 | 2 | 4 | 1 | 8 |
| Vitamins | — | 5 | 2 | — | 7 |
| Transmission in nerve fibers and at synaptic junctions | — | 1 | 2 | 3 | 6 |
| Sensory physiology | 2 | — | — | 4 | 6 |
| Autonomic nervous functions and chemical transmitters | — | 2 | 1 | 3 | 6 |
| Chemotherapy and neuropharmacology | — | 1 | 5 | — | 6 |
| Classical neuroanatomy, neurophysiology, and neurosurgery | 2 | 1 | 1 | — | 4 |
| Insect-borne infections | 2 | 1 | — | — | 3 |
| Tumors | — | 1 | — | 2 | 3 |
| Photo- and fever-therapies | 1 | 1 | — | — | 2 |
| Developmental mechanics | — | 1 | — | — | 1 |
| Insecticides | — | — | 1 | — | 1 |
| Totals | (17) | (25) | (34) | (31) | (107) |

a. Classification of specialties represented by prize-winning research from Nobelstiftelsen (1972, pp. 139–278).

Appendix D

# OFFICIAL OCCUPANTS
# OF THE FORTY-FIRST CHAIR:
# "HONORABLE MENTIONS"
# FOR NOBEL PRIZES

Occupants of the forty-first chair are scientists who have made contributions as significant as the laureates' but who have not been selected for Nobel prizes. In principle, all serious candidates for the prizes would be included among the occupants of the forty-first chair, as would their peers who have been statutorily excluded for one reason or another. In practice, the Foundation's commitment to keeping committee and academy deliberations secret makes this criterion impractical for identifying occupants of the forty-first chair.

The customary silence about candidates for the prizes was partially abandoned in the 1962 edition of *Alfred Nobel: The Man and His Prizes* (and, as I noted, the silence was reinstituted in the 1972 edition of the same volume except where the earliest selections are described). Written by the official historians of the awards, Arne Westgren (for chemistry) and Gorän Liljestrand (for the biomedical sciences), the essays provide information on research the committees judged "prize-worthy" but that were not cited for awards. Thus, the roster of "honorable mentions" for the awards (occupants of the forty-first chair officially designated as the peers of the laureates) includes all those named in the Westgren and Liljestrand papers. In all, they number sixty-nine, of whom thirty-three were alive in 1962, when their names were made public, and who presumably were then contenders for the awards. Of these, fourteen won Nobel prizes between 1963 and 1975. The remaining honorable mentions fall into three categories: scientists now dead and doomed to be permanent honorable mentions, scientists still alive but whose contributions were only temporarily considered prize-worthy, and, finally, living scientists who may still be contenders for the awards. Whether any of these scientists will become laureates will depend on whether their contributions continue to be scientifically consequential, whether more deserving candidates emerge, and, of course, whether these scientists live long enough. Since a number of laureates have been selected since 1962 who were not mentioned in this list, each year that goes by and brings new candidates into consideration decreases the chance that any of those listed here will finally get their prizes.

## "HONORABLE MENTIONS," 1962*

### Permanent Occupants of the Forty-first Chair

**Avery, O. T.**        The Committee found it desirable to wait until more became
(1877–1955)       known about the mechanism of transformation by DNA (p. 281). Ten years later, the verdict was in: "Avery's discovery

---

* Page references refer to the Liljestrand essay in the official history (Nobelstiftelsen, 1962) except where otherwise noted. Scientists who were alive in 1962 but are no longer living are identified by an asterisk.

in 1944 of DNA as carrier of heredity represents one of the most important achievements in genetics and it is to be regretted that he did not receive the Nobel Prize. By the time dissident voices were silenced, he had passed away" (Nobelstiftelsen, 1972, p. 201).

**Bang, O.**  See Ellerman, V.

**Barcroft, J.**  His work on the spleen and its functions was found deserving
(1872–1947)  of a prize in 1933 and again in 1936 (p. 274).

**Barger, G.**  See Harington, C. R.

**Bayliss, W. M.**
(1860–1924)  Their work on secretin was thought to merit a prize by
and  examiners in 1913 and in 1914. The faculty of the Institute
**Starling, E. H.**  suspended the prizes in those years (p. 225).
(1866–1927)

**Berthelot, M.**  "Berthelot's name was suggested almost every year in the
(1827–1907)  early days. If the thermochemical investigations . . . had been honored with a prize, it ought to have been divided [between Berthelot and J. Thomsen, who never was nominated]. . . . In any case, the adjudicators evidently felt that Berthelot should give way to . . . younger men" (Westgren in Nobelstiftelsen, 1962, p. 353).

**Bruce, D.**  "The Nobel Committee's examiner reported that Bruce's work
(1855–1931)  on sleeping-sickness merited a prize . . . but the Committee did not agree with its investigator. . . . The fact that Bruce had had an important share in the investigations of the Malta fever and had, in fact, identified the responsible microbe . . . an achievement that had not been judged worthy of a prize by itself—made no difference since an award for two or more completely independent discoveries was not permissible" (p. 190).

**Cannizzaro, S.**  "Cannizzaro's work was certainly well worth a Nobel Prize,
(1826–1910)  and he was indeed proposed, but strangely enough not until 1907, . . . The Nobel Committee agreed unreservedly that this work was of fundamental importance . . . but such a long time had elapsed since the publication of his results that . . . it [was] impossible to award the prize to him, however much he may have merited it" (Westgren in Nobelstiftelsen, 1962, p. 351).

**Cannon, W. B.**  His work on the emergency function of adrenaline was found
(1871–1945)  to deserve a prize three times—in 1934, 1935, and 1936 (p. 229).

**Cushing, H. W.**  Extensive contributions to neurosurgery also considered to
(1869–1939)  merit a prize by investigators but not by the committee (p. 322).

**Dandy, W. E.**  Work on injecting air through lumbar puncture as diagnosis
(1886–1946)  for localization of tumors found prizeworthy by special investigators but committee rejected the idea (pp. 321–22).

**Dean, H. T.**
(1893–1962)  Their investigations of the use of fluoride to prevent dental
and  caries has been "proposed for a prize but for one reason or
**McKay, F. S.**  another they have not received it . . . their importance is
(1874–1959)  well known and acknowledged" (p. 222).

**d'Herelle, F.**
(1873–1949)
Identification of bacteriophage and studies of it considered to merit an award (p. 210).

**de Kleyn, A.**
See Magnus, R.

**de Vries, H.**
(1848–1935)
Work on mutations considered worthy of prize (one of rediscoverers of Mendel) (p. 249).

**Ellermann, V.**
(no information) and
**Bang, O.**
(1848–1932)
Their 1908 work demonstrating that they could transmit chicken leukemia considered prizeworthy only some time after it was first proposed in 1926, when the Committee was doubtful about its importance, since it did not apply to tumors in general (pp. 206–207).

**Embden, G.**
(1874–1933)
His discovery of lactacidogen was thought to merit the award as early as 1923 but the rule that prevented the division of the award into more than two parts "probably prevented him from receiving a share" (p. 289).

**°Evans, H. M.**
(1882–1974)
Discovery of Vitamin E, the antisterility vitamin, held to deserve a prize (p. 242).

**°Gerard, R.**
(1900–74)
His findings using microelectrodes in neurophysiological investigation "considered for the Nobel Prize" (p. 307–308).

**Gibbs, Willard**
(1839–1903)
Gibbs unquestionably deserved to be awarded a prize but was never nominated. "This is greatly to be regretted, as the name of J. W. Gibbs as the first or possibly the second on the list of Nobel Prize winners for Chemistry would undeniably have been an honorable addition" (Westgren in Nobelstiftelsen, 1962, p. 352).

**Harington, C. R.**
(1897–1972)
and
**Barger, G.**
(1878–1939)
Their synthesis of thyroxin was found to merit a prize (p. 227). E. C. Kendall also mentioned in this context as having done prizeworthy work in isolating thyroxin and determining its structure. He received a prize for his work on cortisone.

**Harrison, R. G.**
(1870–1959)
Committee recommended that he be given a prize for his work on the development of nerve fibers for 1917. The faculty of the Institute, however, decided not to give a prize in that year. Later, in 1933, when the work was reconsidered, opinions diverged (p. 259).

**Kilbourne, F. L.**
See Smith, T.

**Langley, J. N.**
(1852–1925)
"As early as 1901, E. H. Starling was able to claim in a Prize proposal . . . that at least 50 percent of our information in this field consists of Langley's personal discoveries. . . . The significance of these investigations was recognized at once by the physiologists on the Nobel Committee [in 1901, 1912, and 1914] but opposition was voiced by the member representing neurology [F. Lennmalm]" (p. 315).

**Lewis, T.**
(1881–1945)
It was proposed that he share a prize with Einthoven for 1924, which was given for the electrocardiograph but it was thought that too many men were involved in discovering how heart pulses can be used diagnostically (p. 265).
His work on tissue hormones and vascular changes was found to merit a prize in 1935 but he missed out again (p. 273).

**Loeb, J.**
(1859–1924)
His work on parthenogenesis nominated by 100 sponsors from ten countries between 1901 and 1924 but the examiners were doubtful (p. 256).

**Loos, A.**
(no information)

Discovery of ringworm disease and method of transmission disputed when first proposed in 1912. Later when nominated in 1923 it satisfied "all reasonable requirements" but was too old (p. 193).

**Magnus, R.**
(1873–1927)
and
**de Kleyn, A.**
(no information)

Work on nerve reflexes and brain centers "clearly deserved the prize" but Magnus died before final decision made. Later de Kleyn renominated, but Magnus had made the major contribution (p. 311).

**°McCollum, E. V.**
(1879–1967)

His discovery of the growth promoting vitamins A and B in milk considered worthy of prize (p. 240).

**McKay, F. S.**

See Dean, H. T.

**Mellanby, E.**
(1878–1939)

His discovery of Vitamin D and its antirachitic properties considered prizeworthy (p. 240).

**Mendeleev, D. I.**
(1834–1907)

"Had the prize existed in the 1870's and 1880's, Mendeleev would certainly have received it" (Westgren in Nobelstiftelsen, 1962, p. 368).

**Minkowski, O.**

See von Mering, J.

**Neuberg, C.**
(1877–1956)

His work on the intermediate products of fermentation was held to deserve an award several times between 1920 and 1934 (p. 299).

**Quincke, H.**
(1834–1924)

His work on spinal puncture was twice held to deserve a prize—in 1909 and in 1918. The second time it was considered too old to receive an award (pp. 320–21).

**Retzius, G.**
(1842–1919)

Discovery of the importance of hair cells in hearing. His name was brought up many times between 1901 and 1916. But discovery considered too old even though it deserved a prize (pp. 328–29).

**Roux, W.**
(1850–1924)

His work on cell division and half formation found to deserve a prize in 1916 but it was thought to be too old for consideration (p. 257).

**Rubner, M.**
(1854–1932)

His contribution on the role of calories in nutrition "was found by the examiner to deserve a prize" [in 1910] "but for some unknown reason" even though he was renominated it was not resubmitted for special investigation (pp. 284–85).

**°Schick, B.**
(1877–1967)

Test for diphtheria considered worthy of prize (p. 196).

**Smith, T.**
(1859–1934)
and
**Kilbourne, F. L.**
(no information)

Their work on Texas fever found by the Committee to deserve a prize "but the Committee objected possibly because Kilbourne had not been included in the nomination" (p. 189).

**Starling, E. H.**

See Bayliss, W. M. Starling was nominated again in 1926 but his work was considered too old (p. 226).

**Steinach, E.**
(1861–1944)

His work on sex hormones and sex glands considered prizeworthy both in 1930 and in 1938 (p. 235).

**Von Mering, J.**
(1849–1908)

"Proposals for a Nobel Prize were made as early as 1902 and were later repeated, but for some unknown reason they were

and
**Minkowski, O.**
(1858–1931)

never submitted to a special investigation. The discovery [that the removal of the pancreas creates a serious form of diabetes but that a piece of gland inserted under the skin prevents it] would otherwise seem to have met extremely well the testamentary stipulations of Nobel" (p. 232).

## Occupants of the Forty-first Chair as of 1976

**Barr, M. L.**
(1908–    )

Discovery of sex chromatin in 1948 considered prizeworthy (p. 253).

**Cohen, S. S.**
(1917–    )

Work on mode of action of bacteriophages "considered prizeworthy though the number of contributors made it difficult to make a selection for a prize" (p. 211). (N. B.) Delbrück, Hershey, Luria, Lwoff, and Burnet all mentioned in this context and have since received awards.

**Dick, G. F.**
(1881–    )
and
**Dick, G. H.**
(no information)

Test for immunity to scarlet fever considered to merit a prize (p. 196).

**Kalckar, H.**
(1908–    )

Analysis of the sources of galactosemia found to be prizeworthy (pp. 290–91).

**Levan, A.**

See Tjio, J. H.

**Loubatières, A.**
(1912–    )

His work with Jambon on the use of sulphonamides in lowering blood sugar "has been found prize-worthy" (pp. 216–17).

**Magoun, H. W.**
(1907–    )
and
**Moruzzi, G.**
(1910–    )

Their discovery of the arousal reaction held to merit an award (p. 314).

**Selye, H.**
(1907–    )

His analysis of stress reactions is a "prize-worthy achievement" (p. 232).

**Tjio, J. H.**
(1919–    )
and
**Levan, A.**
(1905–    )

Their discovery of the correct number of chromosomes in normal genes was considered prizeworthy (p. 253).

## Temporary Occupants of the Forty-first Chair

The following were once considered for the awards but are no longer serious contenders:

°**Ascheim, S.**
(1878–    )
and
°**Zondek, B.**
(1887–    )

The pregnancy test they developed was considered prizeworthy in 1931 but not in 1937 (pp. 236–37).

Appendix E

# AGE-SPECIFIC ANNUAL RATES
# OF PRODUCTIVITY
# OF LAUREATES AND MATCHED
# SAMPLE OF SCIENTISTS
# WHO SURVIVED TO EACH AGE

**Table E–1. Age-Specific Annual Rates of Productivity of Laureates and Matched Sample of Scientists Who Survived to Each Age**

| Age at Publication | (a) Laureates | N | (b) Matched Sample | N | Ratio a:b |
|---|---|---|---|---|---|
| 20–29 | 1.31 | (41) | .72 | (41) | 1.82 |
| 30–39 | 3.46 | (41) | 1.89 | (41) | 1.83 |
| 40–49 | 4.32 | (38) | 1.55 | (38) | 2.79 |
| 50–59 | 3.63 | (33) | 2.01 | (31)[a] | 1.80 |
| 60+ | 4.04 | (19) | 1.26 | (21)[a] | 3.21 |
| All ages | 3.24 | (41) | 1.48 | (41) | 2.19 |
| Average production | 135.7 | | 62.1 | | 2.20 |

a. Since members of the matched sample were not the same age as the laureates but within 5 years of them, there are small differences in the numbers of each group surviving to each age.

# BIBLIOGRAPHY

**Abelson, P. H.**
1963. A proper accounting. Editorial, *Science* 139: 3549.
1967. Conditions for discovery. In *The search for understanding,* ed. C. Haskins, pp. 27–40. Washington, D.C.: Carnegie Institution.

**Alker, H. R.**
1964. On Measuring Inequality. *Behavioral Science* 9: 207–18.

**Allison, P. D., and Stewart, J. A.**
1974. Productivity differences among scientists: evidence for accumulative advantage. *American Sociological Review* 39, no. 4 (August): 596–606.

**Alvarez, L.**
1962. Adventures in nuclear physics. University of California Faculty Research Lecture at University of California, UCLRL 10476. Mimeo.
1969. Recent developments in particle physics. *Science* 165: 1071–91.

**American Association for the Advancement of Science.**
1975. Interviews (Program in the Public Understanding of Science). Supervised by Norman Metzger. Washington, D.C.

**American Men and Women of Science**
1971–73. 12th ed. New York and London: Jaques Cattell Press and R. R. Bowker Company.

**Amick, D. J.**
1973. Scientific elitism and the information system of science. *Journal of the American Society for Information Science* 24: 317–27.

**Andrade, E. N. Da. C.**
1962. Some reminiscences of Rutherford during his time at Manchester. In *The collected papers of Lord Rutherford of Nelson,* vol. 2, ed. Sir James Chadwick. London: George Allen and Unwin.

**Annan, N. G.**
1955. The intellectual aristocracy. In *Studies in social history: A tribute to G. M. Trevelyan,* ed. J. H. Plumb, pp. 241–87. London: Longmans, Green.

**Asimov, I.**
1960. *The intelligent man's guide to science.* 2 vols. New York: Basic Books.

**Astin, A.**
1963. Undergraduate institutions and the production of scientists. *Science* 141: 334–38.
1965. *Who goes where to college.* Chicago: Science Research Associates.

**Astrachan, A.**
1973. Dogma in the manger. *New Yorker* 49, no. 31: 117 ff.

**Bales, R. F.**
1953. The equilibrium problem in small groups. In *Working papers in the theory of action,* eds. T. Parsons, R. F. Bales, and E. A. Shils, chapter 4. New York: Free Press.

**Barber, B.**
1952. *Science and the social order.* New York: Free Press.
1961. Resistance by scientists to scientific discovery. *Science* 134: 596–602.

**Barron, F.**
1965. The psychology of creativity. In *New directions in psychology*, vol. 2. New York: Holt, Rinehart and Winston.

**Bayer, A. E., and Folger, J.**
1966. Some correlates of a citation measure of productivity in science. *Sociology of Education* 39: 381–90.

**Beadle, G. W.**
1967. Foreword to *Science and imagination: Selected papers of Warren Weaver,* pp. vii–xiv. New York: Basic Books.
1974. Recollections. *Annual Review of Biochemistry* 43: 1–13.

**Beaver, D. de B.**
1966. The American scientific community, 1800–1960: An historical and

statistical study. Unpublished Ph.D. dissertation. New Haven: Yale University.

**Becker, H. S., and Carper, J. W.**
1956a. The elements of identification with an occupation. *American Sociological Review* 21: 341–48.
1956b. The development of identification with an occupation. *American Journal of Sociology* 61: 289–98.

**Ben-David, J.**
1960. Scientific productivity and academic organization in nineteenth century medicine. *American Sociological Review* 31: 451–65.

**Ben-David J., and Collins, R.**
1966. Social factors in the origins of a new science: The case of psychology. *American Sociological Review* 31: 451–65.

**Benzer, S.**
1966. Adventures in the rII region. In *Phage and the origins of molecular biology,* eds. J. Cairns, G. S. Stent, and J. D. Watson, pp. 157–65. Cold Spring Harbor, N.Y.: Cold Spring Harbor Laboratory of Quantitative Biology.

**Berelson, B.**
1960. *Graduate education in the United States.* New York: McGraw-Hill.

**Bergengren, E.**
1962. *Alfred Nobel, the man and his work,* trans., A. Blair. London and New York: T. Nelson.

**Berland, T.**
1962. *The scientific life.* New York: Coward-McCann.

**Bernon, A., ed.**
1964. *Toward the well being of mankind: Fifty years of the Rockefeller Foundation.* Garden City, N.Y.: Doubleday.

**Bernstein, J.**
1962. Profiles: A question of parity. *New Yorker* 38: 49ff.
1975. Profiles—Physicist: I. I. Rabi, I and II. *New Yorker* 51 (13 October): 47–110; (20 October): 47–102.

**Blau, P. M.**
1970. A formal theory of differentiation in organizations. *American Sociological Review* 35: 201–18.
1974. Recruiting faculty and students. *Sociology of Education* 47 (Winter): 93–113.

**Blau, P. M., and Duncan, O. D., with collaboration of Andrea Tyree.**
1967. *The American occupational structure.* New York: John Wiley.

**Boring, E. G.**
1963. Letter. *Science* 142: 622–23.
1964. Cognitive dissonance: Its use in science. *Science* 145: 680–85.

**Boring, E. G., and Boring, M. D.**
1948. Masters and pupils among the American psychologists. *American Journal of Psychology* 61: 527–34.

**Born, M.**
1968. *My life and my views.* New York: Charles Scribner's.
1971. *The Born-Einstein Letters.* Correspondence between Albert Einstein and Max and Hedwig Born, 1916–55. Commentaries by Max Born. Translated from the German by Irene Born. New York: Walker.

**Bottomore, T. B.**
1964. *Elites and society.* London: C. A. Watts.

**Bradbury, W.**
1970. Genius on the prowl. *Life* 69: 57–67.

**Brim, O. G., Jr.**
1968. Adult socialization. *International Encyclopedia of the Social Sciences* 14: 555–62.

**Brown, D. G.**
1967. *The mobile professors.* Washington, D.C.: American Council on Education.

**Brumbaugh, A. J., ed.**
1948. *American universities and colleges.* Washington, D.C.: American Council on Education.

**Bundy, McG.**
1970. Were those the days? *Daedalus* 99: 531–67.

**Bureau of Labor Statistics, U.S. Department of Labor.**
1969. *Handbook of Labor Statistics.* Washington: U.S. Government Printing Office.

**Burton, P. E., and Keebler, R. W.**
1960. "Half-life" of some scientific and technical literatures. *American Documentation* 11: 18–22.

**Bush, G. P., and Hattery, L. H.**
1956. Teamwork and creativity in research. *Administrative Science Quarterly* 1: 361–72.

**Cantacuzene, J.**
1969. Comment obtient-on un prix Nobel scientifique? *Le Monde* (6–12 November), p. 9.

**Caplovitz, D.**
1960. Relations in medical school: A study of professional socialization. Unpublished doctoral dissertation. New York: Columbia University.

**Caplow, T., and McGee, R. J.**
1958. *The academic market place.* New York: Basic Books.

**Caporale, R.**
1964. *Vatican II: Last of the councils.* Baltimore: Helicon Press.

**Carlson, E. A.**
1966. *The gene: A critical history.* Philadelphia: Saunders.

**Cartter, A. M.**
1966. *An assessment of quality in graduate education.* Washington, D.C.: American Council on Education.

**Cattell, R. B.**
1963. The personality and motivation of the researcher from measurements of contemporaries and from biography. In *Scientific creativity: Its recognition and development,* eds. C. W. Taylor and F. Barron, pp. 119–31. New York: John Wiley.

**Chargaff, E.**
1968. Review, *The Double Helix. Science* 159: 1448–49.

**Claessens, D.**
1962. Forschungsteam und Persönlichkeitstruktur. *Kölner Zeitschrift für Soziologie und Sozialpsychologie* 14, no. 3: 487–503.

**Clark, R. W.**
1968. *JBS: The life and work of J. B. S. Haldane.* New York: Coward-McCann.

**Clarke, B. L.**
1964. Multiple authorship trends in scientific papers. *Science* 143: 822–24.

**Cohen, J.**
1964. *Behavior in uncertainty.* London: George Allen and Unwin.

**Cole, J. R.**
1969. The social structure of science. Unpublished dissertation. New York: Columbia University.
1970. Patterns of intellectual influence in scientific research. *Sociology of Education* 43: 377–403.

**Cole, J. R., and Cole, S.**
1971. Measuring the quality of sociological research: Problems in the use of the Science Citation Index. *American Sociologist* 6: 23–39.

1972. The Ortega Hypothesis. *Science* 178 (27 October): 368–75.
1973. *Social stratification in science.* Chicago: University of Chicago Press.
1976. Peer review at National Science Foundation. Report to Committee on Science and Public Policy, National Academy of Sciences.

**Cole, S.**
1970. Professional standing and the reception of scientific discoveries. *American Journal of Sociology* 76: 286–306.
1971. Comparative study of reward systems. Paper presented to American Sociological Association. Denver, Colo.
1972. Age and scientific behavior: A comparative analysis. Paper presented at annual meeting of American Sociological Association. New Orleans, La.

**Coleman, J. S., Campbell, E. Q., Hobson, C. J., McPartland, J., Mood, A. M., Weinfeld, F. B., and York, R. L.**
1966. *Equality of educational opportunity.* Washington, D.C.: U.S. Department of Health, Education and Welfare, Office of Education, U.S. Government Printing Office. OE–38001.

**Coser, L. A.**
1971. *Masters of sociological thought.* New York: Harcourt, Brace, Jovanovich.

**Cournand, A. F., and Zuckerman, H.**
1970. The code of science: Analysis and some reflections on its future. *Studium Generale* 23: 941–62.

**Crane, D.**
1964. The environment of discovery. Unpublished Ph.D. dissertation. New York: Columbia University.
1965. Scientists at major and minor universities: A study of productivity and recognition. *American Sociological Review* 30: 699–714.
1969. Social class origin and academic success: The influence of two stratification systems on academic careers. *Sociology of Education* 42: 1–17.
1970. The academic marketplace revisited: A study of faculty mobility using the Cartter Ratings. *American Journal of Sociology* 75: 953–64.

**Dahl, R.**
1961. *Who governs?* New Haven: Yale University Press.

**Davies, R. E.**
1970. Sheffield and secretions, Krebs and contractions. In *Essays in cell metabolism,* eds. W. Bartley, H. L. Kornberg, and J. R. Quayle, pp. 85–94. London: Wiley-Interscience.

**Davis, K.**
1949. *Human Society.* New York: Macmillan.

**Davisson, C. J.**
1937. Remarks. In *Les prix Nobel en 1937,* ed. Nobelstiftelsen. Stockholm: Imprimerie Royale.

**de Grazia, A.**
1963. The scientific reception system and Dr. Velikovsky. *American Behavioral Scientist* 7: 38–56.

**Delbrück, M.**
1966. A physicist looks at biology. In *Phage and the origins of molecular biology,* eds. J. Cairns, G. Stent, and J. D. Watson, pp. 9–22. Cold Spring Harbor, N.Y.: Cold Spring Harbor Laboratory of Quantitative Biology.

**Dennis, W.**
1956. Age and productivity among scientists. *Science* 123: 724–25.
1958. The age decrement in outstanding scientific contributions: Fact or artifact? *American Psychologist* 13: 457–60.
1966. Creative productivity between the ages of 20 and 80. *Journal of Gerontology* 21: 1–8.

**Dessler, A. J.**
    1970. Nobel Prizes: 1970 awards—physics. *Science* 170: 604–606.
**Dexter, L. A.**
    1970. *Elite and specialized interviewing.* Evanston, Ill.: Northwestern University Press.
**Dobzhansky, T.**
    1973. *Genetic diversity and human equality.* New York: Basic Books.
**Donovan, J. D.**
    1958. The American Catholic hierarchy: A social profile. *American Catholic Sociological Review* 19: 98–113.
**Duijker, H. C. J., and Rokkan, S.**
    1954. Organizational aspects of cross national research. *Journal of Social Issues* 10: 8–24.
**Duncan, O. D., Featherman, D. L., and Duncan, B.**
    1972. *Socioeconomic background and achievement.* New York and London: Seminar Press.
**Durkheim, E.**
    1951. *Suicide.* Trans. J. A. Spaulding and G. Simpson. Glencoe, Ill.: Free Press. (First published 1897.)
**du Vigneaud, V.**
    1952. *A trail of research in sulfur chemistry and metabolism and related fields.* Messenger Lectures. Ithaca, N.Y.: Cornell University Press.
**Eaton, J. W.**
    1951. Social processes in professional teamwork. *American Sociological Review* 16: 707–13.
**Edge, D. O., and Mulkay, M. J.**
    1976. *Astronomy transformed.* New York: John Wiley.
**Edson, L.**
    1967. Two men in search of the quark. *New York Times Magazine* (8 October): p. 54ff.
**Eiduson, B. T.**
    1962. *Scientists: Their psychological world.* New York: Basic Books.
**Eisner, H.**
    1973. University of the FRS. *New Scientist* 25 (January): 197.
**Epstein, C. F.**
    1970. *Woman's place: Options and limits in professional careers.* Berkeley: University of California Press.
**Evlanoff, M., and Fluor, M.**
    1969. *Alfred Nobel: The loneliest millionaire.* Los Angeles: Ward Ritchie Press.
**Faia, M. A.**
    1975. Productivity among scientists: A replication and elaboration. *American Sociological Review* 40: 825–29.
**Farber, E.**
    1953. *Nobel Prize winners in chemistry, 1901–1950.* New York: Henry Schuman.
**Fermi, L.**
    1968. *Illustrious immigrants.* Chicago: University of Chicago Press.
**Feuer, L.**
    1963. *The scientific intellectual.* New York: Basic Books.
**Feynman, R. P.**
    1965. The development of the space-time view of quantum electrodynamics. In *Les Prix Nobel en 1965,* ed. Nobelstiftelsen, pp. 172–191. Stockholm: Imprimerie Royale.
**Fleming, D. H.**
    1954. *William H. Welch and the rise of modern medicine.* Boston: Little, Brown.
    1966. Nobel's hits and errors. *The Atlantic* 218: 53–57.

**Fleming, D. H., and Bailyn, B., eds.**
1969. *The intellectual migration: Europe and America, 1930–1960.* Cambridge: Harvard University Press.

**Frankel, C.**
1971. Equality of opportunity. *Ethics* 81: 191–210.

**Gamow, G.**
1966. *Thirty years that shook physics.* New York: Doubleday.
1970. *My world line.* New York: Viking Press.

**Garfield, E.**
1970a. Citation indexing, historico-bibliography, and the sociology of science. *Proceedings of the Third International Congress of Medical Librarianship,* eds. E. Davis and W. D. Sweeny, pp. 187–204. Amsterdam: Excerpta Medica.
1970b. Citation indexing for studying science. *Nature* 227: 669–71.
1972. Where the action is, was, and will be for first and secondary authors. *Current Contents* (15 March 1972), pp. 5–8.
1973. Current comments. *Current Contents,* no. 40 (October): pp. 5–7.
1974. Current comments. Selecting the all time citation classics: Here are the fifty most cited papers for 1961–1972. *Current Contents,* no. 2 (9 January), pp. 5–8.
1975. Private communication.
1976. 1975 life sciences articles highly cited in 1975. *Current Contents,* no. 15 (12 April), pp. 5–9.

**Garfield, E., and Malin, M. V.**
1968. Can Nobel prizes be predicted? Paper presented at 135th Meeting of American Association for the Advancement of Science. Unpublished.

**Garratt, G. R. M.**
1965. Telegraphy. In *A history of technology,* eds. C. Singer, E. J. Holmyard, A. R. Hall, and T. I. Williams, vol. 4, p. 659 . New York and London: Oxford University Press.

**Garvey, W., and Griffith, B. C.**
1963. Archival journal articles: Their authors and the processes involved in their production. APA-PSIEP no. 7. In *Reports of the American Psychological Association's Project on Scientific Information Exchange in Psychology* 1: 153–86.
1967. Scientific communication as a social system. *Science* 157: 1011–16.

**Garvey, W., Lin, N., and Nelson, C. E.**
1970. Communication in the physical and social sciences. *Science* 170: 1166–73.

**Gaston, J. C.**
1973. *Originality and competition in science.* Chicago: University of Chicago Press.

**Gaston, J. C., Wolinsky, F. D., and Bohleber, L. W.**
1976. Comment on "Social class origin and academic success: The influence of two stratification systems on academic careers." *Sociology of Education* 49: 184–87.

**George K., and George, C. H.**
1955. Roman Catholic sainthood and social status: A statistical and analytic study. *Journal of Religion* 35, no. 5: 85–98.

**Gill, B.**
1976. *Here at the New Yorker.* New York: Random House–Berkley Medallion Publishing Corp.

**Gilman, W.**
1965. *Science: U.S.A.* New York: Viking Press.

**Gilvarry, J., and Ihrig, H. K.**
1959. Group effort in modern physics. *Science* 129: 1277–78.

**Glaser, B. G.**
1963. Variations in the importance of recognition in scientists' careers. *Social Problems* 10: 268–76.
1964. *Organizational scientists: Men in professional careers.* Indianapolis: Bobbs-Merrill.
1965. "Differential association" and the institutional motivation of scientists. *Administrative Science Quarterly* 10: 82–97.

**Goldberger, M. L.**
1969. The Nobel prize: Physics. *Science* 166: 720–22.

**Goode, W. J.**
1966. Family and mobility. In *Class status and power,* 2nd ed., eds. R. Bendix and S. M. Lipset, pp. 582–601. New York: Free Press.

**Goodell, R.**
1977. *The visible scientists.* Boston: Little, Brown.

**Gorden, R.**
1969. *Interviewing: Strategy, technique, and tactics.* Homewood, Ill.: Dorsey.

**Gordon, J. P.**
1964. Research on maser-laser principle wins Nobel Prize in Physics. *Science* 146 (13 November): 897–99.

**Goslin, D. A., ed.**
1969. *Handbook of socialization theory and research.* Chicago: Rand-McNally.

**Gottlieb, D.**
1961. Processes of socialization in American graduate schools. *Social Forces* 40: 124–31.

**Goudsmit, S. A.**
1976. It might as well be spin. *Physics Today* 29 (June): 40–43.

**Gray, G. W.**
1949. The Nobel Prizes. *Scientific American* 181: 11–17.
1961. Which scientists win Nobel Prizes? *Harpers* 222: 78–82.

**Greenberg, D. S.**
1967. The National Academy of Sciences: Profile of an institution II. *Science* 156: 360–64.
1968. *The politics of pure science.* New York: New American Library.

**Griffith, B. C., Jahn, M. J., and Miller, A. J.**
1971. Informal contacts in science: A probabilistic model for communication processes. *Science* 173: 164–66.

**Groenevelt, P. H.**
1971. Onsager's reciprocal relations. *Search* 2 (August): 264–67.

**Gross, N., Mason, W. S., and McEachern, A.**
1958. *Explorations in role analysis.* New York: John Wiley.

**Gross, P. M.**
1973. Editorial. *Science* 180: 1323.

**Groves, L. R.**
1962. *Now it can be told: The story of the Manhattan Project.* New York: Harper.

**Guilford, J. P.**
1962. Factors that aid and hinder creativity. Address at Creative Education Institute at San Jose State College. *Teachers College Record* 63: 380–92.

**Gustafson, T.**
1975. The controversy over peer review. *Science* 190 (12 December): 1060–66.

**Haggerty, P. E.**
1973. Industrial research and development. In *Science and the evolution of public policy,* ed. J. C. Shannon, pp. 189–216. New York: Rockefeller University Press.

**Hagstrom, W.**
1964. Traditional and modern forms of scientific teamwork. *Administrative Science Quarterly* 9: 241–63.

1965. *The scientific community.* New York: Basic Books.

1970. Factors related to the use of different modes of publishing research in four scientific fields. In *Communication among scientists and engineers,* eds. C. Nelson and D. Pollock, pp. 85–124. Lexington: D. C. Heath.

1971. Inputs, outputs and the prestige of university science departments. *Sociology of Education* 44: 375–97.

1974. Competition in science. *American Sociological Review* 39 (February): 1–18.

**Halasz, N.**

1959. *A biography of Alfred Nobel.* New York: Orion Press.

**Hardy, K. P.**

1974. Social origins of American scientists and scholars. *Science* 185 (9 August): 497–505.

**Hargens, L.**

1969. Patterns of mobility of new Ph.D.'s among American academic institutions. *Sociology of Education* 42: 18–37.

**Hargens, L., and Farr, G. M.**

1973. An examination of recent hypotheses about institutional inbreeding. *American Journal of Sociology* 78 (May): 1381–1402.

**Hargens, L., and Hagstrom, W.**

1967. Sponsored and contest mobility of American academic scientists. *Sociology of Education* 40: 23–38.

**Harland, S. C.**

1957. Hermann Muller. In *Nobel Prize winners,* ed. L. J. Ludovici, pp. 185–88. Westport: Associated Booksellers.

**Harmon, L. R.**

1965. *Profiles of Ph.D.'s in the sciences.* Washington, D.C.: National Academy of Sciences–National Research Council, Publication no. 1293.

**Harmon, L. R., and Soldz, H.**

1963. *Doctorate production in United States universities: 1920–1962.* Washington, D.C.: National Academy of Sciences–National Research Council, Publication no. 1142.

**Heard, A.**

1950. Interviewing southern politicians. *American Political Science Review* 44: 886–96.

**Heathcote, N. H. de V.**

1953. *Nobel Prize winners in physics, 1901–1950.* New York: Henry Schuman.

**Heisenberg, W.**

1971. *Physics and beyond: Encounters and conversations.* New York: Harper and Row.

**Hench, P. S.**

1951. Reminiscences of the Nobel Festival. *Proceedings of the staff meetings of the Mayo Clinic* 26, no. 23 (7 November): 424–37.

**Henshel, R. L., and Kennedy, L. W.**

1973. Self-altering prophecies: Consequences for the feasibility of social prediction. *General Systems* 18: 119–26.

**Hess, E. L.**

1970. Origins of molecular biology. *Science* 168: 664–69.

**Hewlett, R. G.**

1961. A pilot study in contemporary scientific history. *Isis* 53: 31–38.

**Hirsch, W.**

1968. *Scientists in American society.* New York: Random House.

**Hodge, R. W.**

1966. Social stratification in science and society. Comments. Meetings of American Sociological Association. Unpublished.

**Hodge, R. W., Siegel, P. M., and Rossi, P.**

1964. Occupational prestige in the United States: 1925–1962. *American Journal of Sociology* 70: 286–302.

**Hoffman, B., with collaboration of Dukas, H.**
1972. *Albert Einstein: Creator and Rebel.* New York: Viking Press.
**Holland, J. L.**
1957. Undergraduate origins of American scientists. *Science* 126: 433–37.
**Holley, I. B.**
1970. Review of "Collected Papers of Robert Goddard." *Science* 170: 522–23.
**Holton, G.**
1974. How to strike scientific gold: Fermi's group and the recapturing of Italy's place in physics. *Minerva* 12 (April): 159–98.
**Homans, G. C.**
1974. *Social behavior: Its elementary forms,* rev. ed. New York: Harcourt, Brace, Jovanovich.
**Houssaye, A.**
1886. Histoire du 41me fauteuil de l'Académie Française. Paris.
**Hughes, H. K.**
1963. Individual group creativity in science. In *Essays on creativity in the sciences,* ed. M. G. Coler, pp. 93–109. New York: New York University Press.
**Hughes, R.**
1928. *American universities and colleges.* Washington, D.C.: American Council on Education. First presented in report to Annual Meeting of Association of American Colleges, 1925.
1934. Report of Committee on Graduate Instruction. *Educational Record* (April): pp. 192–234.
**Hunt, W. H., Crane, W. W., and Wahlke, J. C.**
1964. Interviewing political elites in cross-cultural comparative research. *American Journal of Sociology* 70: 59–68.
**Hunter, F.**
1950. *Community power structure.* Chapel Hill: University of North Carolina Press.
**Hurry, J. B.**
1917. *Poverty and its vicious circles.* London: J. and A. Churchill.
**Huxley, T. H.**
1893–94. *Life and letters of Thomas Henry Huxley,* vol. 1, ed. L. Huxley. New York: D. Appleton.
1900. *Life and letters of Thomas Henry Huxley,* vol. 2, ed. L. Huxley. New York: D. Appleton.
**Hyman, H. H.**
1964. Research design. In *Studying politics abroad,* ed. R. E. Ward, pp. 153–88. Boston: Little, Brown.
1975. Reference individuals and reference idols. In *The idea of social structure,* ed. L. A. Coser, pp. 265–82. New York: Harcourt, Brace, Jovanovich.
**Hyman, H. H., and Sheatsley, P. B.**
1950. The current status of American public opinion. *National Council for Social Studies Yearbook* 21: 11–34.
**Hyman, H. H., and Singer, E., eds.**
1968. *Readings in reference group theory and research.* New York: Free Press.
**Ihde, A. J.**
1969. Theodore William Richards and the atomic weight problem. *Science* 164: 647–51.
**Inhaber, H., and Przednowek, K.**
1976. Quality of research and the Nobel Prizes. *Social Studies of Science* 6 (February): 33–50.
**Institute for Scientific Information.**
1970. Comparative statistical summary. *Science Citation Index.* Philadelphia.
**Janis, I.**
1972. *Victims of Groupthink.* Boston: Houghton-Mifflin.

**Janowitz, M.**
1960. *The professional soldier.* New York: Free Press.
**Jencks, C., and Riesman, D.**
1968. *The academic revolution.* Garden City, N.Y.: Doubleday.
**Jerne, N. K.**
1966. The natural selection theory of antibody formation: Ten years later. In *Phage and the origins of molecular biology,* eds. J. Cairns, G. Stent, and J. D. Watson, pp. 301–12. Cold Spring Harbor, N.Y.: Cold Spring Harbor Laboratory of Quantitative Biology.
**Jevons, F. R.**
1974. The contribution of academic science to industrial growth: A review. *Minerva* 12 (January): 141–43.
**Jewkes, J.**
1958. The sources of invention. *Lloyd's Bank Review New Series,* no. 47, pp. 17–28.
**Jewkes, J., Sawers, D., and Stillerman, R.**
1959. *The sources of invention.* New York: St. Martin's Press.
**Jungk, R.**
1958. *Brighter than a thousand suns: A personal history of the atomic scientists.* Trans. J. Clengh. New York: Harcourt Brace.
**Kaplan, F., comp.**
1939. *Nobel Prize winners: Charts, indexes, sketches.* Chicago: Nobelle.
**Kaplan, N.**
1964. Sociology of science. In *Handbook of modern sociology,* ed. R. E. L. Faris, pp. 852–81. Chicago: Rand McNally.
**Kash, D. E., White, I. L., Reuss, J. W., and Leo, J.**
1972. University affiliation and recognition: National Academy of Sciences. *Science* 175: 1076–84.
**Katz, E., and Lazarsfeld, P. F.**
1955. *Personal influence.* New York: Free Press.
**Kauffmann, G. B.**
1972. Review of Staudinger, *From organic chemistry to macromolecules. Isis* 63: 461–63.
**Kay, W. A.**
1963. Recollections of Rutherford, recorded and annotated by Samuel Devons, *The natural philosopher* 1: 129–55. New York and London: Blaisdell.
**Keller, S.**
1953. The social origins and career lines of three generations of American business leaders. Unpublished Ph.D. dissertation. New York: Columbia University.
1963. *Beyond the ruling class.* New York: Random House.
**Kelley, H. H., and Thibault, J. W.**
1969. Group problem solving. In *Handbook of social psychology,* eds. G. Lindzey and E. Aronson, vol. 4, pp. 1–101. Reading, Mass.: Addison-Wesley.
**Kemble, E. C., Birch, F., and Holton, G.**
1970. Percy Williams Bridgman. *Dictionary of scientific biography,* vol. 2, 457–61.
**Keniston, H.**
1958. *Graduate study and research in the arts and sciences at the University of Pennsylvania.* Philadelphia: University of Pennsylvania Press.
**Kerker, M.**
1976. The Svedberg and molecular reality. *Isis* 67: 190–216.
**Kincaid, H. W., and Bright, M.**
1957. Interviewing the business elite. *American Journal of Sociology* 63: 304–11.
**Knapp, R. H., and Goodrich, H. B.**
1952. *The origins of American scientists: A study made under the direction of*

*a committee of the faculty of Wesleyan University.* Chicago: University of Chicago Press. Rev. edition 1967.

**Knapp, R. H., and Greenbaum, J. J.**
1953. *The younger American scholar: His collegiate origins.* Chicago: University of Chicago Press.

**Kornberg, A.**
1974. *DNA synthesis.* San Francisco: W. H. Freeman.

**Krebs, H.**
1967. The making of a scientist. *Nature* 215: 1441–45.

**Krohn, R. G.**
1971. *The social shaping of science.* Westport, Conn.: Greenwood.

**Kubie, L. S.**
1970. Problems of multi-disciplinary conferences, research teams and journals. *Perspectives in Biology and Medicine* 13: 405–27.

**Kuhn, T. S.**
1962. *The structure of scientific revolutions,* vol. 2, no. 2. Chicago: International Encyclopedia of Unified Science, University of Chicago Press.

**Kuhn, T. S., and Weisskopf, V.**
n.d. Interview on the history of quantum physics. Deposited Philadelphia: American Philosophical Society.

**Lakatos, I.**
1973. History of science and its rational reconstructions. *Boston Studies in the Philosophy of Science* 8: 91–136.

**Larsen, C. W.**
1967. The care and feeding of Nobel Prize winners. Address delivered at Detroit (January 16). Chicago: Office of Public Information, University of Chicago. Mimeo.

**Lasswell, H. D.**
1936. *Politics: Who gets what, when, how.* New York: McGraw-Hill.
1961. Agenda for the study of political elites. In *Political decision-makers,* ed. D. Marvick, pp. 264–87. New York: Free Press.

**Lasswell, H. D., and Lerner, D., eds.**
1965. *World revolutionary elites: Studies in coercive ideological movements.* Cambridge: Massachusetts Institute of Technology Press.

**Lasswell, H. D., Lerner, D., and Rothwell, C. E.**
1952. *The comparative study of elites: An introduction and bibliography.* Stanford, Calif.: Hoover Institute Series, Stanford University Press.

**Lazarsfeld, P. F., Berelson, B., and Gaudet, H.**
1944. *The people's choice.* New York: Duell, Sloane and Pearce.

**Lazarsfeld, P. F., and Thielens, W. T., Jr.**
1958. *The academic mind.* New York: Free Press.

**Lederberg, J.**
1974. Private communication.

**Lederberg, J., Zuckerman, H., and Merton, R. K.**
n.d. Continuity and discontinuity in scientific development: From *schizomycetes* to bacterial sexuality. *Daedalus* (forthcoming).

**Lee, A. McC.**
1951. Individual and organizational research in sociology. *American Sociological Review* 16: 701–707.

**Lehman, H. C.**
1953. *Age and achievement.* Princeton, N.J.: Published for American Philosophical Society by Princeton University Press.

**Lenski, G.**
1961. *The religious factor: A sociological study of religion's impact on politics, economics and family life.* Garden City, N.Y.: Doubleday.

**Lerner, I. M.**
1970. Report on the state of the life sciences. Review of biology and the future of man. *Science* 169: 752–53.

**Levitan, T.**
1960. *The laureates: Jewish winners of the Nobel prize.* New York: Twayne.
**Lin, N., and Nelson, C. E.**
1969. Bibliographic reference patterns in core sociological journals. *American Sociologist* 4: 47–50.
**Lipset, S. M., and Ladd, E. C.**
1971. Jewish academics in the United States: Their achievements, culture and politics. *American Jewish Yearbook,* pp. 89–128.
**Litell, R. J.**
1967. The Nobel establishment: A rare glimpse. *Scientific Research* 11: 48–50.
**Littlewood, J. E.**
1953. *A mathematician's miscellany.* London: Methuen.
**Ludovici, L. J.**
1957. Sir Alexander Fleming. In *Nobel Prize winners,* ed. L. J. Ludovici, pp. 165–77. Westport, Conn.: Associated Booksellers.
**Lukasiewicz, J.**
1966. The handicap race of science—Nobel awards in science and medicine. *American Scientist* 54: 285A–286A.
**Luria, S. E.**
1973. Virology and cancer: Review of *The Molecular Biology of Tumor Viruses. Science* 182: 1338.
**Lurie, E.**
1960. *Louis Agassiz: A life in science.* Chicago: University of Chicago Press.
**Luszki, M. B.**
1958. *Interdisciplinary team research: Methods and problems.* New York: Published for National Training Laboratories by New York University Press.
**Lwoff, A.**
1966. The prophage and I. In *Phage and the origins of molecular biology,* eds. J. Cairns, G. Stent, and J. D. Watson, pp. 88–99. Cold Spring Harbor, N.Y.: Cold Spring Harbor Laboratory of Quantitative Biology.
1968. Review, *The Double Helix. Scientific American* 219 (July): 133–38.
**MacCollum, T., and Taylor, S.**
1938. *The Nobel prize winners 1901–1932.* Zurich: Central European Times.
**MacCracken, J. H.**
1932. *American universities and colleges.* Baltimore: Williams and Wilkins, for American Council on Education.
**Magee, B.**
1973. *Karl Popper.* New York: Viking Press.
**Malinowski, B.**
1961. Introduction in H. I. Hogben, *Law and Order in Polynesia.* Hamden, Conn.: Shoe String Press.
**Manniche, E., and Falk, G.**
1957. Age and the Nobel Prize. *Behavioral Science* 2: 301–307.
**Marsh, C. S.**
1940. *American universities and colleges,* 4th ed. Washington, D.C.: American Council on Education.
**Matthews, D. R.**
1960. *U.S. senators and their world.* Chapel Hill: University of North Carolina Press.
**McElheny, V. K.**
1965. France considers significance of Nobel awards. *Science* 150: 1013–15.
**McWilliams, C.**
1948. *A mask for privilege: Anti-Semitism in the United States.* Boston: Little, Brown.
**Meadows, A. J.**
1972. *Science and controversy.* London: Macmillan.

**Meehan, T.**
1971. What the OTB can learn from Walter Matthau. *New York Times Magazine* (4 July), p. 6ff.

**Menzel, H.**
1958. The flow of information among scientists: Problems, opportunities, and research questions. New York: Bureau of Applied Social Research, Columbia University. Mimeo.

**Merton, R. K.**
1957. Priorities in scientific discovery: A chapter in the sociology of science. *American Sociological Review* 22: 635–59. Reprinted in *The Sociology of Science,* chap. 14.

1959. Notes on problem finding. In *Sociology Today,* eds. R. K. Merton, L. Broom, and L. Cottrell, pp. ix–xxiv. New York: Basic Books.

1960. "Recognition" and "Excellence": Instructive ambiguities. In *Recognition and excellence: Working papers,* ed. A. Yarmolinsky, pp. 297–328. Glencoe, Ill.: Free Press. Reprinted in *The sociology of science,* chap. 19.

1961. Singletons and multiples in scientific discovery. *Proceedings of the American Philosophical Society* 105: 470–86. Reprinted in *The sociology of science,* chap. 16.

1963a. Resistance to the systematic study of multiple discoveries in science. *European Journal of Sociology* 4: 237–82. Reprinted in *The sociology of science,* chap. 19.

1963b. The ambivalence of scientists. *Bulletin of the Johns Hopkins Hospital* 112: 77–97. Reprinted in *The sociology of science,* chap. 18.

1964. Anomie, anomia and social interaction: Contexts of deviant behavior. In *Anomie and deviant behavior,* ed. M. B. Clinard, pp. 213–42. New York: Free Press.

1968a. The self-fulfilling prophecy. Reprinted in *Social theory and social structure,* enl. ed., pp. 475–92. New York: Free Press. (First published 1948.)

1968b. Continuities in the theory of social structure and anomie. In *Social theory and social structure,* enl. ed., pp. 215–48. New York: Free Press.

1968c. With Alice S. Rossi. Contributions to the theory of reference group behavior. In *Social theory and social structure,* enl. ed., pp. 279–334, New York: Free Press.

1968d. Science and the social order. In *Social theory and social structure,* enl. ed., pp. 591–603. New York: Free Press. Reprinted in *The sociology of science,* chap. 12.

1968e. Continuities in the theory of reference groups and social structure. In *Social theory and social structure,* enl. ed., pp. 335–440. New York: Free Press.

1968f. The Matthew Effect in science. *Science* 159: 56–63. Reprinted in *The sociology of science,* chap. 20.

1968g. On the history and systematics of sociological theory. In *Social theory and social structure,* enl. ed., pp. 1–38. New York: Free Press.

1968h. Manifest and latent functions. In *Social theory and social structure,* enl. ed., pp. 73–138. New York: Free Press.

1973. *The sociology of science: Theoretical and empirical investigations,* ed. Norman Storer. Chicago: University of Chicago Press.

1975. The Matthew Effect in science II: Problems in cumulative advantage and distributive justice. William S. Paley Lecture at Cornell Medical School, New York Hospital (September). (Further revised in series of subsequent public lectures.)

**Merton, R. K., and Barber, E.**
1963. Sociological ambivalence. In *Sociological theory, values and social change: Essays in honor of P. A. Sorokin,* ed. E. A. Tiryakian, pp. 91–120. New York: Free Press.

**Merton, R. K., Fiske, M., and Kendall, P.**
1956. *The focused interview.* New York: Free Press.
**Merton, R. K., Reader, G. G., and Kendall, P. eds.**
1957. *The student-physician.* Cambridge: Harvard University Press.
**Merton, R. K., Jahoda, M., and West, P. J. S.**
n.d. Patterns of social life. Unpublished manuscript.
**Michels, R.**
1959. *Political parties.* Trans. E. and C. Paul. New York: Dover. (First published 1915.)
**Miller, S. M.**
1960. Comparative social mobility: A trend report and bibliography. *Current Sociology,* vol. 9, no. 1.
**Mills, C. W.**
1948. *The new men of power.* New York: Harcourt Brace.
1956. *The power elite.* New York: Oxford University Press.
**Monod, J.**
1965. De l'adaptation enzymatique aux transitions allosteriques. In *Les Prix Nobel en 1965,* ed. Nobelstiftelsen, pp. 244–63. Stockholm: Imprimerie Royale.
**Moore, R.**
1961. *The coil of life.* New York: Alfred A. Knopf, Borzoi Books.
**Moulin, L.**
1955. The Nobel prizes for science from 1901–1950—an essay in sociological analysis. *British Journal of Sociology* 6: 246–63.
**Mulkay, M. J.**
1974. Methodology in the sociology of science: Some reflections on the study of radio astronomy. *Social Science Information* 13: 107–19.
1976. The mediating role of the scientific elite. *Social Studies of Science* 6: 395–422.
**Mulkay, M. J., and Edge, D. O.**
1973. Cognitive, technical and social factors in the growth of radio astronomy. *Social Science Information* 12: 25–61.
**Muller, H. J.**
1934. Lenin's doctrines in relation to genetics. In *To the memory of V. I. Lenin,* pp. 565–92. Moscow-Leningrad: Press of Academy of Sciences.
**Mullins, N. C.**
1968. The distribution of social and cultural properties in informal communication networks among biological scientists. *American Sociological Review* 33: 786–97.
**National Institutes of Health Annual Report**
1969. Bethesda, Md.: U.S. Department of Health, Education and Welfare, Public Health Service, National Institutes of Health.
**National Science Board, National Science Foundation**
1975. *Science indicators 1974.* Washington, D.C.: U.S. Government Printing Office. NSF 75–1.
**National Science Foundation**
1969. *American science manpower 1968: A report of the National Register of Scientific and Technical Personnel.* Washington, D.C.: National Science Foundation, 69–38.
1970. *Reviews of data on science resources* (December #19).
1975. *Projections of science and engineering doctorate supply and utilization: 1980 and 1985.* Washington, D.C.: U.S. Government Printing Office. NSF 75–301.
**Needham, J.**
1941. *The Nazi attack on international science.* London: Watts. (Reprinted New York: Arno Press, 1975.)

**Neustatter, H.**
1955. Demographic and other statistical aspects of Anglo-Jewry. In *A minority in Britain,* ed. M. Freedman. London: Ballantine, Mitchell.

**Newman, J. R.**
1956. *The world of mathematics.* 4 vols. New York: Simon and Schuster.

**Nobelstiftelsen**
1951. *Nobel: The man and his prizes.* Norman: University of Oklahoma Press.
1960. *Les prix Nobel en 1960.* Stockholm: Imprimerie Royale.
1962. *Nobel: The man and his prizes.* Amsterdam, London, and New York: Elsevier.
1964. *Les Prix Nobel en 1964.* Stockholm: Imprimerie Royale.
1964–66. *Nobel Lectures in Physiology or Medicine.* 3 vols. Amsterdam and New York: Elsevier.
1964–66a. *Nobel Lectures in Chemistry.* 3 vols. Amsterdam and New York: Elsevier.
1964–67. *Nobel Lectures in Physics.* 3 vols. Amsterdam and New York: Elsevier.
1965. *Les Prix Nobel en 1965.* Stockholm: Imprimerie Royale.
1972. *Nobel: The man and his prizes,* 3rd ed. New York: American Elsevier

**Noonan, J.**
1974. Private communication.

**Northrop, J. H.**
1965. Biochemists, biologists and William of Occam. In *The excitement and fascination of science,* pp. 335–44. Palo Alto: Annual Reviews.

**Olby, R.**
1970. Francis Crick, DNA, and the central dogma. *Daedalus* 99: 938–87.
1974. *The path to the double helix.* Seattle: University of Washington Press.

**Orlans, H.**
1962. *The effects of federal programs on higher education.* Washington, D.C.: Brookings Institution.

**Orth, C. D.**
1959. The optimum climate for industrial research. *Harvard Business Review* 37: 55–64.

**Pardee, A. B., Jacob, F., and Monod, J.**
1959. The genetic control and cytoplasmic expression of "Inducibility" in the synthesis of B-galactosidase by *E. Coli. Journal of Molecular Biology* 1: 165–78.

**Pareto, V.**
1935. *The mind and society,* ed. A. Livingston. 4 vols. Trans. A. Bongiorno and A. Livingston. New York: Dover. (First published as *Tratatto di Sociologia Generale.*)

**Pavel, I., Bonaparte, H., and Sdrobici, D.**
1972. The role of Paulesco in the discovery of insulin. *Israel Journal of Medical Sciences* 8: 488–90. Also published in *Impact of insulin on metabolic pathways,* ed. Eleazar Shafrir. New York: Academic Press.

**Pelz, D. C.**
1963. Relationships between measures of scientific performance and other variables. In *Scientific creativity: Its recognition and development,* eds. C. W. Taylor and F. Barron, pp. 302–10. New York: John Wiley.
1967. Creative tensions in the research and development climate. *Science* 157: 160–65.

**Pelz, D. C., and Andrews, F. M.**
1966. *Scientists in organizations: Productive climates for research and development.* New York: John Wiley.

**Phillips, J. P.**
1955. The individual in chemical research. *Science* 121: 311–12.

**Piel, G.**
1973. The role of graduate education. In *Science and the evolution of public policy,* ed. James A. Shannon, pp. 227–38. New York: Rockefeller University Press.

**Pledge, H. T.**
1939. *Science since 1500.* London: Her Majesty's Stationery Office.

**Polanyi, M.**
1963a. The potential theory of adsorption. *Science* 141: 1010–13.
1963b. Commentary. In *Scientific change: Historical studies in the intellectual, social and technical conditions for scientific discovery and technical invention from antiquity to the present,* ed. A. C. Crombie, pp. 375–80.
1969. My time with x-rays and crystals. In *Knowing and being: Essays by Michael Polanyi,* ed. M. Grene. Chicago: University of Chicago Press.

**Popper, K.**
1970. Normal science and its dangers. In *Criticism and the growth of knowledge,* eds. I. Lakatos and A. Musgrave, pp. 51–58. Cambridge: Cambridge University Press.

**President's Task Force on Science Policy**
1970. *Science and technology: Tools for progress.* Washington, D.C., April.

**Price, D. J. de S.**
1963. *Little science, big science.* New York: Columbia University Press.
1967. Communications in science: The ends—philosophy and forecast. In *Ciba Foundation Symposium on Communication in Science: Documentation and automation,* eds. A. deRueck and J. Knight, pp. 199–209. London: J. and A. Churchill.
1970. Citation measures of hard science and soft science, technology and non-science. In *Communication among scientists and engineers,* eds. C. E. Nelson and D. K. Pollack, pp. 3–22. Lexington, Mass.: Lexington Books.

**Price, D. J. de S., and Beaver, D.**
1966. Collaboration in an invisible college. *American Psychologist* 21: 1011–18.

**Rabi, I. I.**
1960. *My life and times as a physicist.* Claremont, Calif.: Claremont College.

**Ramón y Cajal, S.**
1937. Recollections of my life. *Memoirs of the American Philosophical Society* vol. 8, pt. 1.

**Raven, S.**
1967. The first great woman scientist—and much more. *New York Times Magazine* (3 December), p. 52ff.

**Ravetz, J. R.**
1971. *Scientific knowledge and its social problems.* New York: Oxford University Press.

**Reskin, B. F.**
1976. Sex differences in status attainment in science: The case of the postdoctoral fellowship. *American Sociological Review* 41: 597–612.

**Richards, F. M.**
1972. The 1972 prize for chemistry. *Science* 178 (3 November): 492–93.

**Richardson, S. A., Dohrenwend, B. S., and Klein, D.**
1965. *Interviewing: Its forms and functions.* New York: Basic Books.

**Riesman, D.**
1958a. Some observations on the interviewing in the "teacher apprehension study." In P. Lazarsfeld and W. Thielens, *The academic mind,* pp. 266–370. New York: Free Press.
1958b. Introduction, in D. Lerner, *The passing of traditional society.* New York: Free Press.
1958c. Interviewers, elites and academic freedom. *Social Problems* 6: 115–27.

**Riley, M., Johnson, M., and Foner, A.**
1972. *A theory of age stratification,* vol. 3 of *Aging and society.* New York: Russell Sage Foundation.

**Robertson, D. A., ed.**
1928. *American universities and colleges.* New York: Charles Scribner's Sons for American Council on Education.

**Robinson, J. A.**
1960. Survey interviewing among members of Congress. *Public Opinion Quarterly* 24: 127–38.

**Roe, A.**
1953. *The making of a scientist.* New York: Dodd Mead.
1963. Psychological approaches to creativity in science. In *Essays on creativity in science,* ed. M. A. Coler, pp. 153–82. New York: New York University Press.

**Roose, K. D., and Andersen, C. J.**
1970. *A rating of graduate programs.* Washington, D.C.: American Council on Education.

**Rose, H. and Rose, S.**
1969. *Science and society.* London: Allen Lane—Penguin Press.

**Rosenwaike, I.**
1972. *Population history of New York City.* Syracuse, N.Y.: Syracuse University Press.

**Rozenthal, S., ed.**
1967. *Niels Bohr: His life and his work.* New York: John Wiley; Amsterdam: North Holland.

**Sachtleben, R.**
1958. Nobel prize winners descended from Liebig. *Journal of Chemical Education* 35: 73–75.

**Salomon, J. J.**
1970. *Science et politique.* Paris: Le Seuil.
1972. The mating of knowledge and power. *Impact of Science on Society* 22: 123–32.
1973. *Science and politics.* Trans. Noel Lindsay. Cambridge: M.I.T. Press.

**Samuelson, P. A.**
1972a. Economics in a golden age: A personal memoir. In *The twentieth century sciences: Studies in the biography of ideas,* ed. G. Holton, pp. 155–70. New York: W. W. Norton.
1972b. The 1972 Nobel prize for economic science. *Science* 178 (3 November): 487–89.

**Sayre, A.**
1975. *Rosalind Franklin and DNA.* New York: W. W. Norton.

**Schimanski, F.**
1974. The Nobel experience: The decision-makers. *New Scientist* 64, no. 917 (3 October): 10–13.

**Schmidhauser, J. R.**
1959. The justices of the Supreme Court: A collective portrait. *Midwest Journal of Political Science* 3: 1057.

**Schultz, J.**
1967. Review of Elof Axel Carlson, *The Gene. Science* 157: 296–301.

**Segrè, E.**
1970. *Enrico Fermi: Physicist.* Chicago and London: University of Chicago Press.
n.d. From atoms to antiprotons. 47th Annual Faculty Research Lecture. Berkeley: University of California. Mimeo.

**Selye, H.**
1964. *From dream to discovery: On being a scientist.* New York, Toronto, and London: McGraw-Hill.

**Sewell, W. H., Haller, A. O., and Ohlendorf, G. W.**
1970. The educational and early occupational status attainment process: replication and revision. *American Sociological Review* 35: 1014–27.

**Shaplen, R.**
1958. Annals of science: Adventures of a pacifist: I and II. *New Yorker* 34 (March 15): 51–87 and (March 22): 41–89.

**Shapley, D.**
1972. Nobelists: Piccioni lawsuit raises questions about the 1959 prize. *Science* 176 (30 June): 1405–1406.

**Shatz, A.**
1968. Private communication.

**Shelton, W. R.**
1965. Harold Urey, adventurer. *Science Yearbook,* pp. 348–61. Chicago: World Book Science Annual.

**Sher, W., and Garfield, E.**
1966. New tools for improving and evaluating the effectiveness of research. In *Research program effectiveness,* eds., M. C. Yovitz, D. M. Gifford, R. H. Wilcox, E. Staveley, and H. D. Lemer, pp. 135–46. New York: Gordon and Breach.

**Sherwood, M.**
1974. The Nobel experience: Life at the top. *New Scientist* 64, no. 917 (3 October): 13–17.

**Shockley, W.**
1957. On the statistics of individual variations of productivity in research laboratories. *Proceedings of the IRE* (March), pp. 279–90.
1965. A case: Observations on the development of the transistor. In *The creative organization,* ed. G. Steiner, pp. 130–40. Chicago: University of Chicago Press.

**Siegel, P. M., and Hodge, R. W.**
1969. Social stratification—III: The measurement of social class. *International Encyclopedia of the Sciences,* vol. 15, pp. 316–25.

**Silcock, B.**
1967. Playing the Nobel status game. *The Sunday Times* (12 November).

**Simmel, G.**
1950. Numerical aspects of prominent group members. In *The sociology of Georg Simmel,* ed. and trans. K. Wolff. New York: Free Press.

**Singer, C.**
1960. Science and Judaism. In *The Jews,* ed. L. Finkelstein, vol. 2, pp. 1038–1091. New York: Harper and Brothers.

**Smith, M.**
1958. The trend toward multiple authorship in psychology. *American Psychologist* 13: 596–99.

**Sperber, M. A.**
1974–75. Symbiotic psychosis and the need for fame. *Psychoanalytic Review* 61: 517–34.

**Spilerman, S.**
1970. The causes of racial disturbances: A comparison of alternative explanations. *American Sociological Review* 35: 627–49.

**Staudinger, H.**
1970. *From organic chemistry to macromolecules: A scientific autobiography based on my original papers.* Trans. M. Staudinger. New York: Wiley-Interscience.

**Stein, W. H.**
1972. Autobiography. In *Les prix Nobel en 1972,* ed. Nobelstiftelsen, pp. 125–270. Stockholm: Imprimerie Royale.

**Steinberg, S.**
1974. *The academic melting pot: Catholic and Jews in American higher educa-*

*tion.* Report prepared for Carnegie Commission on Higher Education. New York: McGraw-Hill.

**Stent, G.**
  1969. The Nobel prize for physiology or medicine. *Science* 166: 479–81.
  1970. Remarks at conference on history of biochemistry. American Academy of Arts and Sciences. Newton, Mass.
  1972. Prematurity and uniqueness in scientific discovery. *Scientific American* 227 (December): 84–93.

**Stevenson, L. G.**
  1953. *Nobel prize winners in medicine and physiology.* New York: Schuman.

**Stone, L.**
  1971. Prosopography. *Daedalus* 100: 46–79.

**Storer, N. W.**
  1962. Some sociological aspects of federal science policy. *American Behavioral Scientist* 6: 27–30.
  1966. *The social system of science.* New York: Holt, Rinehart and Winston.
  1967. The hard sciences and the soft: Some sociological observations. *Bulletin of the Medical Library Association* 55: 75–84.

**Streisinger, G.**
  1966. Terminal redundancy, or all's well that ends well. In *Phage and the origins of molecular biology,* eds. J. Cairns, G. Stent, and J. D. Watson, pp. 335–40. Cold Spring Harbor, N.Y.: Cold Spring Harbor Laboratory of Quantitative Biology.

**Stuckey, W.**
  1975. Swedish secret: How to win the Nobel Prize. *Science Digest* 78 (October): 28–36.

**Sturtevant, A. H.**
  1959. Thomas Hunt Morgan, *Biographical Memoirs.* National Academy of Sciences 33: 283–325.
  1965. *A history of genetics.* New York: Harper and Row.

**Sullivan, D.**
  1975. Competition in bio-medical science: Extent, structure, and consequences. *Sociology of Education* 45 (Spring): 223–41.

**Sullivan, W.**
  1966. Who invented the laser? *New York Times* (16 January), p. 7.

**Swatez, G. M.**
  1970. The social organization of a university laboratory. *Minerva* 8: 36–58.

**Szent-Györgyi, A.**
  1963. Lost in the twentieth century. In *Annual reviews of biochemistry,* ed. E. E. Snell, J. M. Luck, F. W. Allen, and G. Mackenzie, vol. 32, pp. 1–14. Palo Alto, Calif.: Annual Reviews.

**Tartakoff, H.**
  1966. The normal personality in our culture and the Nobel Prize complex. In *Psychoanalysis: A general psychology: essays in honor of Heinz Hartmann,* eds. R. M. Loewenstein, L. M. Newman, M. Schur, and A. Solnit, pp. 222–52. New York: International Universities Press Inc.

**Taylor, C., and Barron, F., eds.**
  1963. *Scientific creativity: Its recognition and development.* New York: John Wiley.

**Taylor, C. W., and Ellison, R. L.**
  1967. Biographical predictors of scientific performance. *Science* 155: 1075–80.

**Taylor, D. W., Berry, P. C., and Bloch, C. H.**
  1958. Does group participation when using brainstorming facilitate or inhibit creative thinking? *Administrative Science Quarterly* 3: 23–47.

**Thackray, A. W., and Merton, R. K.**
  1972. On discipline building: The paradoxes of George Sarton. *Isis* 63: 473–95.

**Thistlethwaite, D. L.**
  1963. The college environment as a determinant of research potentiality. In

*Scientific Creativity,* eds. C. W. Taylor and F. Barron, pp. 265–77. New York: John Wiley.

**Thomson, G. P.**
1967. The septuagenarian electron. *Physics Today* 20 (May): 55–61.

**Thomson, S. P.**
1898. *Michael Faraday: His life and work.* New York: Macmillan.

**Tiselius, A.**
1967. On the work of the Nobel foundation: Aspects and experiences. October. Ann Arbor, Mich.: Unpublished speech.

**Turner, R.**
1960. Sponsored and contest mobility and the school system. *American Sociological Review* 25: 163–72.

**Udenfriend, S.**
1970. Von Euler and Axelrod. *Science* 170: 422–23.

**Underhill, R.**
1966. Values and post-college career change. *American Journal of Sociology* 72: 163–72.

**U.S. Bureau of the Census**
1958. Religion reported by the civilian population of the U.S.: March 1957. *Current population reports.* Series P-20, No. 79.
1960. *Historical statistics of the United States: Colonial times to 1957.* Washington, D.C.
1965. *Statistical Abstract of the United States,* 86th ed. Washington, D.C.
1969. *Statistical Abstract of the United States,* 90th ed. Washington, D.C.
1970. *Statistical Abstract of the United States,* 91st ed. Washington, D.C.
1971. *Statistical Abstract of the United States,* 92nd ed. Washington, D.C.
1974. *Statistical Abstract of the United States,* 95th ed. Washington, D.C.
1975. *Statistical Abstract of the United States,* 96th ed. Washington, D.C.

**Utz, W. R.**
1962. Letter. *American Mathematical Society Notices* 9: 196–97.

**van den Haag, E.**
1969. *The Jewish mystique.* New York: Stein and Day.

**Veblen, T.**
1946. *Imperial Germany and the industrial revolution.* New York: Viking Press.

**Verguèse, D.**
1975. Une certaine idée du talent des nations. *Le Monde* 21 (October): 1, 12.

**Visher, S. S.**
1948. *Scientists starred in American men of science, 1903–1943.* Baltimore: Johns Hopkins Press.

**von Humboldt, A.**
1970. On the spirit and the organizational framework of intellectual institutions in Berlin. Trans. E. Shils. *Minerva* 8: 242–49.

**Waksman, S. A.**
1954. *My life with microbes.* New York: Simon and Schuster.

**Wallace, I.**
1968. *The writing of one novel.* New York. Simon and Schuster.

**Walsh, J.**
1973. A conversation with Eugene Wigner. *Science* 181: 527–33.

**Warburg, O.**
1965. Experiments in biochemistry. In *The excitement and fascination of science,* ed. Annual Reviews, pp. 531–44. Palo Alto, Calif.: Annual Reviews.

**Warner L., and Abegglen, J. C.**
1955. *Big business leaders in America.* New York: Harper and Brothers.

**Watson, J. D.**
1966. Growing up in the phage group. In *Phage and the origins of molecular biology,* eds. J. Cairns, G. S. Stent, and J. D. Watson, pp. 239–45. Cold

Spring Harbor, N.Y.: Cold Spring Harbor Laboratory of Quantitative Biology.

1968. *The double helix*. New York: Atheneum.

**Watson, J. D., and Crick, F. H. C.**
1953. Molecular structure of nucleic acids. A structure for deoxyribose nucleic acid. *Nature* 171: 737–38.

**Weber, M.**
1958a. Science as a vocation. In *From Max Weber*, eds. and trans. H. H. Gerth and C. W. Mills. New York: Oxford University Press.

1958b. *The Protestant ethic and the spirit of capitalism*. Trans. T. Parsons. New York: Charles Scribner and Sons. (First published 1904–1905.)

**Weil, S.**
1967. *Le pesanteur et la grâce*. Paris: Union Générale d'Éditions (paper).

**Weinberg, A. M.**
1970. In defense of science. *Science* 167: 141–45.

**Weiner, C.**
1969. A new site for the seminar: The refugees and American physics in the thirties. In *The intellectual migration: Europe and America, 1930–1960*, eds. D. Fleming and B. Bailyn, pp. 152–234. Cambridge: Harvard University Press.

**Weiss, P.**
1971. *Within the gates of science and beyond: Science and its cultural commitments*. New York: Hafner.

**Weyl, N., and Possony, S.**
1963. *The geography of intellect*. Chicago: Henry Regnery.

**Wheeler, L. P.**
1952. *Josiah Willard Gibbs*. New Haven: Yale University Press.

**Whyte, W. H.**
1957. *The organization man*. Garden City, N.Y.: Doubleday Anchor.

**Wilensky, H.**
1956. *Intellectuals in labor unions*. New York: Free Press.

**Williams, G.**
1959. *Virus hunters*. New York: Alfred A. Knopf.

**Williams, L. P.**
1970. Normal science, scientific revolutions, and the history of science. In *Criticism and the growth of knowledge*, eds. I. Lakatos and A. Musgrave, pp. 49–50. Cambridge, Eng.: Cambridge University Press.

**Wilson, L.**
1941. *The academic man*. New York: Oxford University Press.

**Wilson, M.**
1969. How Nobel prizewinners get that way. *Atlantic Monthly* 224: 69–74.

1970. On being a scientist. *Atlantic Monthly* 226: 101–56.

**Wispé, L.**
1965. Some social and psychological correlates of eminence in psychology. *Journal of the History of the Behavioral Sciences* 1: 88–89.

**Wolfle, D.**
1954. *America's resources of specialized talent*. New York: Harper and Brothers.

**Yang, C. N.**
1961. Introductory notes to the article, Are mesons elementary particles? Preliminary draft (May). Mimeo.

1965. Revised introduction to Are mesons elementary particles? In *Enrico Fermi: Collected papers*, ed. E. Segrè, vol. 2, pp. 673–74. Chicago: University of Chicago Press.

**Young, M.**
1959. *The rise of the meritocracy, 1870–2033*. New York: Random House.

**Ziman, J.**
1968. *Public knowledge: An essay concerning the social dimension of science.* Cambridge, Eng.: Cambridge University Press.
1969. Information, communication and knowledge. *Nature* 224: 318–24.
1970. The light of knowledge, new lamps for old. *Aslib Proceedings* 22: 186–200.
n.d. Review of *The Double Helix*. A talk on BBC Radio 3.

**Zimmer, K. G.**
1966. The target theory. In *Phage and the origins of molecular biology,* eds. J. Cairns, G. Stent, and J. D. Watson, pp. 33–42. Cold Spring Harbor, N.Y.: Cold Spring Harbor of Quantitative Biology.

**Zinsser, H.**
1955. *As I remember him*. Boston: Little, Brown.

**Zuckerman, H.**
1965. Nobel laureates: Sociological studies of scientific collaboration. Unpublished Ph.D. dissertation. New York: Columbia University.
1967a. Nobel laureates in science: Patterns of productivity, collaboration and authorship. *American Sociological Review* 32: 391–403.
1967b. The sociology of Nobel prizes. *Scientific American* 217 (November): 25–33.
1968. Patterns of name-ordering among authors of scientific papers: A study of social symbolism and its ambiguity. *American Journal of Sociology* 74: 276–91.
1970. Stratification in American science. *Sociological Inquiry* 40: 235–57.
1972. Interviewing an ultra-elite. *Public Opinion Quarterly* 36: 159–75.

**Zuckerman, H., and Cole, J. R.**
1975. Women in American science. *Minerva* 13, no. 1 (January): 82–102.

**Zuckerman, H., and Merton, R. K.**
1971. Patterns of evaluation in science: Institutionalization, structure and functions of the referee system. *Minerva* 9: 66–100.
1972. Age stratification in science. In *A theory of age stratification,* vol. 3 of *Aging and Society,* eds. M. W. Riley, M. Johnson, and A. Foner, pp. 292–356. New York: Russell Sage Foundation.

# INDEX OF NAMES

Abel, J. J., 120
Abelson, Phillip, 191, 213
Adams, Roger, 104
Adrian, Edgar D., 98n, 288
Alder, Kurt, 286
Alfvén, Hannes, 186–187, 218
Allison, P. D., 7n, 59n, 60n, 219, 236n
Alpert, Harry, 225n
Alvarez, Luis W., 9, 100, 138, 198n, 284
Amick, D. J., 7n
Andersen, C. J., 85n, 89n
Anderson, Carl David, 29, 105, 113, 167, 193, 217, 283
Anfinsen, Christian B., 32, 184, 286–287
Annan, Noel, 27, 97n–98n
Appleton, Sir Edward V., 283
Argo, H. V., 137
Arrhenius, Svante August, 97, 166, 284
Arrow, Kenneth J., 173, 290
Ascheim, S., 300
Asimov, I., 262
Astin, A., 82n
Aston, Francis W., 285
Astrachan, A., 27
Avery, Oswald T., 42, 44, 48
Axelrod, Julius, 32, 73, 91–92, 184, 289
Bailyn, B., 154
Baltimore, David, 289
Bang, O., 297, 298
Banting, Sir Frederick Grant, 55, 287
Bárány, Robert, 287
Barcroft, J., 297
Bardeen, John, 26n, 37, 39, 70, 100, 109, 110, 121, 171n, 177n, 178, 195n, 237n, 283, 284
Barger, G., 297, 298
Barkla, Charles, 282
Barr, M. L., 300
Barton, Derek Harold Richard, 37, 286
Basov, Nikolai G., 166n, 175, 203, 284
Bayer, A. E., 63
Bayliss, W. M., 297
Beadle, George Wells, 32n, 111, 112, 116n, 133–134, 144, 288
Becker, H. S., 124n
Becquerel, Antoine Henri, 35, 97, 282
Ben-David, J., 100n
Benzer, Seymour, 147–148
Berelson, B., 82, 85n, 142, 153n, 155
Bergengren, E., 1n
Bergius, Friedrich, 285
Bergstrom, Sune, 92
Berland, T., 262
Bernon, A., 32
Bernstein, J., 108, 221, 262

Berthelot, M., 297
Berthollet. Claude Louis, 105
Best, Charles, 36, 55
Bethe, Hans Albrecht, 99n, 100, 104n, 155, 199, 200, 215, 284
Bhabha, Homi, 131
Birch, F., 110
Birkhoff, George, 51
Blackett, Patrick M. S., 283
Blau, P. M., 63, 67
Bloch, Felix, 100, 131, 132, 155, 158, 203, 283
Bloch, Konrad E., 289
Blumberg, Baruch S., 289
Bohleber, L. W., 95n
Bohr, Åage N., 96–97, 284
Bohr, Niels, 36, 99, 100, 131, 132, 139–140, 184, 220, 282
Bonaparte, H., 55
Bordet, Jules, 287
Boring, E. G., 100n
Boring, M. D., 100n
Born, Max, 25, 26n, 41, 70, 132–133, 215, 217–218, 283
Bosch, Carl, 285
Bothe, Walther, 283
Bottomore, T. B., 6n, 7n
Bovet, Daniel, 288
Bowen, Catherine Drinker, 34
Bradbury, W., 135
Bragg, Patience, 98n
Bragg, Sir William Henry, 97, 282
Bragg, Sir William L., 97, 98n, 144, 218, 237, 282
Brattain, Walter H., 283
Braun, Karl Ferdinand, 282
Bridgman, Percy William, 29, 110, 283
Brim, O. G., Jr., 123n
Brodie, Bernard B., 92
Brown, D. G., 195, 240
Bruce, D., 297
Brumbaugh, A. J., 157, 170
Buchner, Eduard, 284–285
Bundy, McGeorge, 24–25
Burnet, Sir Macfarlane, 232, 289
Burton, P. E., 149
Butenandt, Adolph, 215n–216n, 285

Calvin, Melvin, 35n, 104n, 127, 138n, 286
Campbell, W. W., 51
Cannizzaro, S., 297
Cannon, W. B., 297
Cantacuzene, J., 26, 88n
Caplovitz, D., 157
Caplow, T., 157
Carlson, E. A., 141
Carper, J. W., 124n
Carrel, Alexis, 155, 192, 196, 287

Cartter, A. M., 85n, 89n
Chadwick, Sir James, 106, 283
Chain, Ernst B., 156, 288
Chamberlain, Owen, 56n, 70, 100, 109, 116, 118, 137, 138n, 167, 283
Chandrasekhar, S., 131
Chargaff, Erwin, 70, 247
Cherenkov, Pavel A., 283
Chew, G. F., 137
Christian, W., 220
Clark, R. W., 156
Claude, Albert, 26n–27n, 217, 289
Cockcroft, Sir John D., 283
Cohen, S. S., 300
Cole, Jonathan, 4n, 7n, 17, 22, 37, 59n, 60n, 62n, 63, 67, 98n, 135n, 193, 199, 238
Cole, Stephen, 4n, 7n, 17, 22, 37, 41, 59n, 62n, 63, 67, 135n, 165, 187, 199, 238, 253
Coleman, J. S., 173
Collins, R., 100n
Compton, Arthur H., 158, 190, 237, 282
Conant, James B., 104
Coolidge, Albert Sprague, 76
Cooper, Leon N., 37, 70, 109, 116, 167, 178, 284
Cori, Carl F., 146, 155, 172, 187, 192–193, 288
Cori, Gerty, 28, 98, 155, 172, 187, 192–193, 288
Cornforth, John Warcup, 287
Coser, L. A., 6
Cotes, Roger, 165
Courant, Richard, 70
Cournand, André F., 4, 146, 155, 178, 192, 233–234, 288
Crane, D., 7n, 95, 153n
Crick, Francis H., 53–54, 164, 183, 184, 192, 205, 224, 289
Curie, Marie, 35, 39, 97, 98, 166n, 177n, 223, 282, 285
Curie, Pierre, 35, 97, 282
Cushing, H. W., 297

Dale, Sir Henry H., 100, 126, 288
Dalén, Nils, 282
Dam, Henrik, 46, 69, 288
Dandy, W. E., 297
Darwin, Charles, 164
Davis, K., 35
Davisson, Clinton, 100, 202–203, 283
Dean, H. T., 297
de Broglie, Louis Victor, 166n, 282
Debye, Peter J. W., 11, 69, 100, 240n, 285
de Duve, Christian, 179, 289
de Gaulle, Charles, 24
Dehlinger, 133

de Kleyn, A., 298, 299
Delbrück, Max, 54, 58, 97n, 100, 109, 121, 126, 131, 147–148, 155, 183–184, 206, 216, 237n, 289
Dennis, Wayne, 165
Dexter, L. A., 256n
d'Herelle, F., 298
de Vries, H., 298
Dick, G. F., 300
Dick, G. H., 300
Diels, Otto, 286
Dirac, Paul A. M., 30, 99, 138, 164, 167, 168, 217, 283
Dobzhansky, Theodosius, 83, 114–115
Doisy, Edward, 30, 46, 288
Domagk, Gerhard, 215n–216n
Dulbecco, Renato, 27n, 113, 289
Duncan, B., 65
Duncan, O. D., 65, 67
Durkheim, Emile, 224
Du Vigneaud, Vincent, 120, 190, 262, 286

Eaton, J. W., 2
Eccles, Sir John Carew, 25, 206, 235, 240n, 289
Eddington, Sir Arthur, 133
Edelman, Gerald M., 187, 289
Edge, D. O., 245
Edinburgh, Duke of, 27
Ehrlich, Paul, 39, 220, 287
Eigen, Manfred, 286
Eijkman, Christiaan, 288
Eilenberg, Samuel, 70
Einstein, Albert, 28, 36, 39, 69, 77n, 132, 164, 213, 240n, 282
Einthoven, Willem, 287
Ellermann, V., 298
Embden, G., 298
Enders, John F., 92, 98, 100, 104n, 109, 166, 193, 288
Ephrussi, Boris, 133
Erlanger, Joseph, 109, 288
Esaki, Leo, 28, 166n, 284
Eschenmoser, Albert, 220n
Evans, Herbert M., 43, 49, 298
Evlanoff, M., 1n

Faia, J. A., 59n, 60n
Falk, G., 217n
Farber, E., 262
Farr, G. M., 153
Farwell, G. W., 137
Featherman, D. L., 65
Fermi, Enrico, 28, 69, 99n, 100, 107–108, 110, 125, 128–129, 130, 137, 146, 154, 184, 212–213, 220, 240n, 283
Feynman, Richard P., 104n, 144, 158, 160, 212, 284
Fibiger, Johannes, 36, 38, 47, 287
Finsen, Niels Ryberg, 287
Fischer, Emil, 35, 39, 105, 128, 220, 284
Fischer, Ernst, 287
Fischer, Hans, 220, 285
Fiske, M., 268n
Fleming, Sir Alexander, 288
Fleming, Donald, 36, 52–53, 88n, 154, 201
Florey, Howard W., 288

Flory, Paul John, 287
Fluor, M., 1n
Folger, J., 63
Fraenkel-Conrat, Heinz, 70
Franck, James, 69, 240n, 282
Frank, Ilya M., 283
Franklin, Rosalind, 205
Friberg, Sten, 47
Friedman, Milton, 290
Frisch, Ragnar, 289

Gabor, Dennis, 176, 284
Gajdusek, D. Carleton, 32, 289
Gamow, G., 164
Garfield, E., 37, 41, 185, 188, 220, 238
Garvey, W., 179
Garwin, R. L., 137
Gasser, Herbert, 100, 109, 116, 118, 288
Gaston, J. C., 7n, 95n
Gay-Lussac, J. L., 105
Gell-Mann, Murray, 37, 38, 158, 167, 184, 284
George, C. H., 64n
George, K., 64n
Gerard, R., 298
Germer, Lester H., 202–203
Giaever, Ivar, 28, 284
Giauque, William Francis, 29, 104n, 286
Gibbs, Josiah Willard, 42, 298
Gill, B., 224n
Gilman, W., 9
Glaesson, 186
Glaser, Barney, 62n
Glaser, Donald A., 7n, 138n, 167, 283
Gödel, Kurt, 70, 158
Goldberger, M. L., 137
Goldhaber, Maurice, 70
Golgi, Camillo, 287
Goodell, R., 235
Goodrich, H. B., 65n–66n, 82n, 85n
Gordon, J. P., 203
Goslin, D. A., 123n
Gosta, Hannes Olof, 284
Goudsmit, Samuel, 43
Granit, Ragnar, 206, 289
Gray, G. W., 26, 88n
Greenberg, D. S., 9, 13, 33
Griffith, B. C., 179
Grignard, François Auguste Victor, 175, 285
Groenevelt, P. H., 186
Gross, P. M., 237n
Guillaume, Charles E., 282
Gullstrand, Allvar, 287
Gustafson, T., 156

Haber, Fritz, 127, 285
Hagstrom, Warren, 7n, 61n, 94n, 114–115, 153n
Hahn, Otto, 213, 286
Halasz, N., 1n
Haldane, J. B. S., 57, 156
Hansen, W. W., 203
Harden, Sir Arthur, 285
Hardy, K. P., 68
Hargens, L., 94n, 152, 153, 161
Harington, Charles Robert, 114–115, 120, 298
Harmon, L. R., 65n, 67, 82n, 88, 89, 158

Harrison, R. G., 298
Hartline, H. K., 206, 289
Hassel, Odd, 286
Haworth, Sir Walter N., 285
Heathcote, N. H. de V., 262
Hecht, Selig, 104
Heisenberg, Werner, 133, 184, 215, 217, 235, 283
Hench, Philip S., 22, 196, 203, 288
Henderson, L. J., 57, 104, 126
Hershey, Alfred, 54, 206, 289
Hertz, Gustav, 282
Herzberg, Gerhard, 37, 286
Hess, E. L., 218
Hess, Viktor F., 69, 240n, 283
Hess, Walter R., 288
Hewish, Antony, 51, 245, 284
Heymans, Corneille, 288
Heyrovský, Jaroslav, 286
Hicks, Sir John, 290
Hill, Archibald V., 98n, 232, 287
Hinshelwood, Sir Cyril, 286
Hodge, R. W., 8, 44, 65n
Hodgkin, Alan Lloyd, 98n, 99, 206, 289
Hodgkin, Dorothy C., 98, 286
Hodgkin, Thomas, 98
Hoffman, Ronald, 220n
Hofstadter, Robert, 283
Holley, Robert W., 32, 171n, 184, 252, 289
Holton, G., 17, 110, 146
Homans, G. C., 147n
Hopkins, Sir Frederick G., 156, 288
Houssay, Bernardo, 237–238, 288
Houssaye, A., 42
Hubble, E. P., 51
Huggins, Charles B., 77, 93n, 206n, 289
Hughes, R., 85n
Huxley, Andrew Fielding, 98n, 206, 289
Hyman, H. H., 269, 277n

Inhaber, H., 219n

Jacob, François, 24, 184, 235, 289
Jahn, M. J., 179
Janis, I., 178n
Janowitz, M., 64n
Jensen, J. Hans, 184, 203, 284
Jevons, F. R., 27
Jewkes, J., 2, 176
Joliot, Frédéric, 97, 220, 285
Joliot-Curie, Irène, 97, 98, 285
Josephson, Brian, 41, 166, 284
Jungk, R., 164

Kac, Mark, 70
Kahneman, Daniel, 222n
Kalckar, H., 300
Kantorovich, Leonid V., 24, 290
Kaplan, F., 262
Kapteyn, Jacobus, 51
Karrer, Paul, 285
Kash, D. E., 88n
Kastler, Alfred, 284
Katz, Bernard, 206, 289
Keebler, R. W., 149

Kekulé, F. A., 105
Keller, S., 6n, 7n, 12, 64, 67
Kelley, H. H., 178n
Kemble, E. C., 110
Kendall, Edward C., 39, 67n, 190, 203, 239, 268n, 288
Kendrew, John Cowdery, 286
Kerker, M., 213
Keynes, John Maynard, 165
Khorana, H. Gobind, 32, 100, 155, 184, 188, 289
Kilbourne, F. L., 298, 299
Knapp, R. H., 65n–66n, 85n
Koch, Robert, 35, 36, 39, 287
Kocher, Emil Theodor, 287
Koopmans, Tjalling C., 290
Kornberg, Arthur, 30–31, 32, 70, 74, 100, 109, 172, 288–289
Kornberg, Thomas, 212n
Kossel, Albrecht, 287
Krebs, Hans Adolf, 104, 124–125, 128, 288
Krogh, August, 287
Kuhn, Richard, 215n–216n, 285
Kuhn, T. S., 132, 139, 164
Kusch, Polykarp, 111, 170, 283
Kuznets, Simon, 290

Ladd, E. C., 68, 69, 71, 72, 73, 76
Lakatos, Imre, 53
Lamb, Willis E., Jr., 170, 241n, 283
Landau, Lev Davidovich, 131, 283
Landsteiner, Karl, 186, 218, 220, 240n, 288
Langley, J. N., 298
Langmuir, Irving, 88, 154, 155, 198n, 285
Larsen, C. W., 29n
Lasswell, Harold D., 7n
Laveran, Charles Louis Alphonse, 287
Lawrence, Ernest O., 30, 99n, 100, 138, 146, 158, 170, 175, 283
Lazarus, D., 137
Lederberg, Joshua, 4, 74, 77n, 109, 116, 118, 144, 158, 166, 167, 174, 185, 235, 238, 288
Lee, Tsung Dao, 26n, 48, 70, 93n, 113, 137, 144, 178, 215, 217, 283
Lehman, H. C., 146, 165, 166
Leloir, Luis Federico, 286
Lenard, Philipp, 282
Lennmalm, F., 298
Leo, J., 88n
Leontief, Wassily, 290
Levan, A., 300
Lewis, G. N., 104, 114–115, 128
Lewis, T., 298
Libby, Willard F., 104n, 138n, 286
Liljestrand, Göran, 38, 206, 296
Lin, N., 179
Lipmann, Fritz Albert, 29, 155, 288
Lippmann, Gabriel, 121, 282
Lipscomb, William N., 287
Lipset, S. M., 68, 69, 71, 72, 73, 76
Litell, R. J., 47, 53

Loeb, Jacques, 114–115, 298
Loewi, Otto, 69, 288
Loos, A., 299
Lorentz, Hendrik Antoon, 35, 282
Lorenz, Gerda, 261n
Lorenz, Konrad Z., 51, 206–207, 245, 289
Loubatières, A., 300
Ludovici, L. J., 262
Lukasiewicz, J., 26
Luria, Salvador E., 54, 111, 112, 113, 121, 155, 186, 206, 237n, 289
Lwoff, André, 24, 247, 289
Lynen, Feodor, 98, 289

MacCollum, T., 262
MacCracken, J. H., 85n
Macleod, John J. R., 36, 55, 287
Magee, B., 25
Magnus, R., 299
Magoun, H. W., 300
Malin, M. V., 37, 185
Malinowski, B., 147n
Manniche, E., 217n
Marconi, Guglielmo, 282
Marsh, C. S., 151n
Martin, Archer J. P., 175, 286
Marvel, Carl Shipp, 120
Matthaei, J. H., 185
Matthews, D. R., 64n, 256n
Mayer, Joseph, 98
Mayer, Maria Goeppert, 28, 98, 99, 133, 155, 184, 192, 203, 284
Mayr, Ernst, 51
McCollum, E. V., 299
McElheny, V. K., 24
McGee, R. J., 157
McKay, F. S., 297, 299
McMillan, Edwin M., 138n, 191, 192, 213, 286
McWilliams, C., 86
Medawar, Sir Peter, 25, 235, 289
Meehan, T., 19
Mellanby, E., 299
Mendeleev, Dmitri Ivanovich, 42, 299
Menzel, H., 179
Merton, Robert, 2, 4, 7n, 12, 34n, 42, 60n, 61n, 62, 67n, 118, 123n, 124n, 135, 165n, 175, 179, 184, 185, 201, 202, 204, 228, 238, 239, 246, 252, 268n
Metchnikoff, Élie, 287
Meyerhof, Otto, 100, 287
Michels, R., 5
Michelson, Albert A., 4, 85n, 86, 88, 121, 154, 155, 282
Miller, A. J., 179
Millikan, Robert Andrews, 30, 49, 105, 237, 282
Mills, C. W., 12
Minkowski, O., 299, 300
Minot, George, 109, 288
Mittag-Leffler, Gosta, 18
Moissan, Henri, 40, 47, 175, 284
Moniz, Antônio E., 288
Monod, Jacques, 24, 25, 184, 185, 289
Moore, Marianne, 224n
Moore, R., 262
Moore, Stanford, 37, 178, 184, 286–287

Morgan, Thomas H., 134, 141–143, 158, 198n, 288
Morrish, A., 137
Moruzzi, G., 300
Mössbauer, Rudolf L., 30, 166, 283
Mosteller, Frederick, 188n
Mottelson, Ben R., 27n, 284
Moulin, L., 88n
Mueller, Paul, 288
Mulkay, M. J., 245, 251
Muller, Hermann Joseph, 30, 99n, 111, 112, 113, 141–143, 288
Mulliken, Robert S., 286
Murphy, William Parry, 109, 116, 118, 196, 288
Myrdal, Gunnar, 290

Natta, Giulio, 286
Needham, J., 25
Néel, Louis Eugène Félix, 284
Nelson, Carnot E., 179
Nernst, Walther, 100, 105, 121, 285
Neuberg, C., 299
Nicolle, Charles, 288
Nirenberg, Marshall W., 32, 38, 184, 185, 289
Nobel, Alfred, 1–2, 17, 18, 22
Noonan, John, 48n
Norrish, Ronald G. W., 286
Northrop, John H., 30, 99, 239–240, 286

Ochoa, Severo, 109, 155, 288
Olby, R., 184
Onnes, Heike Kamerlingh, 282
Onsager, Lars, 43–44, 93n, 186, 192, 218, 286
Oppenheimer, Robert, 114–115
Ostwald, Wilhelm, 105, 285

Packard, M. E., 203
Palade, George E., 289
Pareto, Vilfredo, 5–8, 12
Pasternak, Boris, 216n
Paulesco, N., 55
Pauli, Wolfgang, 30, 69, 119–120, 121, 133, 283
Pauling, Linus, 26n, 100, 104n, 109, 144, 184, 200, 205, 224, 235, 238, 286
Pavel, I., 55
Pavlov, Ivan Petrovich, 35, 179, 287
Perrin, Jean Baptiste, 213, 282
Perutz, Max Ferdinand, 286
Piccioni, Oreste, 56n
Piel, Gerard, 27
Planck, Max, 167, 282
Pledge, H. T., 106
Podgorny, Nikolai V., 24
Poincaré, Henri, 51
Polanyi, Michael, 104, 114–115, 127, 201
Popper, Karl, 25
Porter, George, 286
Porter, Rodney R., 289
Possony, S., 69, 71n
Pound, R. V., 203
Powell, Cecil Frank, 283
Pregl, Fritz, 285
Prelog, Vladimir, 287
Price, D. J. de S., 44, 145, 149, 165, 238

Priestley, J. G., 57
Prokhorov, Alexandr M., 173, 203, 284
Przednowek, K., 219$n$
Purcell, Edward Mills, 110, 203, 283

Quincke, H., 299

Raacke, I. D., 43$n$
Rabi, I. I., 100, 110–111, 119, 121, 170, 221, 230, 262, 283
Rainwater, James, 284
Raman, Sir Chandrasekhara Venkata, 282–283
Ramel, Stig, 18
Ramón y Cajal, Santiago, 214, 223, 231, 287
Ramsay, Sir William, 35, 39, 40, 284
Raven, S., 223
Ravetz, J. R., 13
Rayleigh, Baron, 35, 282
Reader, G. G., 67$n$
Reichstein, Tadeus, 203, 288
Reitz, J. R., 137
Reskin, B. F., 7$n$, 60$n$
Retzius, G., 299
Reuss, J. W., 88$n$
Revelle, Roger, 51
Richards, Dickinson W., Jr., 233–234, 288
Richards, F. M., 178
Richards, Theodore W., 104, 121, 285
Richardson, Owen W., 282
Richet, Charles Robert, 287
Richter, Burton, 284
Robbins, Frederick C., 98–99, 109, 116, 118, 166–167, 193, 288
Robertson, D. A., 85$n$, 157
Robinson, Sir Robert, 286
Rockefeller, David, 31
Roe, Anne, 65$n$
Roentgen, Wilhelm K., 35, 40, 282
Roose, K. D., 85$n$, 89$n$
Rose, H., 13
Rose, S., 13
Rosenbluth, M. N., 137
Rosenwaike, I., 81
Ross, Sir Ronald, 287
Rossi, P., 8, 65$n$
Rous, Francis Peyton, 47, 99, 113, 186, 191, 192, 200, 206, 217, 241$n$, 289
Roux, W., 299
Rubner, M., 299
Rutherford, Ernest, 36, 39, 49, 99$n$, 100, 106, 115, 131, 220, 285
Ružička, Leopold, 285
Ryan, Francis, 167
Ryle, Sir Martin, 51, 245, 284

Sabatier, Paul, 285
Sakharov, Andrei, 24, 216$n$
Salomon, J. J., 13
Samuelson, Paul, 106, 130–131, 173, 290
Sanger, Frederick, 286
Sarton, George, 124$n$
Sartre, Jean-Paul, 216$n$
Sawers, D., 2

Sayre, Ann, 205
Schick, B., 299
Schimanski, F., 19
Schmidhauser, J. R., 64$n$
Schrieffer, John Robert, 37, 70, 85$n$, 109, 116, 118, 166, 167, 178, 284
Schrödinger, Erwin, 100, 283
Schwinger, Julian S., 110–111, 158, 167, 284
Sdrobici, D., 55
Seaborg, Glenn T., 30, 34$n$, 104$n$, 138$n$, 235, 286
Segrè, Emilio, 56$n$, 99$n$, 100, 109, 110, 138$n$, 155, 167, 198$n$, 211–212, 283
Selove, W., 137
Selye, H., 300
Semenov, Nikolai N., 286
Shaplen, R., 1$n$
Shapley, D., 56$n$
Shapley, Harlow, 51
Sheatsley, P. B., 277$n$
Sherrington, Sir Charles S., 40, 156, 217, 288
Sherwood, M., 237, 238$n$, 252
Shockley, William, 60$n$, 241$n$, 283
Siegbahn, Karl M. G., 282
Siegel, P. M., 8, 65$n$
Silcock, B., 88$n$
Simmel, G., 12
Simpson, George Gaylord, 51
Singer, C., 69
Smith, T., 299
Soddy, Frederick, 39, 285
Soldz, H., 82$n$, 88, 89, 158
Solzhenitsyn, Aleksandr, 216$n$
Sommerfeld, Arnold, 99$n$, 104, 133
Sonneborn, Tracey, 111, 112, 113
Spemann, Hans, 288
Stanley, Wendell M., 104$n$, 286
Stark, Johannes, 282
Starling, E. H., 297, 298, 299
Staudinger, Hermann, 186, 218, 286
Stein, William H., 37, 178, 184, 286–287
Steinberg, S., 72
Steinberger, J., 137
Stent, Gunther, 126
Stern, Curt, 70
Stern, Otto, 25, 27$n$, 69, 70, 99$n$, 133, 283
Sternheimer, R. L., 137
Stevenson, L. G., 262
Stewart, J. A., 7$n$, 59$n$, 60$n$, 219$n$, 236$n$
Stillerman, R., 2
Stone, L., 5
Storer, Norman, 7$n$, 61$n$
Strandskov, Herluf, 111
Streisinger, G., 126
Sturtevant, A. H., 142, 143
Sumner, James B., 218, 286
Sutherland, Earl W., 172, 184, 241$n$, 289
Svedberg, Theodor, 213, 285
Synge, Richard, 98$n$, 175, 286
Szent-Györgyi, Albert, 104, 262, 288
Szilard, Leo, 9, 70

Tamm, Igor Y., 24, 283

Tartakoff, H., 20
Tatum, Edward Lawrie, 74, 109, 116$n$, 222, 288
Taylor, S., 262
Teller, Edward, 70, 108, 131, 132, 158
Temin, Howard M., 289
Thackray, A. W., 124$n$
Theiler, Max, 26$n$, 46, 77, 155, 196, 288
Theorell, Hugo, 104, 288
Thibaut, J. W., 178$n$
Thomson, David, 98$n$
Thomson, George P., 97, 202–203, 283
Thomson, Sir Joseph John, 97, 98$n$, 100, 131, 282
Tinbergen, Jan, 97, 289
Tinbergen, Nikolaas, 51, 97, 206–207, 245, 289
Ting, Samuel C. C., 284
Tiselius, Arne, 4, 30, 39–40, 42, 44, 50, 53, 55, 216, 239, 286
Tjio, J. H., 300
Todd, Lord, 286
Tomonaga, Sin-itiro, 284
Torrey, H. C., 203
Townes, Charles H., 170, 175, 284
Turner, Ralph, 132
Tversky, Amos, 22$n$

Udenfriend, S., 92
Uhlenbeck, George, 43
Ulam, Stanislaw, 70
Underhill, R., 124$n$
Urey, Harold Clayton, 29, 104$n$, 192, 236–237, 238$n$, 285

van Vleck, John, 158
van der Waals, Johannes D., 282
van't Hoff, Jacobus Hendricus, 35, 39, 40, 284
Verguèse, D., 26
Virtanen, Artturi, 286
Visher, S. S., 65$n$
von Baeyer, Adolph, 40, 105, 284
von Behring, Emil, 35, 220, 287
von Békèsy, Georg, 24–25, 155, 159$n$, 237$n$, 238, 239, 241$n$, 289
von Euler, Ulf Svante, 37, 55, 97, 289
von Euler-Chelpin, Hans August Simon, 97, 285
von Frisch, Karl, 51, 206–207, 245, 289
von Hayek, Friedrich, 290
von Hevesy, Georg, 285–286
von Jauregg, Julius Wagner, 287
von Laue, Max T. F., 282
von Liebig, Justus, 97$n$, 105
von Mering, J., 299–300
von Neumann, John, 51, 70, 130–131, 155, 158
von Ossietzky, Carl, 215$n$

Waksman, Selman A., 55–56, 77, 262, 288
Wald, George, 100, 104, 235, 289
Wallace, I., 49
Wallach, Otto, 285
Walsh, J., 155

Walton, Ernest T. S., 283
Warburg, Otto H., 38–39, 104, 105, 125, 128, 288
Watson, James D., 53–54, 70, 99n, 100, 111–113, 121, 134–135, 144, 164, 168, 183–184, 191–192, 205, 235, 289
Wattenberg, A., 137
Weaver, Warren, 32n
Weber, M., 252
Weiner, Charles, 69, 154
Weiss, Paul A., 44, 70
Weisskopf, Victor, 70, 114–115, 131–132, 139, 203
Welch, Robert A., 20
Weller, Thomas H., 99, 109, 116, 118, 166–167, 193, 288
Wentzel, Donat, 99
Wentzel, Gregor, 99
Werner, Alfred, 285
Westgren, Arne, 35, 296, 297, 298, 299
Weyl, N., 69, 71n
Wheeler, John, 104, 114–115, 158, 160

Whipple, George H., 288
White, I. L., 88n, 176
Whyte, W. H., 2
Wiegand, Clyde, 167
Wieland, Heinrich O., 98, 212, 285
Wien, Wilhelm, 282
Wigner, Eugene Paul, 30, 99, 104n, 107, 108, 121, 127, 144, 155–156, 158, 199, 200, 235, 283–284
Wilcox, H. A., 137
Wilkins, Maurice H. F., 289
Wilkinson, Geoffrey, 193–194, 287
Williams, L. Pearce, 11
Willstätter, Richard, 285
Wilson, Charles T. R., 175, 282
Wilson, Edmund, 224n
Wilson, L., 157
Wilson, Mitchell, 111
Windaus, Adolf, 39, 285
Wispé, L., 100n
Wolfenstein, L., 137
Wolfle, D., 83
Wolinsky, F. D., 95n

Woodward, Robert Burns, 144, 200, 220n, 286

Yang, Chen Ning, 26n, 48, 70, 93n, 107–108, 113, 128–129, 137, 144, 178, 215, 283
Young, M., 86
Yukawa, Hideki, 30, 283

Zeeman, Pieter, 35, 99n, 282
Zernike, Frits, 175, 283
Ziegler, Karl, 286
Ziman, John, 44, 53–54
Zinsser, H., 104n
Zondeck, B., 300
Zsigmondy, Richard, 285
Zuckerman, H., 4, 7n, 59n, 60n, 98n, 118, 135, 147n, 165n, 176, 179, 193, 201, 221n, 238, 239

# INDEX OF SUBJECTS

Académie des Sciences, 20, 196, 237
Accumulation of advantage in science, 14, 95, 207, 238, 248–254
  additive and multiplicative models, 60–61
  constraints on, 251–254
  definition of, 59–60
  organizations, 149
  processes in, 60–63
Achievement, 254
Age at doctorate, laureates and Academicians, 89, 91, 93–95
Age at first publication, laureates-to-be, 145–146
Age at full professorship, 157–162
Age of masters and apprentices, 118–121
Age at prize, 113–117, 216–218
Age at prize-winning research, 164–169
Age-specific annual rates of productivity, 301
*Alfred Nobel: The Man and His Prizes* (Westgren and Liljestrand), 261, 296
Alfred Nobel Memorial Prize in Economics, 16n
Ambivalence, 137, 180
  in apprenticeships, 138–143
  toward prize, 15, 209–216
American Academy of Arts and Sciences, 57
American Association for the Advancement of Science, 73
American Council on Education, 89n
American laureates, 26, 27, 28, 154
*American Men of Science*, 65n, 145, 158, 159, 199, 261
*American Men and Women of Science*, 9, 10, 238
American Philosophical Society, 34
Apprentices, laureate, 99–121
  age of, relative to masters, 118–121
  ambivalence toward masters, 138–143
  collaboration with masters, 140–143
  conflict with masters, 138–143
  forty-first chair and, 104
  reenactment of master's role, 135–138
  selective recruitment of, 109–111, 115
  self-selection by, 107–109, 111–113, 115, 251
  sharing prize with masters, 116, 118
  socialization of, 122–132

scientific taste, 124, 127–129
  self-confidence and, 129–132
  standards of performance, 124–127
  sponsorship of, 116, 118, 132–135
Arches of Science Medal, 33
Ascription, 254
Atoms for Peace Award, 199
Awards
  post-prize, 236–238
  pre-prize, 196–200

Barnard Medal, 22
Bell Laboratories, 170, 171, 241
*Biological Abstracts*, 46
British laureates, 26, 27
Brown University, 171, 241n

California, University of
  at Berkeley, 28–29, 30, 31, 84, 87n, 88, 90, 155, 170, 171, 240, 241
  at Los Angeles, 85n, 241n
  at San Diego, 241n
California Institute of Technology, 29, 30, 84, 87n, 89n, 90, 93, 111, 171, 193
Carnegie Institution, 159n, 171
Case Institute of Technology, 30, 85, 93, 155, 171
Chicago, University of, 29n, 30, 84, 87n, 88, 89n, 90, 93, 170, 171, 241
Citation analysis
  pre-prize, 37–38, 148–149, 187–189
  prize-winning research, 184–187
Collaboration
  master and apprentice, 140–143
  post-prize relations, 232–234
  prize-winning research, 176–178
College of the City of New York, 30–31, 85
Columbia University, 30, 84, 88, 89n, 90, 155, 170, 171, 240, 241
Communication, informal, 108, 155
Comparative Statistical Summary, I. S. I., 185
Conflict in apprenticeships, 138–143
Cornell University, 30, 84, 89n, 90, 155, 171, 199, 241
Cresson Medal, 22

Dartmouth College, 84, 87n
Discoveries
  multiple independent, 202–205, 247
  premature, 185–186, 192
  serendipitous, 211

Doctoral origins of laureates, 86–95
Draper Medal, 199

Earlham College, 85n
Elites
  biological and social, 6
  governing, 12
  Pareto and, 5–8
  scientific, 8–13
  segmental, 12
  self-perpetuation of, 106
  social selection of, 7
  socialization for: *see* Socialization in science
  socioeconomic origins of, 63–64
  sponsorship by, 132–135
  strategic, 13
  ultra-elite, 11–13, 17, 71
Enrico Fermi Award, 199
Evocative environments, 131, 172–173
Evokers of excellence, 125–127

Federation of American Scientists, 23
Field Medals, 21, 22, 33
Florida, University of, 85n
Forty-first chair
  definition of, 42
  occupants of, 42–50, 140, 143, 158, 161, 188, 221, 244–245, 254
    masters and apprentices, 104
    1976, 300
    permanent, 296–300
    temporary, 205–207, 300
    vs. laureates, 214
  phenomenon of, 244, 245
  selection of laureates and, 44–50
Franklin Institute of Philadelphia, 22
Franklin Medal, 22, 199
French laureates, 26
Frick Chemical Laboratory, 30
Furman University, 85n, 93

Gatekeeping in science, 62, 139–140, 200
General Electric Company, 171, 241n
George Washington University, 90
German laureates, 25–26
German science, Nazi effect on, 70–71
Guggenheim Foundation, 32

Harvard University, 29, 31, 84, 87n, 88, 89n, 90, 111, 155, 170, 171, 192, 193, 194, 199, 240, 241, 257, 258
History of science, heroic view of, 230

Honorific awards, 196–200

Illinois, University of, 84, 87n, 89n, 90, 171
Inbreeding, academic, 152–154
Indiana University, 90, 171, 241n
Influentials in science, 201
Institute for Advanced Study, 149, 157, 159n, 171
Institute for Scientific Information, 185
Institutional mobility after the prize, 239–241
Intellectual property, 179
International Mathematical Union, 21
Interview guide, 267–269
Interviewing, 256–279
  procedures of, 3–4, 256–270
    appointments, 259
    attitudes toward prospect of being interviewed, 260
    initial phase, 256–258
    interview guide, 267–269
    letters, 258–259
    preparation, 261–267
    tape recorder, 269–270
  techniques of, 270–278
    images of interviewer, 275–277
    technical language, 271
    top elites, 277–278
    types of questions, 271–275

Jewish laureates
  overrepresentation of, 71–78
  undergraduate origins, 86–87
Johns Hopkins University, 84, 87n, 88, 89n, 90, 155
Joint prize-winners, 232–233
Justice, distributive, 232, 251, 252

Kenyon College, 85n
Kinship ties, 96–99

Lafayette College, 85, 93
Lasker award, 32n, 199
Laureates, 11–13
  accumulation of advantage and: see Accumulation of advantage in science
  age at doctorate, 89, 91, 93–95
  age at full professorship, 157–162
  age at prize, 216–218
  American, 26, 27, 28, 154
  apprentices: see Apprentices, laureate
  British, 26, 27
  emigrés, 69–71
  foreign born, 69–71, 192
  French, 26
  geographic origins, 81
  German, 25–26
  graduate origins, 86–95
  influence of research, 40–42
  Jewish, 71–78, 86–87
  kinship ties, 96–99
  masters: see Masters, laureate
  mobility of: see Mobility
  national distribution of, 25–27

non-academic employment, 157
organizational claims to, 29–33
political advocacy by, 23–24, 235–236
productivity of: see Productivity
promotion rates, 157–162
religious origins of, 68–69, 78–82
saints and, 48n
in science, 1901–76, 282–290
scientific excellence of, 35, 37–42
selection of, 17–19
  controversy over, 56
  forty-first chair and, 44–50
  socioeconomic origins of, 63–68, 78–80, 93–95
Soviet, 27
undergraduate origins, 82–86
vs. occupants of forty-first chair, 214
visibility of, 183, 222, 235
women, 98, 192–193
Laureates-to-be, 184
  age at first publication, 145–146
  collaborative publication, 146–149
  first jobs, 149–157
  foreign born, 154
  impact of research, 37–38
  productivity of, 145–149
  recognition of: see Reward system of science
  role performance, 145–146
Ledlie Prize, 33

Massachusetts Institute of Technology, 84, 87n, 88n, 89n, 90, 93, 241
Masters, laureate, 99–121
  age at prize and, 113–117
  age of, relative to apprentices, 118–121
  ambivalence toward apprentices, 138–143
  collaboration with apprentices, 140–143
  conflict with apprentices, 138–143
  forty-first chair and, 104
  process of mutual search and, 107–113
  reenactment of roles of, 135–138
  as role models, 125
  selective recruitment by, 109–111, 115
  sharing prize with apprentices, 116, 118
  socialization by, 122–132
    scientific taste, 124, 127–129
    self-confidence and, 129–132
    standards of performance, 124–127
  sponsorship by, 116, 118, 132–135
Matthew Effect, 34, 59n, 62–63, 140–141, 228, 251, 252

Max Planck Medal, 199
Mayo Clinic, 74, 159n, 171, 241
Mayo Foundation, 190
Metallurgical Laboratory of the Manhattan Project, 30
Michelson Prize, 30
Michigan, University of, 85, 87n, 89n, 90, 171
Michigan College of Mining, 85n
Michigan State College, 85n, 90
Minnesota, University of, 90, 156, 190
Missouri, University of, 85n
Mobility
  consequences of, 224–227, 229–230
  post-prize, 239–241
  pre-prize, 194–195
  social, 132
  upward, 144–162
Morrison Prize, 199

Name-ordering of authors, 180–183
National Academy of Sciences, 10, 14, 20, 33, 63, 82, 88, 89, 115, 159, 187–188, 237, 249
  election to, 196–198
  women in, 193
National Institutes of Health, 32, 74, 170, 171, 241
National Register of Scientific and Technical Personnel, 9, 10
National Research Council, 10
National Science Board, 8, 26, 27
National Science Foundation, 9, 10, 20, 63, 156
Nebraska, University of, 85n
Neglect in science, 187
New York University, 85n, 171, 241n
Nobel Committee, 56, 261
Nobel Foundation, 2, 4, 18, 26n, 33, 35, 261
Nobel Institute, 19
Nobel prize: see Prize
"Nobel Prize Complex," 19–20
Nobelstiftelsen: see Nobel Foundation
Noblesse oblige, 15, 207, 228, 238
  prize-winning research and, 179–183, 188
  productivity of laureates-to-be and, 147–148
  resources for research and, 253
Nominations, consensus on, 39–40
Normative structure of science, 175–176, 209–210, 214
Norwegian Storting, 2

Oberlin College, 84, 87n, 93
Obliteration by incorporation, 185
Observability, 155–156
Oregon State University, 85n
Organizational concentration, 160–161, 251

first jobs, 151–152, 156
post-prize laureates, 240
prize research, 170–172

Pennsylvania, University of, 90
Pittsburgh, University of, 90
Post-prize
  awards, 236–238
  collegial relations, 231–232
  mobility, 239–241
  productivity, 242
  recognition, 238–239
  relations with collaborators, 232–234
  research: see Post-prize research
  visibility of laureates, 232
Post-prize research
  productivity, 221–230
  regression effects on, 222
  significance of, 219–221
Premature discoveries, 185–186, 192
Pre-prize
  citation analysis, 37–38, 148–149, 187–189
  mobility, 194–195
  recognition, 189–202
Presidential Medal of Merit, 199
President's Science Advisory Committee (PSAC), 13
President's Task Force on Science Policy, 20
Princeton University, 30, 88, 89n, 90, 155–156, 171, 241
Prix Balzan, 20, 23
Prix Nobel, Les, 261
Prize
  ambivalence toward, 15, 209–216
  annual number of winners, 54
  delay in award, 115–116, 172, 214–215, 218
  distribution of, 3, 12
  double winners of, 26n, 39, 220
  efficacy of, 243–248
  field limitations on, 245
  fields covered by, 18, 48, 50–52
  honorarium for, 17, 19, 20–21
  impact on science, 230, 246–248
  joint winners, 232–233
  number of nominees for, 44–46
  prestige of, 19–23, 35–36
  refused, 215–216
  rules governing, 18–19
  scientific work after, 218–220
  selection of winners: see Selection
  shared, 116, 118
  symbolic uses of, 14, 23–35
  timing of, 216–218
  ultra-elite of science and, 11–13
  visibility of, 16–17, 20, 21–22
Prize research
  controversy over, 219
  errors in, 212–213
  selection of, 39, 52–54, 210–214

Prize-winning research, 163–207
  age at, 164–169
  claims to, 178–183
  collaborative and individual, 176–178
  distribution by specialties, 174
  foci of, 173–189
  impact of, 183–189
  interval between prize and, 218
  multiple independent discoveries in, 202–207
  recognition following: see Reward system of science
  site of, 170–172
  specialty and year of award, 292–293
Productivity
  age-specific annual rates of, 301
  authentic and spurious changes in, 227–229
  laureates-to-be, 145–149
  long-run changes in, 229–230
  post-prize, 242
  short-run changes in, 221–227
Promotions, 157–162
  following prize-winning research, 189–195
  timeliness of, 195
Purdue University, 85n

Recognition
  delayed, 187, 198, 214–216
  post-prize, 238–239
  pre-prize, 189–202
Reference group behavior, 130
Research Corporation, 32
Reward, timeliness of, 162
Reward system of science, 62, 161–162, 189–202, 250, 254
  honorific awards, 196–200
  positions of influence and authority, 200–202
  promotion, 189–195
  slippage in, 191
Rochester, University of, 90, 171, 241n
Rockefeller Foundation, 32, 170, 171, 240, 241
Rockefeller Institute, 30, 74, 149–151, 157, 159n, 170, 171, 192, 240
Rockefeller University, 29, 31, 240, 241
Roles
  redefinition of, with prize, 234–236
  reenactment of, 15, 135–138, 179, 228
Royal Caroline Medico-Surgical Institute, 2, 18–19, 20
Royal Society of London, 17, 20, 196, 199, 237
Royal Swedish Academy of Sciences, 2, 18–19, 20
Rumford Premium, 57
Rutgers University, 85n, 171, 241

Salk Institute, 241n
Science Citation Index, 37, 41, 148–149, 185, 188, 220

Scientific American, 33
Scientific growth, 173–174
  forty-first chair and, 44
  heroic theory of, 247
Scientific taste, 124, 127–129
Scientists
  occupational prestige, 8
  socioeconomic origins compared to laureates', 66–68
  stratification of, 3, 8–13, 248
Selection, 17–19
  controversy over, 56
  forty-first chair and, 44–50
  successive approximations in, 48–50
Selective recruitments, 109–111, 115
Self-fulfilling prophesies, 250
Self-selection, 95, 107–109, 111–113, 115, 251
Serendipity, 211
Social mobility, 132
Social selection, 95, 111, 156
Socialization in science, 96, 118, 122–132
  scientific taste, 124, 127–129
  self-confidence and, 129–132
  standards of performance, 124–127
Socioeconomic origins of laureates, 63–68, 78–80, 93–95
Socioeconomic status, first jobs and, 152–153
South Dakota, University of, 85n
Specialties, 174, 292–293
Sponsorship, 116, 118, 132–135
St. Louis University, 30, 171, 241n
Stanford University, 29, 89n, 155, 171, 173, 241
Stevens Institute of Technology, 155
Stratification system in science, 3, 8–13, 248
Success
  subversive effects of, 252
  wrecked-by-success syndrome, 252
Swarthmore College, 84, 87n, 93
Swedish Academy at Stockholm, 2
Swedish Central Bank, 16n

Taste, scientific, 124, 127–129
Texas, University of, 171
Treatise on General Sociology (Pareto), 5–7
"Truffle dogs" in science, 110
Tyler, John and Alice, Ecology Award, 20

Ultra-elite of science, 11–13, 17, 71
United States Naval Academy, 85n
United States Plant, Nutrition, & Soil Lab., 171
Ursinus College, 85n, 93

Vanderbilt University, 85n, 155, 241n
Visibility
  of apprentices, 135
  of authors, 182

Visibility (*cont.*)
  of laureates, 183, 222, 235
  of prize, 16–17, 20, 21–22
Washburn University of Topeka, 85*n*
Washington University (St. Louis), 30, 90, 155, 170, 171, 172, 190, 241

Western Reserve University, 241*n*
Whitman College, 85*n*
*Who's Who*, 261
Willard Gibbs Medal, 238*n*
Wisconsin, University of, 85, 87*n*, 89*n*, 90, 155, 171, 241
Women laureates, 98, 192–193

Wooster, College of, 85*n*, 93

Yale University, 84, 87*n*, 88, 89*n*, 90, 171